Engineering for Resolution of the Energy–Environment Dilemma

Committee on Power Plant Siting
NATIONAL ACADEMY OF ENGINEERING

NATIONAL ACADEMY OF ENGINEERING
WASHINGTON, D.C. 1972

This study has been supported by the
NATIONAL SCIENCE FOUNDATION and ATOMIC ENERGY COMMISSION
under Contract NSF-C310, T.O. 204
and by the electric utility industry through the
REGIONAL ELECTRIC RELIABILITY COUNCILS.
The full technical reports of the Working Groups of the
Committee on Power Plant Siting are contained in the main report,
ENGINEERING FOR RESOLUTION OF THE ENERGY-ENVIRONMENT DILEMMA.

Available from

Printing and Publishing Office
National Academy of Sciences
2101 Constitution Avenue, N.W.
Washington, D.C. 20418

Library of Congress Catalog Card Number 79-186370
ISBN 0-309-01943-5

Printed in the United States of America

THE NATIONAL ACADEMY OF ENGINEERING

 The National Academy of Engineering was established in December, 1964. The Academy is independent and autonomous in its organization and election of members, and shares in the responsibility given the National Academy of Sciences under its congressional act of incorporation to advise the federal government, upon request, in all areas of science and engineering.

The National Academy of Engineering, aware of its responsibilities to the government, the engineering community, and the nation as a whole, is pledged:

1. To provide means of assessing the constantly changing needs of the nation and the technical resources that can and should be applied to them; to sponsor programs aimed at meeting these needs; and to encourage such engineering research as may be advisable in the national interest.

2. To explore means for promoting cooperation in engineering in the United States and abroad, with a view to securing concentration on problems significant to society and encouraging research and development aimed at meeting them.

3. To advise the Congress and the executive branch of the government, whenever called upon by any department or agency thereof, on matters of national import pertinent to engineering.

4. To cooperate with the National Academy of Sciences on matters involving both science and engineering.

5. To serve the nation in other respects in connection with significant problems in engineering and technology.

6. To recognize in an appropriate manner outstanding contributions to the nation by leading engineers.

Committee on Power Plant Siting
National Academy of Engineering
Washington, D.C.
November 15, 1971

Mr. Clarence H. Linder
President
National Academy of Engineering
Washington, D.C.

Dear Mr. Linder:

Submitted herewith is the final report of the Committee on Power Plant Siting of the National Academy of Engineering. This report does not offer complete or final solutions to power plant siting questions, but we believe it does contain a body of material which can be useful in the further development of solutions, and also in the Academy's future consideration of wider energy questions.

The results of the Committee's efforts appear herein in two sections: the Steering Group report in Part I, the reports of the Working Groups in Part II (the latter appear as condensations in the Summary volume).

The conclusions and recommendations of the Steering Group, which appear in Chapter 3 of Part I, are directed both to short range and long range possibilities for mitigation or resolution of power plant siting difficulties. The reports of the Working Groups, which appear in Part II, assemble and to some extent develop a considerable body of more specialized material which should be relevant and useful to the siting process and which has not previously been available in any one place.

As you know, in forming the committees, we put great emphasis in getting a wide diversity of viewpoints. Participants were consciously chosen to represent environmental groups and electric power utilities as well as the many relevant disciplines and fields. It is therefore gratifying to affirm that each of the group reports contained herein has the broad backing from each member of the group which prepared it, if not necessarily in every case the specific approval of each detail by each participant.

The reports contained herein have also had the benefit of comments by the Committee on Public Engineering Policy and the Project Committee of the Academy. However, the final responsibility for the Steering Group report and each Working Group report rests with the authoring group rather than with individuals who have had an opportunity to comment.

Although the prime responsibility of the Committee on Power Plant Siting of the National Academy of Engineering was rather narrowly defined by its title and sponsorship, it was clearly necessary from the beginning to chart the program within a much broader context. One reason for this is, of course, that electric power plants are only parts of larger systems for producing, distributing, and consuming electricity. Even more important is the fact that extremely significant parts of the solutions to power plant siting problems must be found by procedures which lie considerably beyond generally accepted boundaries of engineering. Not only did the Committee have to explore domains such as meteorology, hydrology, marine biology, radiation physics, and economics; it also had to seek beachheads in areas of law, governmental organization, public policy, and politics.

Clearly, a definitive and final contribution to our subject was impossible within the scope of our resources of time and funding. Nevertheless, we view the work with some satisfaction and also with a feeling of urgency. The satisfaction stems from the substantial consensus we have reached on approaches to mitigating and resolving power plant siting conflicts and on a considerable body of background material. The urgency stems, of course, from the fact that the energy–environment conflict is getting worse and that much must be done before it can get better.

We are not suggesting, however, that the Committee on Power Plant Siting should be perpetuated. On the contrary, we believe that our proper task will be complete with the publication of this report and with the completion of the Second Forum on Power Plant Siting which we propose herein. We do, however, believe most emphatically that the subject of power plant siting should not be dropped by the National Academy of Engineering, but rather incorporated within the context of a much broader program on the total energy problem, including the efficiency of energy uses and the optimum utilization of all energy sources.

Respectfully and sincerely yours,

W. DEMING LEWIS, *Chairman*
Committee on Power Plant Siting

PREFACE

Conflicts between environmentalists and electric power utilities over power plant siting issues have been publicized widely in recent months; they are examples of difficulties which can develop in the utilization of technology with changing value judgments regarding the side effects which accompany the benefits desired. An important obligation of the engineering community is to work to reduce undesirable side effects to satisfactory levels in new designs and to contribute to the resolution of existing conflicts. This report of the Committee on Power Plant Siting of the National Academy of Engineering deals with the urgent and precedent-setting problem of power plant siting, an area in which the need for conflict resolution is urgent.

The study which is the basis for the report was initiated by the Academy to provide a factual basis for decision making in dealing with the complex problems associated with power plant siting. The effort was focused primarily on approaches which might become useful during the next 20 years. The study involved individuals from organizations which could be expected to have sharply differing positions on the issues involved. As is evident from the list of participants, the members of the Steering Group and Working Groups include those who have been identified professionally with different sides of the questions treated as well as those whose positions might well be classed as neutral. It is particularly significant and gratifying therefore to note that the report contains a substantial number of conclusions and recommendations on which all or almost all participants agreed.

The report proposes in broad outline a procedure of mitigating power plant siting conflicts, which, if implemented, could become effective in the relatively near future. There are also identified research and development projects which if pursued might contribute to minimizing the present conflict to a substantial degree within a decade or so. Finally, considerable information directly relevant to the energy–environment crisis is included in the reports of the Working Groups. Since this material was previously unavailable in a convenient single source, its assembly should be valuable for future studies.

A fundamental premise which motivated the program of the Committee on Power Plant Siting is that sound engineering can do much to mitigate the conflict between the demands of society for electric energy and its simultaneous desire to preserve the quality of the environment. The results of the study give credence to the validity of that assertion.

The success of this study is due to the substantial contributions of time and effort by many individuals. The National Academy of Engineering is grateful for the excellent response of all those involved. Particular note should be made of the leadership given to the program by Dr. W. Deming Lewis and Dr. Denis M. Robinson, Chairman and Vice-Chairman, respectively, of the Committee on Power Plant Siting.

The program would not have been possible without the financial support of the Atomic Energy Commission, the National Science Foundation, and the electric utility industry as a whole, acting through the nine Regional Electric Reliability Councils. This support is gratefully acknowledged.

CLARENCE H. LINDER, *President*
National Academy of Engineering

COMMITTEE ON POWER PLANT SITING

STEERING GROUP

DR. W. DEMING LEWIS, *Chairman* *; President, Lehigh University

DR. DENIS M. ROBINSON, *Vice Chairman* *; Chairman of the Board, High Voltage Engineering Corp.; Chairman, Marine Biological Laboratory

DR. STANLEY I. AUERBACH, Division Director, Ecological Sciences Division, Oak Ridge National Laboratories

PROFESSOR DAVID F. CAVERS, President, Council on Law Related Studies

MR. ROLAND C. CLEMENT,* Vice President, National Audubon Society

DR. MERRIL EISENBUD, A. J. Lanza Laboratory, New York University

MR. S. DAVID FREEMAN, Director, Study of National Energy Policy, Twentieth Century Fund

MR. WILLIAM R. GOULD,* Senior Vice President, Southern California Edison Company

DR. HELMUT E. LANDSBERG,* Research Professor, Institute for Fluid Dynamics & Applied Mathematics, University of Maryland

PROFESSOR PHILIP H. LEWIS, JR., Chairman, Department of Landscape Architecture, University of Wisconsin

MR. THEODORE J. NAGEL,* Vice President, System Planning, American Electric Power Service Corp.

DR. SAM H. SCHURR, Director, Energy and Minerals Division, Resources for the Future

MR. LELAN F. SILLIN, JR., President and Chairman, Northeast Utilities

MR. EDWIN H. SNYDER, Director and Consultant, Public Service Electric & Gas Company

MR. JOSEPH C. SWIDLER, Chairman, New York State Public Service Commission

MR. FRANK M. WARREN, President, Portland General Electric Company

MR. FREDERICK H. WARREN, Advisor on Environmental Quality, Federal Power Commission

MR. G. O. WESSENAUER, Consultant, Chattanooga, Tennessee

Staff of the Committee on Power Plant Siting

MR. E. D. EATON, *Executive Secretary*
MRS. LEILA VON MEISTER, *Staff Assistant*
MR. DANIEL P. SHEER, *Staff Assistant* (April 1–August 31, 1971)
MRS. MAXINE BROWN
MRS. ELLEN DAVIS

* Members of Committee on Power Plant Siting Executive Committee.

WORKING GROUPS

I(a) Environmental Protection: Power Plant Siting and Air Quality—Engineering Considerations *

DR. GLENN R. HILST, *Chairman;* Vice President, Environmental Research, Aeronautical Research Associates of Princeton, Inc., Princeton, New Jersey

MR. WILLARD A. CRANDALL, Consolidated Edison Company, New York, New York

ROBERT FRANK, M.D., University of Washington, Seattle, Washington

DR. FRANCOIS FRENKIEL, Naval Ship Research and Development Center, Washington, D.C.

DR. SHELDON K. FRIEDLANDER, California Institute of Technology, Pasadena, California

DR. FRANK GIFFORD, National Oceanic and Atmospheric Administration, Oak Ridge, Tennessee

MR. JOHN HEALY, Los Alamos Scientific Laboratory, Los Alamos, New Mexico

DR. JAY S. JACOBSON, Boyce Thompson Institute, Yonkers, New York

DR. CARL B. MOYER, Aerotherm Corporation, Mountain View, California

MR. MAYNARD E. SMITH, Brookhaven National Laboratory, Long Island, New York

I(b) Environmental Protection: Water

DR. JOSEPH A. MIHURSKY, *Chairman;* Natural Resources Institute, University of Maryland, Prince Frederick, Maryland

MR. JAMES ADAMS, Pacific Gas and Electric Company, Emeryville, California

MR. BRUNO BRODFELD, Stone and Webster Corporation, Boston, Massachusetts

DR. CHARLES COUTANT, Oak Ridge National Laboratories, Oak Ridge, Tennessee

DR. DONALD HARLEMAN, Massachusetts Institute of Technology, Cambridge, Massachusetts

DR. ROBERT HOLTON, U.S. Atomic Energy Commission, Washington, D.C.

MR. ROBERT T. JASKE, Pacific Northwest Laboratory, Richland, Washington

DR. LOREN JENSEN, Johns Hopkins University, Baltimore, Maryland

DR. PETER A. KRENKEL, Vanderbilt University, Nashville, Tennessee

DR. EDWARD RANEY, Cornell University, Ithaca, New York

DR. MARTIN ROESSLER, University of Miami, Miami, Florida

DR. KIRK STRAWN, Texas A&M University, College Station, Texas

DR. C. M. TARZWELL, National Marine Water Quality Laboratory, West Kingston, Rhode Island

I(c) Environmental Protection: Radiological Aspects of Power Plants and Their Fuel Cycles †

MR. CHARLES L. WEAVER, *Chairman;* Acting Director, Division of Surveillance and Inspection, Environmental Protection Agency, Rockville, Maryland

MR. JOSEPH J. DINUNNO, U.S. Atomic Energy Commission, Washington, D.C.

MR. J. M. GERUSKY, Pennsylvania Department of Health, Harrisburg, Pennsylvania

DR. MORTON I. GOLDMAN, NUS Corporation, Rockville, Maryland

DR. EDWARD P. RADFORD, Johns Hopkins University, Baltimore, Maryland

MR. JAMES M. SMITH, General Electric Company, San Jose, California

---

* Dr. Robert N. Rickles, Commissioner, Department of Air Resources, New York City, was initially named to this Working Group. He resigned August 9, 1971, because of disagreement with the point of view of the study.

† Dr. Dean Abrahamson, Director of Environmental Studies, University of Minnesota, was initially a member of this Working Group. He resigned April 22, 1971, citing lack of time to participate in the Working Group activities.

DR. JAMES H. WRIGHT, Westinghouse Electric Corporation, Pittsburgh, Pennsylvania

MR. CARL M. UNRUH, Battelle Memorial Institute, Richland, Washington

I(d) Environmental Protection: Aesthetics and Land Use

PROFESSOR BRUCE H. MURRAY, *Chairman;* Department of Landscape Architecture, University of Wisconsin, Madison, Wisconsin

MR. G. S. BINGHAM, Bonneville Power Administration, Portland, Oregon

MR. FRANK BURGGRAF, JR., New York Public Service Commission, Albany, New York

PROFESSOR F. STUART CHAPIN, University of North Carolina, Chapel Hill, North Carolina

PROFESSOR PHILIP H. LEWIS, JR., University of Wisconsin, Madison, Wisconsin

PROFESSOR DANIEL MANDELKER, Washington University, St. Louis, Missouri

MR. T. A. PHILLIPS, Federal Power Commission, Washington, D.C.

MR. J. E. SCHUMANN, Pacific Gas and Electric Company, San Francisco, California

II Systems Approach to Site Selection

MR. WILLIAM R. GOULD, *Chairman;* Senior Vice President, Southern California Edison Company, Rosemead, California

MR. THEODORE J. NAGEL, *Vice Chairman;* Vice President, System Planning, American Electric Power Service Corporation, New York, New York

SUBGROUPS

Technical and Economic Aspects of Siting

MR. HARVEY F. BRUSH, *Chairman;* Bechtel Corporation, Los Angeles, California

DR. JAMES FLANNERY, Department of the Interior, Washington, D.C.

MR. RUBLE A. THOMAS, Southern Services, Inc., Birmingham, Alabama

DR. JAMES WRIGHT, Westinghouse Environmental Systems Department, Pittsburgh, Pennsylvania

Environmental Aspects of Siting

PROFESSOR LESTER LEES, *Chairman;* California Institute of Technology, Pasadena, California

MR. PETER BORELLI, Sierra Club, New York, New York

MR. ROLAND COMSTOCK, Northern States Power Company, Minneapolis, Minnesota

MR. WILLIAM DOUB, U.S. Atomic Energy Commission, Washington, D.C.

Licensing and Regulatory Restraints

MR. JOHN CONWAY, *Chairman;* Consolidated Edison Company, New York, New York

MR. PAUL CLIFTON, California Resources Agency, Sacramento, California

MR. FRANCIS P. COTTER, Westinghouse Electric Corporation, Washington, D.C.

MR. J. FREDERICK WEINHOLD, Office of Science and Technology, Washington, D.C.

III Energy and Economic Growth

MR. ABRAHAM GERBER, *Chairman;* Vice President, National Economic Research Associates, Inc., New York, New York

MR. MARTIN BERNSTEIN, National Economic Research Associates, Inc., New York, New York

MR. BERT J. BLEWITT, Public Service Electric and Gas Company, Newark, New Jersey

MR. HENRY CAULFIELD, Colorado State University, Fort Collins, Colorado
MR. GERALD GAMBS, Ford Bacon and Davis, Inc., New York, New York
MR. WILLIAM R. HUGHES, Charles River Associates, Inc., Cambridge, Massachusetts
MR. RENE H. MALES, Commonwealth Edison Company, Chicago, Illinois
MR. WILLIAM R. NEW, Tennessee Valley Authority, Chattanooga, Tennessee
MR. MILTON F. SEARL, Resources for the Future, Inc., Washington, D.C.
DR. CARLOS STERN, University of Connecticut, Storrs, Connecticut

CONTENTS

II REPORTS OF THE WORKING GROUPS

I
REPORT OF THE STEERING GROUP

This report presents the consensus of the Steering Group and agreement on major conclusions and recommendations. However, not all of the details represent the specific views of all members.

1 INTRODUCTION: THE ENERGY–ENVIRONMENT DILEMMA

As this report is being written, in the summer of 1971, it is clear that the energy–environment dilemma is a problem which is urgently upon us now and must be resolved by utilizing wise and comprehensive planning and good management. In some parts of the United States, New York City for example, this problem has reached major proportions. It seems likely to get worse before it will get better. It threatens to spread over wider areas unless a general solution can be found. Because of the long time required for the planning, siting, and construction of new generating facilities, this problem cannot be resolved overnight. We will be fortunate to observe a substantial mitigation in less than a decade. Meanwhile we must act to avoid letting matters get worse than they have already.

Americans, both privately and through their corporate and public enterprises, have had a voracious appetite for energy in all available forms. We use coal, natural gas, petroleum, water power, and increasing amounts of nuclear power in our homes, our businesses, our factories, and our transportation systems. The demand for electric power for familiar uses is expanding, and further new uses emerge each year. For example, urban mass transit systems, material recycling processes, and environmental protection measures are expected to become significantly greater consumers of electric energy. Abundant and cheap energy has become one of the principal foundations of our uniquely prosperous and technological civilization. About a quarter of what we use (in the year 1970, approximately 17 quadrillion Btu's out of a total of 69 quadrillion) is used to generate electricity.

The demand for energy has been supplied adequately but not always without adverse environmental impacts. As energy requirements have grown and populations have become denser, the contaminants from many stacks and vehicles have become part of the environment of tens of millions of people. Heated water discharges have changed the ecosystems of rivers and lakes, and hydroelectric reservoirs have altered the landscape. Transmission lines have crossed previously untouched scenery. The development of nuclear power has been accompanied by the warming of rivers and by the fear of radiation.

The electric utilities, government-owned and investor-owned, are managed generally with a spirit of public service. Their responsibility to provide energy is legally enforceable if the customer demand is reasonable. The investor-owned electric utilities are regulated regarding rates, financing, accounting, safety, customer service, and in other areas by the Federal Power Commission and in most cases state utility commissions. As a general rule, public power agencies are regulated to a certain degree by the state itself and by their governing boards, which are responsible to the public. All private utilities and some public agencies must also comply with the regulations of numerous government units, including air and water control agencies, zoning commissions, city councils, the SEC, FTC, and FCC, to name but a few. Nevertheless, the utilities have found themselves to be principal targets of public, and in particular, conservationist and antipollution concern.

Whether they are government-owned or investor-owned and regulated by commissions in open proceed-

3

ings, the electric utilities are peculiarly sensitive in response to public opinion. They are large and highly visible targets because of the logic and history of their development and because they are rooted in their franchise areas. Some of their plants are located in densely populated urban areas and produce highly visible smoke plumes. Modern generating plants, even though they can technologically be located many miles from a city, are large and have great potential impact upon the environment. Understandably, they often draw the attention of those whose concern is the preservation of the natural environment. When this occurs, it is clear that many methods are available within our system of government and court review by which construction and operation of a new power plant can be stopped or greatly delayed by determined opponents.

Conservationists who would preserve natural values and environmentalists who would prevent undesirable by-products of change have a common interest in power plant siting. In most of America's past, the conservationist groups have contained a small and dedicated minority of Americans. There were the Thoreaus, the Audubons, the John Muirs, the Theodore Roosevelts, the Gifford Pinchots. The environment was rarely an issue. Most of us were too busy building, too busy finding and utilizing the resources to worry much about what happened to the natural beauty and habitability of our relatively unspoiled continent. But in the last few years, a new awareness of conservationist and environmentalist issues has overtaken us. We see that we shall become choked in the by-products of our mechanized society's processes unless we recognize and act upon the knowledge that the control of these by-products is as much a necessity as are the processes themselves. The issues involved have moved so close to many of us that we must recognize their relevance to our lives and in the last analysis perhaps to our survival.

Yet, apparently, we cannot deny the need for electric power. The average person in our mechanized society demands directly and indirectly a growing portion of the resources of this land, including electricity. A bountiful supply of electrical energy is an indispensable part of our standard of living. The growth of population is an exponential pressure in the predicted rate of growth of electric energy consumption.

The dilemma caused by the American requirement for more electric power with the simultaneous hope that the environment will be protected by limiting construction of new electric generating facilities is only a fractional part of a much larger and growing contradiction between man's natural desire to improve his living standard by using the resources of this planet and his desire to conserve its resources. But the particular conflict to which we address our attention in this report is

sharper, more immediate, and more publicized than others of a similar nature.

Even without regard to environmental concerns, there has been difficulty in some areas in meeting the requirements for more and more electric power. Some electric utilities have been hard pressed to cope with the steep growth in the needs of industry and of individuals who enjoy such things as electric heat in the winter and air conditioning in the summer. Increasingly, curtailments, or the threat of curtailments, of electric service have faced some critical areas of the country. Though the great Northeast blackout of November 9, 1965, was itself not due to power shortage, it was one of the events that drew attention to problems of system stability and reliability, and to the dependence of the public on electric power supply. Concern with diminishing power generating reserves and with the problems—financial, technical, and political—of meeting steeply increasing demands for electric power both reliably and economically followed.

In the 1960's in some areas of this country, public criticism of plans for generating capacity expansion was intensifying. Environmentalists were increasingly successful in delaying or blocking altogether plans for the augmentation of bulk power supply. A substantial body of thought and political effort emerged, directed at stopping power plant construction altogether, in the sincere belief that this was the only way in which environmental standards could be met. Although one can respect the motivation of those using this approach, the analysis of Working Group III, Energy and Economic Growth (p. 40), makes it clear that such an approach will not provide the solution society is seeking in the near future. The most conservative growth in energy will need substantial additions to the power supply. One has to plan accordingly; more power can be used to clean up the environment itself.

The utilities in the most besieged areas—New York City being the prototype—found themselves confronted with sharp deficiencies in peak power generating capacity and no apparent ways out. The *New York Times,* in the years 1970 and 1971, while frequently referring to the problems of power shortage, has also made many references to the activities of the groups which were blocking the construction of new power generating capacity.

Neither side in this impasse was evil, malicious, or pursuing selfish aims. The conservationists were merely trying to restore some values that had been lost and to avoid further losses. The utilities were simply trying to do what they had always done successfully—to meet their legally imposed responsibility and supply the demand for power. The great majority of each group was unquestionably trying with dedication to discharge a responsible obligation to society. The fault, if any, was not with the intent of either party; it was that society as a whole had not yet faced up to the dimensions of the problem. Its normal methods of conflict resolution, such as recourse to the adversary processes of the courts, were insufficient to resolve the basic problem.

As is commonly the case in a conflict situation, a few members of each side chose to characterize the opposite side as evil, or at least cynical. Many facile but inadequate solutions were generated with good intent but without adequate consideration of all of the factors. Responsible and thoughtful people met in many arenas to discuss and to seek resolution of the conflict. Federal and state legislatures became deeply concerned with the problem, environmental agencies were established, the Office of Science and Technology of the President conducted studies which formed the basis for legislative proposals. This process still continues. A selection of written reports of these efforts, some of them heroic, is listed on p. 19.

Meanwhile, the problem is getting worse. The energy–environment dilemma which is faced so painfully by New York City now seems likely to spread to many other parts of the nation. The southwestern United States, for example, may be in for even greater difficulty than New York City. In that region, nuclear plants proposed for the coastal regions have come under criticism because of the complicating hazard of earthquakes. Fossil-fueled plants proposed for desert regions, such as at Four Corners, are resisted because of their potential impact on the existing environment. New York City's past ability to import power from neighboring areas with reserve capacity will not be easily available to the Southwest because of the great distances involved and the resultant lack of facilities to provide assistance from outside areas.

We clearly do not have to wait a generation, a decade, or even a year. The energy–environment dilemma is with us now and is growing. Perhaps it is only a precursor of many worse problems which man is bringing on himself as he increasingly utilizes the resources of the planet. We prefer to believe that it is only a symptom of a disease which can be mitigated in the shorter run and hopefully cured in the longer run by adequate and prompt treatment. That belief has led us to undertake this study.

THE PROGRAM ON POWER PLANT SITING

ORIGIN AND SCOPE OF THE PROGRAM

The program on power plant siting was established as part of the National Academy of Engineering's effort to respond to the total energy problem of the nation. In 1970, the National Academy of Engineering Committee on Engineering Aspects of Environmental Quality * commissioned a modest effort to define the scope and to outline the conduct of a power plant siting study. This preliminary planning phase was conducted at Massachusetts Institute of Technology during the summer of 1970 by a group which included young and highly motivated scientists, engineers, and one lawyer. The report of this group contained a compilation of much material relevant to the problem and a good bibliography; it laid down in general outline, if not in precise detail and timing, the program that has actually been pursued by the National Academy of Engineering.

Power plant siting was accepted initially as the scope of the study because the siting option provides the most immediate and obvious recourse of the designer of a new generating facility who wishes to protect or control the environment. Environmental control can be facilitated simply by siting a new power plant in an area where its undesirable influences are least evident and where space and distance will dilute or attenuate objectionable by-products.

Although the scope of the work was initially confined to siting, it soon became clear that the energy–environment dilemma could not be resolved by siting techniques

alone. An effective program could not be carried out properly without a wide-ranging consideration of many factors and the participation of individuals with many viewpoints and expert knowledge in many fields.

ORGANIZATION OF THE PROGRAM AND COMMITTEE ON POWER PLANT SITING

Overall planning and coordination of the program was performed by a Steering Group. This group was selected to the extent possible to be representative of environmentalist and conservation interests, of utility viewpoints, and to include "independent" but expert individuals who could help in various ways to advance the work of the committee. The Steering Group contained representatives of government, of private industry, and of the universities. Its members shared a common goal; to do what they could to see that the electrical energy–environmental crisis would be resolved in the most expeditious and statesmanlike manner possible.

A number of more specialized working groups were established by the Steering Group to enable the program to cover the wide range of technical and specialized knowledge relevant to power plant siting. Each of these groups was established with the same criteria of balance and of motivation as the Steering Group itself.

Since environmental factors affecting power plant siting choices are the least well established and understood elements which must enter the site selection process, a set of four working groups was selected to consider these environmental factors from several points of view. These groups were concerned respectively with Air, Water, Radiation, and Appearance and Land Use.

---

* Renamed in 1971 "The Committee on Environmental Engineering."

A "Systems Engineering" working group was also established to investigate specifically the possibilities for a rational and systematic approach to the siting of new electric power generating facilities. "Systems engineering," while rarely and certainly not in this instance an exactly prescribed process, is an essential part of comprehensive engineering decision making and design. Here the words "systems engineering" have been used to describe the procedure whereby all pertinent input into a problem is accumulated and assigned its proper priority or hierarchy. After giving due weight to that priority or hierarchy, a solution to the problem is produced representing an optimum arrangement or a balance of all pertinent factors. It is a process that requires the careful analysis of technical, environmental, and governmental and public approval requirements. It also requires a creative synthesis of combinations of technological resources to meet these requirements, the careful analysis of these combinations to make sure that they do indeed meet requirements, and the selection of an alternative which provides the optimum arrangement.

A proper responsibility for any committee is to determine whether it really has a problem. The suggestion has been made in a number of quarters that there need be no environmental–energy conflict, that we should simply diminish the rate of growth of consumption of electric energy. This is a serious suggestion and indeed must be considered seriously and balanced with the possible benefits which higher rates of growth might bring. The question which the Economics and Energy Growth working group had to face is whether it might be possible to diminish the siting problem substantially by arranging matters so that fewer new sites would be necessary. As the report of that group makes clear, the limiting concept of "zero growth" appears neither politically nor economically feasible. Otherwise the remainder of the program might not have been necessary.

The Steering and Working Groups together have come to be known as the Committee on Power Plant Siting (COPPS).

A number of individuals not involved specifically with any committee activity also took part in the program at one time or another, particularly in connection with the Forum which was conducted in Washington for several days during March, 1971. (See the Appendix.)

The program was staffed by a small but dedicated full-time group in the Academy offices in Washington, augmented by the general staff of the National Academy of Engineering.

Fundamental Premise of the Program

The power plant siting program was self-generated by the National Academy of Engineering in response to a need expressed by the Committee on Engineering Aspects of Environmental Quality. The fundamental premise which motivated the program is that good engineering can do much to mitigate the conflict, more apparent than real, between society's demands for electric energy and its wish to preserve the environment. Engineers tend to feel that the present dilemma is not an inherent one and that much can be done toward its resolution by good engineering. The engineering process, to be successful, must, of course, use as inputs all of the requirements which must be met, both environmental and in terms of power supply and distribution. Expert knowledge from biologists, ecologists, health scientists, and many others must be used in the engineering process.

Since the imposition of new design constraints based upon environmental requirements may eliminate certain economical design options, the engineering path to the resolution of the conflict will probably result in a more expensive design than one which simply cuts the Gordian knot by a negotiated compromise between sacrifice of environmental objectives and sacrifice of power capacity. In the short run such compromises may indeed be necessary, but in the longer run the American public has always been willing to pay for what it wants. As technology advances, environmental impacts will have been greatly reduced in comparison with the additional expenditure required.

In undertaking this program, the Committee on Power Plant Siting recognizes that engineering alone can certainly not develop all of the answers. Not all conflicts can be resolved by technology alone. Furthermore, much research and development will be needed to augment the abilities of engineers to contribute in this field. Many other professions and processes, including those of law, education, public discussion, and politics will also be required. It is nevertheless clear that the process of resolution of the conflict must obviously be consistent with the possibilities of nature and economics, in other words, on matters intrinsic to engineering.

The Major Thrusts of the Program

The major thrusts of the program related the following efforts to

1. Identify those conflicts, issues, or questions in power plant siting where good engineering can help to
   a. Avoid adverse effects on the environment by proper and timely consideration of environmental impact
   b. Take due account of environmental science and value judgments in the engineering process

2. Appraise the adequacy of existing or proposed approaches to power plant siting by giving attention to

   a. Existing methods of setting environmental standards

   b. Technological methods for controlling environmental impacts

   c. Promising methods of systems analysis and synthesis

3. Identify those areas where further study, research or development seem likely to improve methods for power plant siting

These tasks were selected as appropriate by the participants in view of the constraints imposed upon such a program by short time and the intermittent activities of committee members, who are selected on the basis of their familiarity with areas relevant to the study. Programs of this nature must succeed primarily by an exchange of ideas and by arriving at consensus in a relatively short time. This type of committee activity has in the past proved effective in defining and appraising problems, evaluating alternative solutions, and determining desirable directions for future work.

The Forum

As part of the program on power plant siting a forum was held on March 16–19, 1971. This forum served both as an end in itself and as an input to the work of the remainder of the program. More explicitly, the purposes of the forum were as follows:

1. To give high visibility to current authoritative appraisals of existing and prospective factors and methods in power plant siting including relevant environmental standards, design techniques, economic factors, legal and licensing procedures, and system design methods.

2. To identify areas of agreement and disagreement established on approaches to power plant siting.

3. To encourage thought and discussion that could lead to further improvements.

4. To provide inputs for subsequent deliberations on power plant siting and environmental problems.

A brief account of this forum is contained in the Appendix.

Funding of the Program

The program was supported by generous grants from the National Science Foundation, from the Atomic Energy Commission, and from the government- and investor-owned electric utilities. Conservationists and environmental groups supported the program liberally through participation in the work of the program, as did indeed the other sponsors, but were, as organizations supported by gifts, unable to make financial contributions to the program.

Reports of the Program

It was decided that the results of the program would be reported in two and possibly three forms. In one form the reports of the steering committee, the working groups, and a number of relevant appendixes will be bound together. In another briefer volume, the report of the steering committee and the summary and conclusions of each of the working groups will be bound together with a limited amount of supplementary material. Finally, if suitable funding can be found, a second forum will be conducted during which the major participants in the program will discuss their conclusions with a widely representative group of interested individuals.

3   CONCLUSIONS AND RECOMMENDATIONS

This chapter contains the conclusions and recommendations of the Steering Group of the Committee on Power Plant Siting. These conclusions and recommendations were formulated by the Executive Committee of the Steering Group and were based upon inputs received during the Forum held on March 16–19, 1971, upon a substantial amount of published background material, some of which is referred to in the bibliography, and of course upon the individual knowledge and experience of the Steering Group members.

GENERAL

Conclusions

1. There is a substantial and growing crisis in the supply of electric power in the United States. This crisis results to a considerable extent from the conflict between society's requirements for more electric power and society's requirements for environmental protection and resource conservation. This conflict is not confined to electric power, but the electric power dilemma is sharper, more immediate, and more publicized than other aspects of this conflict.

2. While maintaining engineering integrity and economic feasibility, power supply decisions must accord environmental objectives coordinate importance and, as a minimum, the construction and operation of generation and transmission facilities should make every effort to avoid substantial irreversible damage to the environment. The National Environmental Policy Act (NEPA) and the recent Calvert Cliffs * decision of the courts have the effect of requiring adherence to these principles by the federal agencies involved in power plant siting.

3. Processes and information now being used to resolve the electric power conflict are inadequate.

    a. As is discussed in the report of Working Group III, stopping or substantially delaying construction of new power generating facilities is an inadequate solution. Power supplies are already insufficient in some areas, and society's demands are increasing.

    b. Approval for the construction and operation of new generating facilities normally requires too many steps, and licenses and permits from too many separate agencies, and therefore too much time. Approval for transmission meets similar obstacles in many areas.

    c. The adversary process in the courts often requires too much time and in addition is inadequate to cope with the complexities of the engineering and ecological problems involved.

    d. Lack of effective channels for consideration of public interest aspects of facility development and lack of orderly public decision-making are prejudicial to—sometimes foreclosing—effective effort by environmentalists in behalf of their concerns, while at the same time absence of those channels shackles utilities with serious delays, uncertainties, and costly design revisions.

    e. There has not been enough research and development toward better technological methods for supplying adequate power without unacceptable adverse

* ——— U.S. App. D.C. ——— Nos. 24839 and 24871 decided July 23, 1971.

effects on the environment or to provide a basis for rational setting of environmental standards. This lack arises partly because the power industry is highly fragmented, partly because investor-owned utilities have in the past not always been permitted to recover the costs of research and development, and partly because widespread intense concern about the environment is relatively new.

4. It is incumbent upon society to find out how to accommodate growth in electric power supply without substantial adverse environmental effects. There are approaches to resolving this problem. They may be divided roughly into the short and long range.

a. The short-range approaches can do much to mitigate but certainly not completely solve the problem during the next decade or so. Such approaches require a restructuring of the processes by which decisions concerning new generating capacity are made, including the decisions concerning power plant siting, and also a larger commitment to the development and application of technological fixes.

b. A considerably expanded and adequately coordinated research and development program can, in due course, lead to a much more complete resolution of conflict. As is demonstrated in the COPPS Working Group reports, there is now a substantial fund of science and technology applicable to power plant siting problems. However, much of this material must be better organized for effective application, and there also is great need to enlarge basic knowledge. Furthermore, because the time from the availability of proven new technology to the actual operation of generating facilities which utilize it is normally about a decade, research and development cannot be expected to produce many short-term improvements.

5. The short-term amelioration of the conflict will cost much money and will not completely avoid damages to the environment. With application of new technology, proper long-range planning and coordination, including the internalization of all costs of production, the concurrent activities of energy production and environmental goals can come closer to reality. The additional cost will necessarily be high, yet they must be economically feasible in relation to benefits (see recommendation 8, p. 12).

6. Good and comprehensive engineering can contribute much to the development of better methods of easing the energy–environment crisis. However, planning electric facilities and systems has to deal with the factors affecting engineering integrity, safety, dependability and efficiency, environmental constraints, economic and financial considerations, and all of these in relation to public preferences. Planning is greatly complicated by the long lead time for providing capacity to meet future loads, by changing technology, and by

possible changes in the magnitude and character of demands for electric service. Furthermore, as is demonstrated in the report of Working Group II, many other processes must also be involved, including those of law, politics, education, and public information. Engineering properly includes collaboration and interaction with those other professions in dealing with the many-faceted problems of power plant siting. Practicing engineers acquire competence for that through experience, and appropriate training to that end is now generally included in the curricula of engineering colleges.

Recommendations

1. Major changes must be made in society's handling of the power plant process.

2. A new certifying system is required to take all interests into account and to arrive at acceptable power plant siting decisions in a timely fashion. This subject is discussed further in the section on Procedures and Organization, below.

3. A substantially larger and well-planned and coordinated research and development program should be undertaken. (See section on Research and Development.)

4. The power plant siting process must involve comprehensive engineering. Such engineering must take into account not only the possibilities of more advanced technology, but also constraints based upon the best available information about possible environmental effects. Engineering, which includes a consideration of economic factors, must also be integrated in decision-making with the processes of law and public participation. A primary responsibility of the engineer is to synthesize viable solutions to the problems of producing new electric power capacity without unacceptable environmental impacts.

5. Because of the significant engineering component involved in the resolution of the energy–environment dilemma, the National Academy of Engineering should take continuing interest in the problems of power plant siting as part of the Academy's concern for the total energy problem in the context of its public responsibilities. (See p. 17.)

PROCEDURES AND ORGANIZATION

Conclusions

1. There are many shortcomings in present procedures and organization related to obtaining approval for the siting, construction, and operation of new electric power generating facilities. Beyond the technological and economic difficulties which electric power utilities have faced for some time, the following aspects are involved:

a. There is a multiplicity of mutually independent agencies which must approve each new application.

b. State and federal standards are technically inadequate as a basis for planning and design. Environmental concerns and standards based upon them are changing rapidly, with acquisition of new information.

c. In many areas there is disagreement in value judgments. There is no accepted method for resolving these disagreements.

d. It takes years (from three to twenty) to bring new technology from conception to application.

f. At present, utility planners and even regulatory agencies have only very inadequate means for ascertaining the specific nature of public opposition until after initiation of adversary intervention. This deficiency can greatly increase utility costs and forfeit environmental protection. Administrative procedures, if overly judicialized, are neither fast enough to settle the difficulties in a timely fashion nor competent to bring about a resolution of the technical difficulties. They can result in a settlement or a compromise between two opposing views, but they are not capable of meeting the requirements of both sides simultaneously.

2. Timely resolution of power plant siting problems can be effected only if all relevant factors, technological, economic, environmental, and public opinion, are considered in an integrated fashion from the beginning.

Recommendations

1. Agencies should be established to act with finality, subject to due process of law, on power plant siting issues through unified procedures that give comprehensive considerations to all factors relevant to the public interest and to the views of all concerned parties. This certification function should, as nearly as possible, be a "one-stop" procedure, and it should be exercised at the lowest governmental level that has adequate jurisdiction and that can marshall the required technical and other competence. In order to expedite decisions, the certifying agencies should specify as fully as possible the character and extent of the supportive information to be presented in advance by proponents and opponents.*

---

* Presentations made by panelists of Working Group II: Systems Approach to Power Plant Siting, at the National Academy of Engineering Committee on Power Plant Siting Forum, March 17, 1971, Washington, D.C.: "California's Approach to Siting Thermal Power Plants," by Paul L. Clifton, Chairman, Power Plant Siting Committee, California Resources Agency; "Power System Planning and Licensing of Thermal Plants," by Joseph C. Swidler, Chairman, New York State Public Service Commission; "Remarks on the Environmental Approach to Power Plant Siting," by Peter Borelli, Eastern Representative, Sierra Club; "Northern States Power Company's Approach to Open Planning," by Robert Poppe, Department of Environmental Affairs, Northern States Power Company, Minneapolis.

2. The establishment of such agencies will clearly require national and probably state legislative and administrative action. Specific recommendations concerning such action are beyond the proper scope of this study. However, it is clear that the new agencies would have certain characteristics.

a. They must achieve public acceptance and trust.

b. They must be in a position to understand and respond appropriately to the views and interests of all parties who are concerned with power plant decisions.

c. They must apply existing environmental standards, and have competence to understand ecological, technological, and economic factors relevant to power plant siting decisions.

3. It is clear that the agencies described above must be highly competent in a variety of fields. It is therefore essential that they be provided with adequate and diversified staffs. These agencies must clearly be competent in business, economics, power plant technology, environmental science, and standards. They must have, at the top level in each agency at least, the ability to make judicious decisions based upon a complex of arguments and factors, some of which will be intangible and nonquantifiable. In creating new agencies, the existing resources and competence should be utilized, and efficiency enhanced by eliminating overlapping.

4. To the maximum extent possible, the activities of the new agencies should be designed to employ processes of cooperation, systematic problem solving, and consensus seeking rather than adversary processes. It follows that the staff of the agencies must encourage communication among all parties interested in a new power plant on an informal but open basis for an adequate period prior to formal hearings and the formal decision. We recognize that these procedures in some instances may be impractical and that circumstances may be such that formal adversary proceedings may have to be introduced at an early stage.

5. Power plant siting decisions should be settled at the lowest governmental level that has adequate scope, namely, at the state level wherever possible. Since power supply problems and their environmental impact are often multistate in character, serious thought must be given concerning what agency would have jurisdiction in multistate cases. There are various precedents for the cooperative operation of agencies in affected states through consortia, compacts, and other mechanisms.

6. The utilities should be charged with responsibility for proposing solutions to each new requirement for power. Under this responsibility it will be up to the utilities to demonstrate that new power generating capacity is required; to consider alternative means by which this need can be met, and to propose one of them; to estimate the impact of the proposed new facility

upon the environment; and to demonstrate that this environmental impact is in conformity with standards.

7. From recommendation (6), it follows that the systems engineering activities of the power producing utilities must include considerations on environmental impact and environmental control in addition to the power producing technology and economics which have been necessary since the beginning of the electric power industry. (See report of Working Group II.)

8. Agencies responsible for regulating utilities' finances should acknowledge environmental protection expenditures as proper utility costs. The costs of environmental surveys, monitoring, and research as well as the costs of avoiding or ameliorating environmental damage are necessary costs of electric power, and they should be internalized in the rates for service. Such costs should be related to benefits, and be consistent with what the public is able and willing to pay.

9. The electrical power producing industry should continue to expand its processes of public information, and to the extent possible, of open discussion. The utilities should also expand their practices of working informally with the regulatory bodies and interested persons to include a substantial amount of discussion about factors relevant to a new proposed power plant site prior to the formal application and hearing.

10. We recommend that the process of environmental standard setting be strengthened by the institution of a network of monitoring stations. They should be established at power plant sites prior to construction and should be continued throughout operation at an appropriate level.

11. A national bank of information about environmental standards, environmental impact, and environmental management should be established (possibly under the auspices of a national coordinating group for research and development as suggested later).

12. Environmental standards and constraints placed upon new power plants should not be altered once commitments to major expenditures have been made, except by common consent of all concerned, or unless acute dangers can be demonstrated.

13. Legislative procedures should establish objectives or criteria. Standard setting should generally be an administrative procedure.

14. Environmental criteria and standards set upon power plants should specify the end results to be obtained but should not limit options by regulating how these results should be reached. In furtherance of this, present and prospective environmental standards should be examined to assure their technical validity and feasibility. (See recommendation 4, p. 18.)

15. Since informed public opinion will assist problem solving, programs of public information and education

about problems and progress in this area should be encouraged.

16. Control of environmental impacts which can have a damaging effect on human health should have the highest priority.

17. In making decisions which involve the acceptance of some environmental impact, substantial irreversible damage should be regarded as unacceptable. Where there is doubt concerning irreversible environmental effects, a conservative approach should be taken. This does not mean that extravagant claims of environmental hazard should be taken at face value.

18. As understanding of environmental effects and their control expands, increasing efforts at establishing rational bases for trade-offs between environmental impacts and dollars should be made. Evaluation of benefits should provide estimates of the costs of successive increments of power supply or environmental protection.

19. Because of the expectation that new technology and understanding will improve the ability to reduce adverse environmental impacts, engineering decisions should be made so that paths to more beneficial options in the future will be kept open.

20. Finally, since the power plant siting problem is only one part of a complex of many problems involving energy and the environment, we urge interested agencies, including the National Academy of Engineering, to continue the search for practical and rational solutions to energy and environmental problems.

PRIORITY RESEARCH AND DEVELOPMENT

The reports of the Working Groups provide much information with regard to needed research and development, and, based on that, this section presents general conclusions and recommendations regarding R&D organization and funding together with a concise summary of those research and development topics which appear to COPPS to have significant bearing on the power plant siting problem.

Character of Priority R&D Topics

We believe that R&D is needed over the whole range of knowledge to be used in setting standards for electric utilities and in making engineering decisions to minimize impact on the environment. Because we must rely on the scientific and engineering knowledge we already have for standards about to be set and for siting decisions of the next five years, the following recommendations are directed to problems that extend beyond that time span. It is urgent that research and development on these priority subjects be intensified or commenced now to provide for the future, particularly for the period five to twenty years from now.

Along with improvement of generation and transmission technology, R&D is needed to help preserve or improve the quality of air and water (river, lake and ocean), and the beauty/comeliness/human quality of landscape, seascape, and townscape. The range of such research and development is very great:

In size—from the breeder reactor to the water temperature tolerance of salmon fingerlings; from millions of gallons of warm water per minute which must be successfully mixed into ocean, lake, or river, to millionths of a gram of radioactive element which must be prevented from escaping into the surroundings.

In cost—from about $300 million per year now being spent for the breeder reactors urgently needed in the mid-1980's, to a score of grants for biological studies, each of which would be funded at 1/10,000 of that annual rate.

For our interdisciplinary Steering Committee to recommend an order of priorities over the whole range would have been difficult in the time available. There are complex interrelationships among the unknowns. For example, we don't know enough about the effect on human lungs and respiratory tract of SO_2 combined with the particles from fossil fuel stations. Nor do we know enough about the long-term effect on humans, animals, and plants of the radioactive emissions permitted nuclear power stations by the present AEC standards. If we knew more about these two subjects, future standards for fossil fuel and nuclear stations could be determined with greater wisdom. The costs of meeting these standards would be one factor affecting the choice of site and type of fuel. We need information on both these effluents, and it is therefore not useful to rate one as of higher priority than the other. While it is difficult to suggest R&D priorities for such widely separated fields, within each of the major research problems we need to decide where to place greater emphasis.

The effort and dollars to be spent in getting much of this information is truly very small in relation to the damage that could be caused to the quality of life by having the future standards too slack or the huge unnecessary cost to the electricity user of having them too tight. The whole budget for hundreds of such needed investigations disappears in comparison with the long-term cost to the public at large of making the wrong choices based on inadequate information. Thus, liberal funding without rigid priorities will be well repaid.

When we pass to the other end of the range of R&D to be done—breeder reactor, nuclear fusion, magnetohydrodynamics, and other methods—the dollar amounts are large for each program, and some hard choices

have already been made. But here again the information on which the choice is made is likely to be incomplete. Unexpected difficulties arising from engineering experience and even from unknowns in basic science have to be faced after years of work. We need a breeder reactor in less than twenty years—we want this to be a safe, reliable reactor with minimum environmental impact. The choice nationally and after years of work has been to give maximum backing to the liquid sodium fast breeder. To be ready in time to alleviate the fuel and energy problems, it needs all the money it is getting annually and probably more. There are problems and delays in developing this fast breeder, and we should be prepared for more of these in the next fifteen years.

To what extent can we—should we—double up on this development and fund a parallel major development of one of the alternative breeder reactor types? The savings realizable by having the freedom to choose the more desirable type could far outweigh the extra cost of the dual program. The cost of having *no* workable breeder would be high indeed.

The Committee on Power Plant Siting and its Working Groups, deeply conscious of responsibilities to the quality of life, believe in the necessity and the payoff for human health, happiness, and spirit of the R&D it recommends. Much of this R&D is already being done by innumerable groups in government, universities, and industry. Cooperation and coordination is vital to avoid waste of our manpower and science resources. In the present mood and practice of reduced funding, particularly in many university programs, we speak out emphatically in favor of government, foundation and industry grants to highly qualified investigators capable of doing excellent work in any and all of the R&D lines which will assist in wiser power plant siting decisions.

Organization, Coordination, and Funding

Conclusions

1. Development of the needed competence in electric power generation and transmission and related environmental protection requires an enlarged program of research and development that will:

● Extend and deepen knowledge of fundamental processes;

● Organize relevant scientific and technical knowledge and other information to facilitate its availability and use; and

● Develop applications of scientific and technical advances to the stage of reduction to practice.

2. The existing research and development in this field is inadequate because:

● The environmental crisis has only recently been accorded the priority which it now has;

• Costs of research and development performed by the investor-owned electric utilities have in some cases in the past been difficult to recover through the rates the utilities were permitted to charge; and

• The electric power utilities have been highly fragmented partly as a result of governmental policy.

3. Relevant research is now being done by the government, particularly the Atomic Energy Commission, by the power equipment manufacturers, and in the environmental fields, by a variety of institutions. However, it is not coordinated, and the results are not available in any one place or easily accessible.

4. Funding of research and development in this general area is inadequate.

Recommendations

1. The plurality of effort and enterprise that is a significant strength of this country should be preserved in the research and development programs relevant to bulk electric power supply, including power plant siting. No competent organization, whether government, investor-owned, or academic, should be excluded as a possible agent for the performance of needed research and development.

2. An organization should be established which has the responsibility to coordinate, but not the authority to control, the research and development programs. This organization could be similar to that described by the Electric Research Council * with public and government representatives. In making this recommendation we recognize that other coordinating mechanisms may be necessary but urge strongly that overall coordination be accomplished.

3. Basic functions of the above organization in R&D would be to:

• Secure agreement on the priority of R&D needs;

• Secure and disseminate information about ongoing and projected R&D programs;

• Identify unfilled gaps and arrange for filling them.

4. Relevant research about the environment should be coordinated and supported with the rest of the program as discussed below.

5. The responsibility for coordinating the research and development program should carry with it the responsibility for assuring that information relevant to the energy–environmental crisis be stored, organized, and made readily accessible.

6. Since generation of electric power is by nature not competitive in the commercial sense, COPPS recom-

mends that the results of the research and development programs and the resources of the information bank recommended in (5) above be made available to all who have any legitimate need. This would include the investor-owned and publicly owned utility industry, the suppliers of fuel and equipment, federal, state and local government agencies responsible for energy and environmental matters, and environmentalist and conservation groups.

7. Funding of the research, development, and information handling activities above identified should be accomplished as a combination of:

• Support at an appropriately increased level by those federal and other public agencies concerned with energy or the environment.

• Increase in direct funding by the utilities. This should be encouraged by the passage of legislation which permits these utilities to recover such expense.

Priority R&D—General Topics

A greatly expanded research and development program should be mounted to provide the scientific and technical basis for improvement of practice in the near term, and also to contribute to major advances in the longer term.

Near-Term Subjects

1. Reduction of SO_2, NO_x, and particulate matter from fossil-fuel stacks.

2. Gasification of coal as a means of reducing sulfur and making the vast coal resources available in a clean form.

3. Increased knowledge of pollution impacts on public health.

4. Management of rejected heat.

5. Increased knowledge of aquatic ecosystem tolerances.

6. Development of the breeder reactor, including greater support to an alternative to the liquid sodium fast breeder.

7. Comprehensive baseline surveys of the receiving media should be made for proposed power plants. Predictive environmental models should be constructed, and monitoring should be continued through several operational years.

8. Mechanisms should be established for improved exchange of scientific and technical information, including a monitoring system of air, water, food supply, and public health, plus systematic analysis of the data collected. This should be of national and later international extent.

---

* Electric Research Council. *Electric Utilities Industry Research and Development Goals Through the Year 2000.* Report of the Research and Development Task Force of the Electric Research Council, June 1971. (ERC Pub. #1-71) pp. 177 *ff.*

9. Development of more economical methods of high-voltage underground transmission.

Longer-Term Subjects

1. Development of offshore siting of nuclear power plants.

2. Investigation of new approaches to power plant siting, including underground generation and totally enclosed switching stations.

3. Research and development of nuclear fusion technology.

4. Research on new methods of energy conversion.

5. Investigation of the interaction of power plant siting with broad national energy and land use planning policy aims.

Priority R&D—Specific Topics

Specific R&D recommendations made both by the Steering and Working Groups are listed under the following headings:

- Those pertaining to general technology and environmental impacts
- Those concerned with effluents into the air, including radiation
- Those concerned with effluents into the water (rivers, lakes, groundwater, ocean) and including power station radiation that ends up in water
- Fuel resources and their best uses
- Use of the land, the sea and river coasts and the aesthetics of buildings, towers, transmission lines, and the noise associated therewith
- Energy and economic growth

In separating the recommendations under these headings, the Steering Group has been impressed by the number of interrelated subjects. For example, there is both conflict and synergism between the best use of fuel and the preservation of clean air and clean water.

General Technology and Environmental Impacts

1. Development of more efficient generating processes, including gas turbine topping cycles; reduction of water transit time in condensers; improvement of boiler furnace design and combustion control including fluidized-bed boiler furnaces, better engineering surfaces to reduce heavy metals in discharge water, and development of environmentally inert processes and materials such as nonfouling surfaces.

2. Before each major new power facility is constructed, reconnaissance surveys and qualitative descriptions of the biotic community and the physical environment should be made, and after it is operating, environmental and ecological monitoring should be continued for an appropriate time; this should lead to composition of ecological maps for judging the suitability of sites and the probable effect of power facilities on the environment. [See also recommendation (3) under Effluents into the Water, p. 16.]

3. Data acquisition, research, and development work are necessary in order to make it possible to feed a complex water problem into a computer for solution and also be able to retrieve all significant information; this should include modeling research to determine the thermal residual of highly diluted effluents with very low temperature difference; *viz.* in large lakes with low natural circulation.

4. Research should consider as a system the thermal discharge to the ecosphere as a whole.

5. Research should investigate the role of vegetation in recycling elements and as temporary sinks for pollutants; this should include the basis for differences in plant susceptibility to stack emissions and possible genetic influences on biological systems of repeated exposures over many generations.

Effluents into the Air

1. Determination of whether the "safe" long-term radioactive dose for the human population has a threshold, and whether there is an adverse effect which is linear from zero dose. (This may affect nuclear plant siting.)

2. Investigation of the health effects of oxides of nitrogen, ozone and its satellites.

3. Investigation of the effects on human performance and behavior of carbon monoxide in air, particularly the importance of possible temporary impairment of function as a result of carbon monoxide in urban atmospheres especially in relation to use of gas turbines or gas turbine topping cycles.

4. Investigation of the necessary standard for SO + particulates and for other air-polluting combinations that appear to have greater effect on human respiratory tissue than has either one alone.

5. More information is needed regarding the emission performance of power generating gas turbines, particularly with respect to oxides of nitrogen and aldehydes.

6. More data should be secured on stack emissions and their effects, including the types of emissions and fluctuations in time and with distance, the chemical and physical alteration of pollutants after emission, and the effect of rain, condensation, gravitational fall, and radioactive decay on the residence times in the atmosphere of critical air pollutants.

7. Meteorological data should be secured regarding:
- Wind and temperature characteristics at research and power sites, and regional dispersion and sinks;
- Behavior of cooling-tower plumes;
- Local weather and climate modification by atmospheric pollutants from power plants, the effect of cooling-tower plumes on stack height effectiveness, and effects on regional climatology of water vapor and heat releases.

8. At some sites, we need R&D to meet economically the recent SO_2 standards with available fuels and to deal with the sulfur-loaded absorbents used in stack removal of SO_2.

9. Improved fossil-fuel generation technology is needed for
- reduction of NO_x at existing and future plants, including lower-temperature combustion, wet and dry scrubbers, boiler modifications, and additives; and
- reduction of particulates (some of the present electrostatic precipitator methods are not working well with low-sulfur coal, and few existing precipitators are effective on the smaller particles of a few microns diameter and less).

10. Improved nuclear generation technology is needed for control of radioactive waste releases during plant operation and during fuel reprocessing and for protection from accidents.

11. Development of improved condenser cooling systems, especially more flexible supplementary cooling systems for low stream flow situations, including alternatives such as dry cooling towers and related heat transfer methods, and reduction of cooling tower drift, blowdown, and mist, for both fresh and saline water.

12. Improved removal of sulfur prior to combustion, by coal, lignite, and shale ore gasification and other means, and also control of particulates (fly ash), SO, and NO_x emissions from gas turbines (minimizing smoke and reducing NO_x conflict to some extent).

Effluents into the Water

1. Data are needed regarding
- How much heating of bodies of water can be tolerated by ecosystems and by economic and amenity uses;
- Effect of thermal plumes on fish, plankton, and other marine organisms;
- The impact of the thermal, chemical, and biocide content of cooling tower blowdown.

2. Better understanding is needed regarding
- Life cycles and migrations of estuarine and offshore fauna;
- Behavioral responses of mobile aquatic species to thermal and chemical loadings;

- The significance of daily and seasonal fluctuations of temperature and dissolved oxygen, especially in relation to maintenance of populations of desired species.

3. Regional ecological surveys should be made of major bodies of water to determine the species composition, disposition, and ecology of the biota, including migration pathways, spawning grounds, etc., and the commercial and recreational uses over the wide areas considered for siting.

4. Investigation should be made of the feasibility of systems of canals of large size, like elongated cooling ponds, to take the warm water from a town site to a point where it can be discharged into a larger body of water with ecological safety.

5. Improved design should be developed for intake devices and systems to minimize the number of organisms drawn into the circulating and makeup water of all types of power plants, and revolving screens for water intake structures should be improved.

6. Methods should be developed for disposal of condensate from geothermal systems which do not recycle all of the water used in the plant steam cycle.

Fuel Resources and Their Best Use

1. Further improvements should be sought in present methods of using the heat energy in our fuels, including gas turbine topping cycles, increased working temperatures, etc.

2. The liquid metal fast breeder reactor, at present receiving top priority, should be supplemented with more backing for one of the possible alternatives. Success with either, leading to a sufficient number of breeder reactors in power plant operation, would extend the useful burning of our uranium fuels (by a factor of 25 or more), greatly reduce the problem of storing radioactive material, lessen the thermal waste heat into the ecosphere (compared with the lower temperature nonbreeders of equal electric output), and reduce the dependence on fossil fuels and thus avoid increase of chemicals and ash added to the ecosphere.

3. Development of fusion reactor technology requires longer-term R&D, but that is well justified because the heavy water in the ocean is an almost unlimited energy source, and furthermore, it is expected that the reactor will be proven to be the cleanest source of power—with minimal chemical or radioactive effluents.

4. Exotic energy sources should be investigated: geothermal energy with further development and study could become a useful supplementary source in some regions; and solar, tidal, and wind energy are available if we know how to harness and convert them with greater efficiency. In some regions of the world this is

already being done; in others, it may become practical and necessary.

5. Beneficial uses of waste heat: Although the amount of heat leaving the steam plants is very great, its temperature is little above that of the cooling water intake while the volume of water is enormous (e.g., 360,000 gallons per minute for a single power plant of modern size). Therefore use of this heat by pumping it through a multitude of ever-dividing pipes has not been practicable. Nevertheless, the various working groups have recommended investigation of possibilities for salvaging rejected heat and developing them as actively as seems promising of feasibility. Some possible applications suggested for investigation are back pressure steam for industrial processing and absorption air conditioning, extension of agricultural growing season, elimination of fog pockets and ice patches on highways, and supplemental residential heating. The feasibility of such applications might be enhanced if considered in the context of communities designed around a power station. Controlled thermal discharges into water might be a positive commercial advantage to some fish and wildlife species. In combination with icebreakers, thermal discharges might extend winter navigation. The country's air-conditioning needs are greatest just when the rivers are lowest and are not fully available for direct power station cooling. How can we use the low-quality heat energy which must be extracted from the steam at the low pressure end of the turbine to cool the surrounding homes and offices?

Land Use, Aesthetics, and Noise

1. Metal-enclosed compact gas-insulated switching stations, underground nuclear power sites, and underground transmission of power can significantly enlarge multiple uses of power facility sites as well as improve their appearance. Mine mouth or underground coal gasification would protect urban landscapes.

2. Modeling and other methods should be developed to give full weight to the best long-term public interest in power facility sites, multiple use of them, and the aesthetics of plants, substations, transmission lines, along with public safety, and use of fuel, water, air and land resources, including mining and storage of fuel.

Energy and Economic Growth

1. Economic studies are needed to provide better understanding of the economic impact of higher energy prices, price elasticity, nonprice effects on demand, and the environmental effects of the shifts in consumption patterns.

2. Studies are needed also to ascertain the opportunities for more efficient energy utilization.

3. We need research in the social sciences on methods to assure a useful feedback in national attitudes and values for rational decision-making, and to provide public understanding of the consequences of rising power needs.

Final Comment

Our R&D recommendations would not be complete without reference to the parallel effort of the R&D Goals Task Force of the Electric Research Council, which first met in October 1970 and whose report was released in June 1971.* The people who have worked on and been responsible for the findings of the ERC report have formulated a proposed program of comprehensive scope. They have knowledgeably faced the important but difficult task of assigning relative priorities within the fields of R&D, and they have recommended dollar expenditure for each suggested program for each year in detail for the next five years and in outline through the end of the century. We commend that Task Force and its sponsors for making a major contribution to resolution of the electric energy–environmental quality dilemma.

Even a cursory reading of the report of the Working Groups of the NAE Committee on Power Plant Siting will leave an unbiased reader in no doubt of the existence of many and very serious problems, but equally with the strong conviction that the intellectual and scientific resources of this country are well able to find satisfactory solutions, provided these resources can be put to work on the problems soon and in a continuing manner.

FUTURE NATIONAL ACADEMY OF ENGINEERING ACTIVITIES

Conclusions

1. Electric energy is only one form of energy, and power plant siting is thus only part of the larger whole. These wider questions deserve further exploration by the Academy.

2. The scope, magnitude, and duration of this program has left many questions unexplored and some others inadequately answered. Nevertheless, COPPS feels that the time has come for a recapitulation and report of its progress and an end of its work.

---

* *Electric Utilities Industry Research and Development Goals Through the Year 2000. Op. cit.*

Recommendations

In view of the broader implications of this program and the urgency of the energy–environmental crisis in the area of power plant siting, we recommend that

1. The results, conclusions, and recommendations achieved to date be published and widely disseminated.

2. If adequate funding can be found, a second Forum be held early in 1972. A purpose of this Forum would be to present the output of the program to legislators, officials of regulating agencies, utility representatives, environmentalists, the press, and, in fact, representatives of all institutions which have active interest in the field. Another purpose of this Forum would be to discuss how one-stop agencies can be established and to explore interdisciplinary cooperation for this effort. Attention should also be given to the operating characteristics of such new agencies.

3. The Committee on Power Plant Siting should be terminated after submission of its report.

4. The National Academy of Engineering should consider examination of the scientific bases of environmental standards and their usability for application.

5. The National Academies of Engineering and of Sciences, in fulfillment of their public responsibilities, should:

a. Continue to be responsive to requests from government for advice in the field of energy supplies and their relation to environmental quality.

b. Continue to provide leadership to the scientific and engineering professions in enhancing capability to deal with problems in that field.

c. Consider how best to contribute to development under credible auspices of public understanding of those problems through valid and readily comprehensible information relevant to the public concerns.

d. Seek to cooperate with professionals of other disciplines in developing specific proposals for implementing a consolidated certification procedure.

SELECTED BIBLIOGRAPHY

GENERAL REFERENCES

Abrahamson, Dean E. *Environmental Cost of Electric Power: a Scientists' Institute for Public Information Workbook.* New York: Scientists' Institute for Public Information, 1970. 36 pp.

Arthur D. Little, Inc. *Energy Policy Issues for the United States During the Seventies.* Prepared for the National Energy Forum. Sponsored by the National Division, The United States National Committee, World Energy Conference. Cambridge, Massachusetts, July, 1971. 78 pp.

Booz, Allen & Hamilton, Inc. *An Inventory of Energy Research: A Report Prepared for the National Science Foundation.* Washington, D.C., October 15, 1971. 450 pp.

Calvert Cliffs' Coordinating Committee et al., *Petitioners,* vs. *United States Atomic Energy Commission and United States of America, Respondents.* ——— U.S. App. D.C. ——— Nos. 24839 and 24871, decided July 23, 1971.

Electric Research Council. *Electric Utilities Industry Research and Development Goals Through the Year 2000.* Report of the Research and Development Task Force of the Electric Research Council. Pub. #1-71. June, 1971. 176 pp.

Electric Utility Industry Task Force on Environment. *The Electric Utility Industry and the Environment, a Report to the Citizens Advisory Committee on Recreation and Natural Beauty.* Washington, D.C.: The Citizens Advisory Committee on Environmental Quality, 1968. 105 pp.

Environmental Quality Laboratory, California Institute of Technology. *People, Power, Pollution: Environmental and Public Interest Aspects of Electric Power Plant Siting.* September 1, 1971.

Federal Power Commission. *National Power Survey: A Report by the Federal Power Commission, 1964.* Washington, D.C.: U.S. Government Printing Office, 1964. 2 Vol. 719 pp.

Federal Power Commission. *Prevention of Power Failures: Vol. II, Reliability of Electric Bulk Power Supply.* Washington, D.C.: U.S. Government Printing Office, June 1967. 138 pp.

Office of Science and Technology Energy Policy Staff. *Considerations Affecting Steam Power Plant Site Selection.* Washington, D.C.: U.S. Government Printing Office, December 1968. 132 pp.

Office of Science and Technology Energy Policy Staff. *Electric Power and the Environment.* Washington, D.C.: U.S. Government Printing Office, August 1970. 71 pp.

Plant Siting Task Force of the Edison Electric Institute Committee on Environment. *Major Electric Power Facilities and the Environment.* New York: Edison Electric Institute, February 1970. 31 pp.

Resources for the Future. *Research Needs: Consumption, Production, Technology, Environmental Effects, Policy Issues.* A report to the National Science Foundation written in cooperation with the MIT Environmental Laboratory, October 1971.

Risser, Hubert E. "Power and the Environment, A Potential Crisis in Energy Supply." *Environmental Geology Notes.* No. 40, December 1970. Urbana, Illinois: Illinois State Geological Survey. 47 pp.

Schurr, S., and Netschert, B. *Energy and the American Economy 1950-1975.* A report published for Resources for the Future. Baltimore: Johns Hopkins Press, 1960. 774 pp.

Scientific American. Issue on Energy and Power. Vol. 224, No. 3. September, 1971.

Sillin, Lelan F., Jr. *A Project for Prometheus.* Hartford, Connecticut: Public Affairs Department, Northeast Utilities, 1971 (address delivered at the Engineering Foundation Conference, Andover, New Hampshire, August 16, 1971).

Singer, Dr. S. Fred. "Human Energy Production as a Process in the Biosphere." *Scientific American.* Vol. 223, September 1970. p. 174–190.

U.S. Congress, House, Committee on Interstate and Foreign Commerce. *Power Plant Siting and Environmental Protection.* Hearings before the Subcommittee on Communications and Power. 92nd Congress. 1st Session. Washington, D.C.: U.S. Government Printing Office, 1971.

U.S. Congress, Joint Economic Committee. *The Economy, Energy and the Environment.* A report prepared by the Environmental Policy Division of the Legislative Reference Service. Washington, D.C.: U.S. Government Printing Office, September 1, 1970. 131 pp.

U.S. Congress, Senate, Committee on Government Operations. *The Intergovernmental Coordination of Power Development and Environmental Protection Act.* Hearings before the Subcommittee on Intergovernmental Operations. 91st Congress. 2nd Session. Washington, D.C.: U.S. Government Printing Office, 1970. 882 pp.

WATER

Clark, J. "Thermal Pollution and Aquatic Life." *Scientific American.* Vol. 220, March 1969. p. 18

Eisenbud, Merril, and George Gleason. *Electric Power and Thermal Discharges.* New York: Gordon and Breach, 1969. 423 pp.

U.S. Department of the Interior, Federal Water Pollution Control Administration. *Industrial Waste Guide on Thermal Pollution.* Corvallis: Oregon Department of the Interior Federal Water Pollution Control Administration, Pacific Northwest Water Laboratory, September 1968. 112 pp.

AIR

"Air Pollution: Power Magazine Special Report." New York: *Power Magazine,* n.d. 47 pp.

Nadler, Allen A., *et al. Air Pollution, A Scientists' Institute for Public Information Workbook.* New York: Scientists' Institute for Public Information, 1970. 28 pp.

TRANSMISSION

U.S. Department of the Interior. *Environmental Criteria for Electric Transmission Systems.* Washington, D.C.: U.S. Government Printing Office, 1970. 56 pp.

RADIATION

Clopton, John C. [ed]. *Environmental Radioactivity: A Symposium Held at the Johns Hopkins University, January 19-20, 1970.* Baltimore: Department of Geography and Environmental Engineering, The Johns Hopkins University, 1970. 272 pp.

Nelkin, Dorothy. *Nuclear Power and Its Critics: The Cayuga Lake Controversy.* Ithaca, New York: Cornell University Press (Science, Technology, and Society Series), 1971.

U.S. Congress, Joint Committee on Atomic Energy. *Selected Materials on Environmental Effects of Producing Electric Power.* 91st Congress. 1st Session. Washington, D.C.: U.S. Government Printing Office, August 1969. 553 pp.

OTHER REFERENCES

Federal Power Commission. *1970 National Power Survey.* Washington, D.C.: U.S. Government Printing Office, in press.

U.S. Congress, Senate, Committee on Interior and Insular Affairs. *A National Fuels and Energy Policy Study.* Hearings and Committee Prints pursuant to S. Res. 92-45. 92nd Congress, 1st Session. Washington, D.C.: U.S. Government Printing Office, 1971.

APPENDIX

FIRST FORUM ON POWER PLANT SITING*
March 16 and 17, 1971
Washington, D. C.

SUMMARY OF EVENTS

On March 16 and 17, 1971, the National Academy of Engineering Committee on Power Plant Siting (COPPS) convened the First Forum on Power Plant Siting, bringing together some 225 men and women knowledgeable and active in electric energy–environmental quality problems. Participants represented a broad range of interests and included utility executives, lawyers, environmental and social scientists, legislators, regulatory officials, and officers of environmental protection and other citizen organizations from the United States, Canada, and other nations.

The Forum identified substantial areas of agreement as well as unresolved major issues, some of them intensely controversial, concerning power plant siting. Proven or promising approaches to resolving some of the problems and priority research needs for improving siting procedures were also delineated. Such approaches encompass a wide range of subjects, from economics and facility design to environmental standards and licensing procedures—and much of the research, it was agreed, should be multidisciplinary.

Interaction of highly qualified and experienced individuals characterized the Forum and yielded valuable information, ideas, and perspectives for the Working Groups concerned with the technical aspects of the COPPS report. These Forum inputs were especially useful to the Steering and Working Groups (which met for intensive working conferences for three days after

* Held under the auspices of the National Academy of Engineering Committee on Power Plant Siting in the Auditorium of the National Academy of Sciences building, Washington, D.C.

the Plenary Sessions) in focusing attention on the unresolved issues and the principal contending arguments that should be treated in the reports.

HIGHLIGHTS OF THE FORUM PROGRAM

In his welcoming address, NAE President Clarence Linder briefly discussed the committment of the National Academy of Engineering to contribute to resolution of urgent public issues that have strong engineering content. One aspect of this process is to provide rational approaches that can be the basis for responsible discussion and decision. To that end, President Linder affirmed, the report of the Committee on Power Plant Siting would recommend procedures for taking account of and evaluating the diverse factors involved, giving close attention to providing usable methodology for siting decisions.

A panel on "Environmental Concerns and National Energy Needs," organized by Mr. Frank M. Warren, President of Portland Electric Company and a member of COPPS Steering Group, offered the industry view of power plant siting problems and described the difficulties that utilities face in meeting environmental standards. Speakers cited the difficulties inherent in increasing capacity at a rate of 6–7% per year and the need to keep old facilities on line. The discussion dealt with the loss of capacity and diseconomies that resulted from retrofitting old equipment, the lack of conclusive data to justify some standards—standards for which proven equipment is often not available—and long delays in courts caused by environmental interveners that compound the problems. The panel called for a one-stop agency to replace the up to fifty agencies now having review over siting decisions and for more industry input into the setting of environmental standards so as to insure that there will be consideration of the technology available for compliance. Sharp challenges to some of these points were made during subsequent floor discussion.

"Protection of Environmental Quality" was the subject of an address by Mr. David Sive, an attorney noted for his representation of environmental groups. Mr. Sive outlined the rationale for utilization of judicial procedures in siting disputes. Utilities, he said, are not suited to consider, much less determine, many of the values inherent in decisions of where and whether to site a new power plant. He argued that the courts can and should make judgments on social values, citing the recent court decision barring completion of a new fossil fuel generating facility in the Los Angeles Basin on the grounds that the air quality of the basin was of paramount concern, rather than the public desire for more electricity. Mr. Sive pointed out the merit of adversary proceedings in exposing all relevant facts. However, he also advocated a one-stop licensing agency for power plant siting, provided that the agency is created so as to adequately protect environmental interests.

The luncheon speaker, Representative John P. Saylor (Pennsylvania, 22nd District), expressed congressional concern over inadequate investigation of sites for the approximately 250 new power stations projected as needed by 1990. Mr. Saylor also questioned the adequacy of present nuclear power plant design—an inadequacy evidenced by the requirement for federal insurance protection under the Price-Anderson amendment. Mr. Saylor concluded with a reminder that electric power generation and transmission is interstate in character, and so its regulation is a congressional concern.

Dr. William C. Ackermann (Chief, Illinois State Water Survey) led the Panel on "Energy–Environment Concerns," which assessed the relative magnitude of the environmental and economic impacts of large generating stations. These include both intense local effects of power generation and its more subtle, but highly significant, regional, national, and global implications. Major points discussed by the panel included the possible severity of local heat dissipation problems, the prospects for the several types of fuel supply, the need and prospects for new technology in the utilization and processing of fossil fuels, and the basis, in terms of new and expanding usage, of projections of future growth in the consumption of electric power. There were also com-

ments on the advantages, both to the public and to the utility, of early notice of proposed locations of power plants, and the difficulties inherent in trying to internalize the unknown or undefined environmental and aesthetic costs of generating electricity.

The third panel presented "Engineering Approaches to Power Plant Siting" and was chaired by Dr. Denis M. Robinson (Chairman, Marine Biological Laboratory at Woods Hole, and Vice-Chairman of COPPS). Some of the "technological fixes" which could be applied to electric generation and transmission to lessen their adverse environmental impacts were described. The political as well as scientific and engineering considerations essential for good engineering design were also discussed. One member, Dr. Glenn R. Hilst (Chairman, COPPS Working Group on Air Quality), stated that to design power plants intelligently, two things must first be decided: how much electricity and what environmental quality is desired. In summarizing the panel's discussion, Dr. Robinson concluded that although generation technology will not undergo radical change in the next 15 years, technology will be available to ameliorate its environmental effects, and that the hope for the future of electric power lies in new modes of electric generation and transmission.

In an after-dinner address, Commissioner James T. Ramey of the United States Atomic Energy Commission emphasized the view of the AEC of the necessity of considering all factors in evaluating the environmental impact of any means of generating electricity. He indicated that when waste disposal, mining, and transport of fuels are taken into account, as well as the plant itself, comparison of environmental effects often will be favorable to nuclear generation.

In a series of separate panel sessions during the second morning, Forum participants met with the six Working Groups. Diverse speakers dealt with a wide variety of topics related to each Working Group report. For example, Mr. Robert Poppe of the Northern States Power Company presented his company's open planning approach; Mr. Joseph Swidler and Mr. Paul Clifton (chairmen of the New York State Public Service Commission and the California Power Plant Siting Committee, respectively), explained the transmission and the generation siting procedures of their states to one group; Dr. Joseph Mihursky, Chairman of the Water Working Group, explained his development and use of thermobiotic predictive models for the Chesapeake Bay to a different session; Mr. Harvey Brush of the Bechtel Corporation described the future prospects for containment of radioactive wastes at another panel; and the Chairman of the Department of Landscape Architecture at the University of Wisconsin, Dr. Philip Lewis, described his team approach to land use planning. In all, about 25 papers were presented and discussed that morning.

In the final plenary meeting of the Forum, the chairman of each of the six Working Groups summarized his morning session, stating the objectives of his group in preparing its report. It was noted that while there was disagreement in some spheres within each Working Group, there was also significant concurrence. The general consensus, for example, was that a multidisciplinary approach was needed in order to consider adequately all the aspects of a power plant siting problem. The use of mathematical models and other new tools to predict effects was highly recommended by all groups, as was the internalization of the environmental costs of electric generation.

MAJOR POINTS AND CONCLUSIONS

Dr. W. Deming Lewis, COPPS Chairman, summarized the Forum as follows: Participants agreed that improved advance planning procedures are needed to assure proper power plant location and design. There was discussion of a national energy policy, but no agreement was reached. Although there was general recognition of the need to incorporate technological, ecological, and social preference information into the planning and the decision-making processes, no consensus developed regarding suitable organizational arrangements, especially as concerns citizen involvement in planning. The general opinion was that a one-stop decision-making body would be preferable to current decision-making procedures, that such a body must represent all interests, and that it must have at its disposal new means of evaluating the tradeoffs between power generation and environmental quality.

Participants also concurred on the need for greatly enlarged research to find or develop new and better tools and technology for gathering and using information in site planning and decision making. Hope was expressed that use of mathematical models can help us understand the functioning and mechanisms of the environment, that graphic techniques can help visualization of changes in aesthetic values, and that means for measuring public opinion to help comprehension of human environment will be developed. Dr. Robinson expressed the consensus of the Forum when he said ". . . most, if not all, of our power plant pollution problems, unlike many more stubborn problems that assail us in the political and social world, can be fixed, technologically. . . ."

Speakers at the Forum showed that, although generation and transmission technology will not change radically in the next 15 years, technological fixes are or will be available to alleviate the pollution problems caused by generation of electricity during that period. However, it was asserted that the hope for the future lies in new generation technologies, in breeder reactors, in fusion power, and in magnetohydrodynamics.

Forum participants also recognized an obligation to improve the environment and the necessity of bringing all relevant information to bear on decisions as to the kind of environment we want. There was emphasis on the importance of making this type of political decision before good engineering design can begin.

A major contribution of the Forum toward resolving power plant siting problems grew out of the interchange of ideas among the Forum participants. Environmentalists and utility executives, government officials and university professors became acquainted and engaged in discussion. A consequence of this interplay is increased awareness and understanding of diverse viewpoints, an essential for accommodation and collaboration.

FORUM REGISTRANTS

Dr. Dean E. Abrahamson, University of Minnesota, Minneapolis, Minnesota
Dr. William C. Ackermann, Illinois State Water Survey, Urbana, Illinois
Mr. James Adams, Pacific Gas and Electric Company, Emeryville, California
Mr. Rolf Andreasson, Embassy of Sweden, Washington, D.C.
Mr. Charles Arrington, Jr., New York Environmental Conservation Dept., Albany, New York
Mr. Gerald G. Bachman, Omaha Public Power District, Omaha, Nebraska
Mr. Walter H. Bailey, National Research Council, Washington, D.C.
Mr. John Barnby, IITRI, Washington, D.C.
Mr. H. C. Barnes, American Electric Power Service Corporation, New York, New York
Mr. Batiste, Washington, D.C.
Dr. Clifford K. Beck, Atomic Energy Commission, Washington, D.C.
Mr. George Beiser, Battelle Memorial Institute, Washington, D.C.
Mr. Martin Bernstein, National Economic Research Associates, Inc., New York, New York
Lt. Gen. A. W. Betts (Ret.), Southwest Research Institute, San Antonio, Texas
Mr. Franz W. Beyer, Duke Power Company, Charlotte, North Carolina
Mr. G. S. Bingham, Bonneville Power Administration, Portland, Oregon
Mrs. Kerstin B. Binns, National Academy of Engineering, Washington, D.C.
Mr. Julius Bleiweis, Northeast Power Co-ordinating Council
Mr. Bert J. Blewitt, Public Service Electric and Gas Company, Newark, New Jersey
Mr. John J. Boland, Johns Hopkins University, Baltimore, Maryland
Mr. Kurt Borchardt, House Committee on Interstate and Foreign Commerce, Washington, D.C.
Mr. Peter Borrelli, Sierra Club, New York, New York
Mr. Bruno Brodfeld, Stone and Webster Corporation, Boston, Massachusetts
Mr. Harry K. Borne, U.K. Scientific Mission, Washington, D.C.
Mr. Harvey F. Brush, Bechtel Corporation, Los Angeles, California
Mr. W. A. Bulter, Environmental Defense Fund, Washington, D.C.
Mr. R. J. Burger, National Academy of Engineering, Washington, D.C.
Mr. Frank Burggraf, Jr., New York Public Service Commission, Albany, New York
Mr. George S. Campbell, University of Connecticut, Storrs, Connecticut
Mr. Wayne A. Carbiener, Battelle Memorial Institute, Columbus, Ohio
Mr. R. A. Carpenter, Library of Congress, Washington, D.C.
Mr. J. A. Casazza, Public Service Electric and Gas Company, Newark, New Jersey
Mr. Edson G. Case, Atomic Energy Commission, Washington, D.C.
Mr. Henry P. Caulfield, Jr., Colorado State University, Fort Collins, Colorado
Mr. David F. Cavers, Harvard Law School, Cambridge, Massachusetts
Dr. Bernard Chew, Atomics International, North American Rockwell, Canoga Park, California
Mr. Roland C. Clement, National Audubon Society, New York, New York
Mr. Paul Clifton, Power Plant Siting Committee, Resources Agency, State of California, Sacramento, California
Mr. John Conway, Consolidated Edison Company, New York, New York
Mr. S. D. Cornell, National Academy of Sciences–National Academy of Engineering, Washington, D.C.
Mr. Francis P. Cotter, Westinghouse Electric Corporation, Washington, D.C.
Mr. Charles C. Coutant, Oak Ridge National Laboratory, Oak Ridge, Tennessee
Mr. Willard A. Crandall, Consolidated Edison Company, New York, New York
Mr. W. Donham Crawford, Edison Electric Company, New York, New York
Mr. Robert S. Currie, Southern California Edison Company, Los Angeles, California
Mr. R. K. Davis, George Washington University, Washington, D.C.
Mr. Harold L. Deloney, Louisiana Power and Light Company, New Orleans, Louisiana
Dr. Jesse C. Denton, National Science Foundation, Washington, D.C.
Mr. Joseph J. DiNunno, Atomic Energy Commission, Washington, D.C.
Dr. Warren H. Donnelly, Library of Congress, Washington, D.C.
Mr. R. Michael Dowe, Booz, Allen Applied Research, Inc., Bethesda, Maryland
Mr. Daniel Dreyfus, Senate Interior Committee, Washington, D.C.
Miss Diana Drisko, George Washington University, Washington, D.C.
Professor Leonard Dworsky, Cornell University, Ithaca, New York
Mr. Mahlon Easterling, California Institute of Technology, Pasadena, California
Mr. E. D. Eaton, National Academy of Engineering, Washington, D.C.
Dr. Merril Eisenbud, New York University Medical Center, New York, New York
Mr. Charles Elkins, Environmental Protection Agency, Washington, D.C.
Mr. Howard C. Elmore, Chelan County Public Utility District No. 1, Wenatchee, Washington
Mr. James Flannery, Department of the Interior, Washington, D.C.

Dr. Robert Frank, University of Washington, Seattle, Washington
Mr. Gene Frederikson, General Accounting Division, Washington, D.C.
Mr. S. David Freeman, Twentieth Century Fund, Washington, D.C.
Mr. Alex Fremling, Atomic Energy Commission, Washington, D.C.
Dr. Francois N. Frenkiel, Naval Ship Research and Development Center, Washington, D.C.
Mr. Sheldon K. Friedlander, California Institute of Technology, Pasadena, California
Mr. Alexander Gakner, Federal Power Commission, Washington, D.C.
Mr. Gerard C. Gambs, Ford, Bacon & Davis, Inc., New York, New York
Dr. F. E. Gartrell, Tennessee Valley Authority, Chattanooga, Tennessee
Mr. Austin Gavin, Pennsylvania Power and Light Company, Allentown, Pennsylvania
Mr. Donald P. Geesaman, Lawrence Radiation Laboratory, Livermore, California
Mr. Abraham Gerber, National Economic Research Associates, Inc., New York, New York
Mr. Thomas M. Gerusky, Pennsylvania Department of Environmental Resources, Harrisburg, Pennsylvania
Dr. John C. Geyer, Johns Hopkins University, Baltimore, Maryland
Dr. Frank Gifford, Oak Ridge National Laboratory, Oak Ridge, Tennessee
Mr. J. E. Gilleland, Tennessee Valley Authority, Chattanooga, Tennessee
Mr. George Gleason, Atomic Industrial Forum, New York, New York
Dr. Morton I. Goldman, NUS Corporation, Rockville, Maryland
Mr. Robert L. Goodell, Delaware River Basin Commission, New Jersey
Mr. Richard L. Gordon, Pennsylvania State University, University Park, Pennsylvania
Mr. Richard Gore, Pennsylvania State University, University Park, Pennsylvania
Mr. William R. Gould, Southern California Edison Company, Rosemead, California
Mr. Halvorsen, New York, New York
Mr. Steve H. Hanks, Johns Hopkins University, Baltimore, Maryland
Dr. Donald Harleman, Massachusetts Institute of Technology, Cambridge, Massachusetts
Professor Robert Hartmann, University of Wisconsin, Madison, Wisconsin
Dr. V. S. Hastings, Commonwealth Edison Company, Chicago, Illinois
Mr. John W. Healy, University of California, Los Alamos, New Mexico
Dr. James Hester, New York City Government, Housing and Development Administration, New York, New York
Miss Frances Hightower, National Academy of Engineering, Washington, D.C.
Dr. Glenn R. Hilst, Aeronautical Research Associates of Princeton, Inc., Princeton, New Jersey
Mr. Fred Hittman, Hittman Corporation, Columbia, Maryland
Mr. Larry Hobart, American Public Power Association, Washington, D.C.
Dr. Bart Hoglund, Argonne National Laboratory, Argonne, Illinois
Dr. Frederick A. L. Holloway, Standard Oil Company (N.J.), New York, New York
Dr. Robert Holton, Atomic Energy Commission, Washington, D.C.
Mr. William R. Hughes, Charles River Associates, Inc., Cambridge, Massachusetts
Mr. Phineas Indritz, House Conservation and Natural Resources Subcommittee, Washington, D.C.
Dr. Jay S. Jacobson, Boyce Thompson Institute, Yonkers, New York
Mr. Robert T. Jaske, Battelle Northwest, Richland, Washington
Dr. Loren D. Jensen, Johns Hopkins University, Baltimore, Maryland
Mr. William Jordan, Environmental Protection Agency, Washington, D.C.
Mr. Drexel Journey, Federal Power Commission, Washington, D.C.
Mr. Wally Judd, Delaware
Mr. J. O. Kadel, The Marley Company, Kansas City, Missouri
Mr. Kanterovitz, Cambridge, Massachusetts
Mr. R. Kasper, George Washington University, Washington, D.C.
Dr. Rita D. Kaunitz, Clean Air Commission, State of Connecticut, Westport, Connecticut
Dr. Guy J. Kelnhofer, Jr., State Planning Agency, St. Paul, Minnesota
Professor C. E. Kindsvater, Georgia Institute of Technology, Atlanta, Georgia
Mr. Tor Kolflat, Sargent and Lindy, Chicago, Illinois
Mrs. Carolyn Konheim, Department of Air Resources, City of New York, New York, New York
Mr. Wallace Kornack, Atomic Energy Commission, Washington, D.C.
Dr. Peter Krenkel, Vanderbilt University, Nashville, Tennessee
Dr. Helmut Landsberg, University of Maryland, College Park, Maryland
Dr. John A. Laurmann, National Academy of Sciences–National Academy of Engineering, Washington, D.C.
Mr. Byron Lee, Jr., Commonwealth Edison Company, Chicago, Illinois
Professor Lester Lees, California Institute of Technology, Pasadena, California
Mr. C. Leggett, Washington, D.C.

Professor Philip H. Lewis, University of Wisconsin, Madison, Wisconsin
Dr. W. Deming Lewis, Lehigh University, Bethlehem, Pennsylvania
Mr. Clarence H. Linder, National Academy of Engineering, Washington, D.C.
Mr. Robert Loftness, Department of State, Washington, D.C.
Mr. Lee R. Love, Duquesne Light Company, Pittsburgh, Pennsylvania
Dr. James J. MacKenzie, Massachusetts and National Audubon Societies, Lincoln, Massachusetts
Mr. Rene H. Males, Commonwealth Edison Company, Chicago, Illinois
Mr. Bernard Manowitz, Brookhaven National Laboratory, Upton, New York
Mr. Walter J. Matthews, Public Service Company of Indiana, Plainfield, Indiana
Mr. William Matuszeski, Council on Environmental Quality, Washington, D.C.
Mr. F. J. McAlary, Allegheny Power Service Corporation, New York, New York
Mr. John C. McCaine, Hazelton Laboratories, Inc., Falls Church, Virginia
Mr. Michael McClintock, University of Wisconsin, Madison, Wisconsin
Mr. Frank A. McCrackin, Southern California Edison Company, Los Angeles, California
Mr. Wayne A. McRae, IONICS, Inc., Watertown, Massachusetts
Mr. Charles F. Meyer, General Electric Company, Santa Barbara, California
Dr. Joseph A. Mihursky, Natural Resources Institute, Prince Frederick, Maryland
Mr. Thomas A. Miskimen, American Electric Power Corporation, New York, New York
Dr. Carl B. Moyer, Aerotherm Corporation, Mountain View, California
Dr. J. H. Mulligan, Jr., National Academy of Engineering, Washington, D.C.
Dr. Dorothy A. Muncy, Consulting City Planner, Arlington, Virginia
Professor Bruce H. Murray, University of Wisconsin, Madison, Wisconsin
Mr. Paul D. Myers, Southern California Edison Company, Rosemead, California
Mr. Theodore J. Nagel, American Electric Power Service Corporation, New York, New York
Dr. Joseph Nemec, Booz Allen and Hamilton, Cleveland, Ohio
Mr. William R. New, Tennessee Valley Authority, Chattanooga, Tennessee
Professor Bernard Niemann, University of Wisconsin, Madison, Wisconsin
Dr. Elburt F. Osborn, Department of the Interior, Washington, D.C.
Mr. Jim Peck, Hazelton Laboratories, Inc., Falls Church, Virginia
Mr. Harry Perry, Library of Congress, Washington, D.C.
Mr. Michael Pertschuk, Senate Commerce Committee, Washington, D.C.
Mr. T. A. Phillips, Federal Power Commission, Washington, D.C.
Mr. Robert H. Poppe, Northern States Power Company, Minneapolis, Minnesota
Mr. Louis B. Potter, American Bar Association, Chicago, Illinois
Mr. Hubert Q. Pray, Mississippi Power and Light Company, Jackson, Mississippi
Dr. Donald W. Pritchard, Johns Hopkins University, Baltimore, Maryland
Mr. Nathaniel Pulsifer, Ipswich, Massachusetts
Dr. Edward Radford, Johns Hopkins University, Baltimore, Maryland
The Honorable James T. Ramey, Atomic Energy Commission, Washington, D.C.
Mr. William T. Reid, Battelle Memorial Institute, Columbus, Ohio
Mr. Vic Reinemer, Office of the Honorable Lee Metcalf, Washington, D.C.
Professor Arnold W. Reitze, George Washington University, Washington, D.C.
Mr. Robert S. Restall, New England River Basin Commission, Boston, Massachusetts
Mr. T. Reynolds, Ontario Hydro-Electric Power Commission, Toronto, Ontario, Canada
The Honorable Robert Rickles, New York City Air Pollution Control Commission, New York, New York
Mr. Maurice Rifkin, Mitre Corporation, McLean, Virginia
Dr. Denis M. Robinson, High Voltage Engineering Corporation, Burlington, Massachusetts; Marine Biological Laboratory, Woods Hole, Massachusetts
Mr. W. M. Roe, Watertown, Massachusetts
Dr. Martin Roessler, University of Miami, Miami, Florida
Professor Albert J. Rosenthal, Columbia University, New York, New York
Mr. A. David Rossin, Argonne National Laboratory, Argonne, Illinois
Mr. Eric Rubin, Public Service Commission, Albany, New York
Mr. Stephen G. Salay, Department of the Interior, Washington, D.C.
Mr. Richard H. Sandler, Office of the Honorable Philip A. Hart, Washington, D.C.
The Honorable John P. Saylor, House of Representatives, Washington, D.C.
Mr. E. A. Schultz, Illinois Power Company, Decatur, Illinois
Dr. Sam H. Schurr, Resources for the Future, Inc., Washington, D.C.
Mr. Milton F. Searl, Office of Science and Technology, Executive Office of the President, Washington, D.C.
Mr. Daniel P. Sheer, Johns Hopkins University, Baltimore, Maryland
Mr. H. G. Simens, Bechtel Corporation, San Francisco, California
Mr. David Sive, Winer, Neuberger and Sive, New York, New York

Mr. James M. Smith, General Electric Company, San Jose, California
Mr. Maynard E. Smith, Brookhaven National Laboratory, Upton, Long Island, New York
Mr. Waldo E. Smith
Mr. Edwin H. Snyder, Public Service Electric and Gas Company, Newark, New Jersey
The Honorable Louis J. Sparvero, Public Utilities Commission, Harrisburg, Pennsylvania
Professor Arthur M. Squires, City University of New York, New York, New York
Dr. Chauncey Starr, University of California, Los Angeles, California
Mr. Stergilun, Franklin Institute, Philadelphia, Pennsylvania
Dr. Carlos Stern, University of Connecticut, Storrs, Connecticut
Mr. D. H. Sterrett, Duke Power Company
Mr. James Stout, Federal Power Commission, Washington, D.C.
Dr. Kirk Strawn, Texas A&M University, College Station, Texas
Mr. A. Sunda, Federal Power Commission, Washington, D.C.
The Honorable Joseph C. Swidler, New York State Public Service Commission, Albany,
   New York
Miss Ann Szymkowicz, University of Maryland, Prince Frederick, Maryland
Dr. C. M. Tarzwell, National Marine Water Quality Laboratory, West Kingston, Rhode Island
Mr. L. L. Teel, National Academy of Engineering, Washington, D.C.
Mr. Michael S. Terpilak, Environmental Protection Agency, Rockville, Maryland
Mr. Frank Thomas, Federal Power Commission, Washington, D.C.
Mr. Ruble A. Thomas, Southern Services, Inc., Birmingham, Alabama
Mr. C. A. Thompson, Baltimore, Maryland
Mr. George W. Toman, Bonneville Power Administration, Department of the Interior,
   Washington, D.C.
Mr. George Tomlinson, Federal Power Commission, Washington, D.C.
Mr. Thomas W. Trice, Mid-Atlantic Area Council, Baltimore, Maryland
Mr. John Trump, Massachusetts Institute of Technology, Cambridge, Massachusetts
Mr. Richard C. Tucker, National Water Commission, Arlington, Virginia
Mr. E. Winslow Turner, Senate Subcommittee on Intergovernmental Relations, Washington,
   D.C.
Mr. Carl M. Unruh, Battelle Memorial Institute, Richland, Washington
Mr. John F. Vogt, Jr., New Orleans Public Service, Inc., New Orleans, Louisiana
Mr. John Volkert, Public Service Commission, Albany, New York
Mr. Frank M. Warren, Portland General Electric Company, Portland, Oregon
Mr. Frederick H. Warren, Federal Power Commission, Washington, D.C.
Dr. William D. Watson, Jr., University of Minnesota, Minneapolis, Minnesota
Mr. Charles L. Weaver, Environmental Protection Agency, Rockville, Maryland
Mr. J. Frederick Weinhold, Office of Science and Technology, Executive Office of the President,
   Washington, D.C.
Mr. J. Robert Welsh, Southwest Power Pool, Shreveport, Louisiana
Mr. G. O. Wessenauer, Chattanooga, Tennessee
Dr. Alfred T. Whatley, Western Interstate Nuclear Board, Lakewood, Colorado
Mr. John P. White, Arkansas Power and Light Company, Little Rock, Arkansas
Mr. Lee C. White, Semer, White & Jacobsen, Washington, D.C.
Dr. Merrill Whitman, United States Atomic Energy Commission, Washington, D.C.
Professor Robert Williams, University of Michigan, Ann Arbor, Michigan
Dr. Daniel Williard, University of Wisconsin, Madison, Wisconsin
Dr. Mason Willrich, University of Virginia, Charlottesville, Virginia
Mr. Conrad Wirth, Washington, D.C.
Professor Warren Witzig, Pennsylvania State University, University Park, Pennsylvania
Mr. James C. Woodruff, Michigan Public Service Utilities Commission, Lansing, Michigan
Dr. James Wright, Westinghouse Electric Corporation, Pittsburgh, Pennsylvania
Mr. Kurt Yager, Mitre Corporation, McLean, Virginia
Mr. Marvin M. Yarosh, Oak Ridge National Laboratory, Oak Ridge, Tennessee
Mr. H. J. Young, Edison Electric Institute, New York, New York
Dr. A. Zucker, National Academy of Sciences—National Academy of Engineering, Washington,
   D.C.

COPPS STAFF

Mr. E. D. Eaton, *Executive Secretary*
Mrs. Maxine Brown
Mrs. Leila von Meister

II
REPORTS OF THE WORKING GROUPS

The reports of the Working Groups, which were substantially autonomous in defining their subjects and the manner of their treatment, should be regarded as supportive material published on the responsibility of each Working Group with the benefit of comments from the Steering Group, but not necessarily endorsed by the Steering Group or the National Academy of Engineering.

Working Group I(a)

Environmental Protection: Power Plant Siting and Air Quality—Engineering Considerations

INTRODUCTION

The following report was prepared as a contribution to the considerations of the National Academy of Engineering Committee on Power Plant Siting (COPPS) regarding engineering contributions to resolution of the problems of electrical power generation and its environmental impacts. As its contribution to this timely task, Working Group I(a): Air concentrated on a delineation of the causes and effects of air pollutants from fossil-fueled power plants and the meteorological and engineering factors which enter into the specification of those feasible engineering alternatives which could help minimize environmental impacts while maintaining electrical energy production at or near societal demand levels. In this first iteration, it has not been possible to proceed to exact specification of these alternatives or their evaluation. That important step, which should follow as soon as possible, requires significant interchanges with the other Working Groups, as well as manpower, time, and resources which were not available in this initial activity.

Having made this first step, this Working Group recommends that the analyses begun here be continued. The potential for costly and tragic errors due to continued *ad hoc* solutions directed to simplistic criteria and too narrow objectives must be avoided. Thoughtful and objective assessments of alternatives, before actions are committed, provide the best hope for satisfactory resolution of these complex problems.

Among the multiple facets of the current problems of adequate electrical power generation *and* maintenance of an equitable environment, air quality is probably the most prominent. This appears to be true for two reasons:

(1) the trend towards large-capacity, centralized power generation plants makes them highly visible and concentrates their atmospheric pollution emissions, and (2) the atmosphere is by far the least fettered of the geophysical fluids and, therefore, can and does carry pollutants indiscriminately to land and coastal areas where man and the things he holds valuable reside. By contrast, rivers and lakes generally stay within their banks and, while the deterioration of the environment they represent is also serious to human welfare, the locale of the problems and those things affected by environmental changes are much more clearly defined.

This is not to prejudge the relative importance of environmental impacts which electrical power generation presents via the atmosphere, the surface waters, and the land surfaces—this task has been reserved for another working group. Rather we wish to emphasize, within our own restricted purview, the ubiquity and indiscriminate spread of pollutants which have been released to the atmosphere. Once the pollutants are released, we are totally dependent upon uncontrolled natural processes of transport, dilution, and removal of these materials. As a result, human control of the environmental effects of airborne materials emanating from power plants must be exercised at the source; the alternative of cleaning that air which we want to use for breathing, plant respiration, and the like, is not a viable substitute, except perhaps for the most extreme personal situations.

To help COPPS in its assessment of engineering alternatives in power plant siting and environmental quality, we accepted the task of providing basic information on

(1) the physical and biological effects of power plant emissions and (2) the scientific and engineering bases for alternatives in siting, design, and operation. This information can be used for several purposes: (1) to ensure that previous and current procedures are evaluated in the light of currently available information in engineering, meteorology, chemistry, and the health and environmental sciences, (2) to provide guidelines for future power sources and sites, and (3) to obtain a more equitable balance between energy demands and environmental quality.

Our charter clearly prescribed the consideration of engineering alternatives for power generation, alternatives which may provide the basis for rational assessment of the trade-offs between electrical power production and environmental quality. However, throughout our deliberations the Working Group has recognized that electrical power generation is only one of the contributors to environmental problems, albeit a prominent one. In order to focus on our assigned task, we have isolated power generation, but with a clear recognition that we are looking at the *relative* contribution of power generation to environmental quality problems. *The larger problem of total environmental quality requires consideration of all facets of the allocation of a finite resource among competing activities.* In this case, the resource is the variable and finite capacity of the environment for assimilation of wastes. The Working Group had neither the resources nor the time to address this larger problem. Someone must do so, however, or we shall surely develop inequities much larger than those which arose out of an earlier naive assumption that this resource was a free good of infinite (or at least inexhaustible) extent.

In approaching its own facet of this problem this Working Group has recognized that there are two dependent variables and two independent variables in the equations relating engineering alternatives to power production and environmental quality. Stated in question form, the dependent variables are

1. What level of power production is required?
2. What level of environmental quality must be maintained?

It is a tenet of the present NAE study that these questions, and the variables they address, are not independent of one another. At some point, it seems clear, the cost of additional power generation capacity, with controls necessary to maintain environmental quality, will exceed the perceived worth of the additional electricity. Or, conversely, the denial to some segment of society of further access to electrical power will represent a larger perceived disbenefit than the

environmental "worth" which would be threatened by that increment of power generation. As its part of the inputs necessary to analyses of these trade-offs, this Working Group has focused attention on the details of environmental impacts associated with electrical power generation. These materials are presented in Chapter 2 of this report. Because these impacts frequently involve intangible aspects of man's relationship to the world around him, we have not attempted to assign tangible dollar values to them. Such value judgments must be assigned by social criteria and should be derived in the context of similar judgments for other facets of this problem.

In describing environmental impacts ascribable to electric power production, we have chosen three main topics: (1) the effect of chemically active, toxic, and radioactive materials on human health, (2) the effects of these materials on plants, animals, inanimate objects, and geophysical processes, and (3) the effects of sensible and latent heat, and of water vapor emissions. These choices automatically focused attention on stationary power plants using conventional fossil fuels (including the gas turbine) or nuclear fission energy sources. The impact of hydroelectric systems on the *air* environment was judged to be negligible. Future energy sources, such as fusion, fuel cells, and solar energy, were also recognized as being essentially pollution-free.

Given this choice of stationary, fossil-, or nuclear-fueled power generation facilities, all of which present a sizable environmental impact potential, we identified two independent variables which could be manipulated, within limits, to achieve power generation goals *and* environmental quality criteria. These are

1. The geographical location of the power plant or multiple plants
2. The process control of effluents from these plants

Although these variables are also subject to nonatmospheric considerations, e.g., transmission facilities and fuel supplies, all of the engineering alternatives must eventually be incorporated into these two variables, since the environmental impact is completely determined by *how much* material is released to the atmosphere and *where* it is released.

The meteorological and engineering alternatives which enter into choices of plant siting and emissions control are presented in Chapter 3 of this report.

The Working Group has considered the problems of formulation and enforcement of regulations and standards which serve to implement social choices between productivity and environmental quality. At the moment, actions in this arena are strongly slanted towards restoring or preserving environmental quality. However, we are much more impressed by the obvious state of flux of these activities—the nearly annual major amendments of the Federal Clean Air Act are probably the best measure of how rapidly the processes of regulation and standard setting are changing. Given this state of affairs, we believe the study to which we are contributing is most timely and, we hope, helpful. Major specific recommendations regarding regulations and standards must also arise out of the broader analyses to which we are contributing.

Finally, we must note with all the vigor at our command that there are areas of ignorance, error, and uncertainty in our knowledge of environmental impacts and power generation which are detrimental to an objective assessment of these complex problems. We have noted these throughout the text of this report. We do not believe that all decisions and actions must be held in abeyance until these gaps are filled. But, given the ever-increasing complexity and magnitude of these problems, we are convinced that a vigorous research program, designed to overcome these gaps, is essential. One of the most valuable products of studies of the kind undertaken by COPPS is the identification of areas in which ignorance, error, and uncertainty do preclude adequate definition and evaluation of pressing and future technologically oriented social problems. We can think of no stronger rationale for research.

THE NATURE AND SOURCES OF AIR POLLUTION FROM POWER PLANTS

POLLUTANT FORMATION AND EMISSION RATES

Table 1 provides a listing of the types and present annual emission rates for air pollutants resulting from the operation of a typical modern 1,000-MW(e) fossil-fueled steam power plant of conventional design. The pollutant emissions from the operation of a 1,000-MW(e) installation of gas turbines (burning gas, oil, or jet fuel) are listed in Table B-1, Appendix B. For our present purposes, the source or method of formation of each of the air pollutants listed and other materials emitted into the atmosphere is of primary concern.

Particulates

With respect to coal-fired boilers, the source of particulates (fly ash) emitted into the atmosphere is the ash content of the fuel burned. Boiler design and method of operation determine to some small degree the extent that a portion of ash in the fuel may drop out in the boiler as "bottom ash" and not appear as particulates in the combustion gases emitted from the boiler. The "bottom ash" may represent on the order of 20% of the total ash content of the coal burned.

In the case of oil, the particulates originate from two sources—the ash content of the oil burned and chemical reactions occurring in the combustion process.

Natural gas has no "ash content" which contributes particulate matter to the atmosphere. The particulates emitted as the result of burning gas in a steam boiler or gas turbine result from chemical reactions occurring in the combustion process.

Oxides of Sulfur

This pollutant, in all cases, has as its source the sulfur content inherent in the fuel burned. Unless control systems are employed to remove oxides of sulfur from the combustion gases, essentially all of these oxides (primarily SO_2 plus small quantities of SO_3) pass into the atmosphere. The amount of these materials that is retained by the "bottom ash" or the fly ash removed by such gas cleaning devices as electrostatic precipitators is extremely small in respect to the total quantity involved.

TABLE 1   Air Pollutant Emissions from a Typical 1,000-MW(e) Conventional Power Plant [a, b]

Pollutant-	Annual Release (10^6 lbs)		
	Coal [c]	Oil [d]	Gas [e]
Particulates	9.9	1.6	1.02
Oxides of sulfur	306.0	116.0	0.027
Oxides of nitrogen	46.0	47.8	26.6
Carbon monoxide	0.460	0.0184	Negligible
Hydrocarbons	1.150	1.47	Negligible

[a] Source: Joint Committee on Atomic Energy, "Selected Materials on Environmental Effects of Producing Electric Power." 91st Congress, First Session. U.S. Government Printing Office, August, 1969: Tables II, III, IV (pp. 125–126).
[b] Based on normal average heat rates, load factors, and fuel properties.
[c] Burning 2.3 × 10^6 ton/year. Assuming 3.5% sulfur content of which 15% remains in the ash, and a 9% ash content with 97.5% fly ash removal efficiency.
[d] Burning 460 × 10^6 gallons/year. Assuming 1.6% sulfur content and 0.05% ash content.
[e] Burning 68 × 10^9 scf/year.

Oxides of Nitrogen

Oxides of nitrogen, primarily NO plus smaller quantities of NO_2 (frequently lumped together and referred to as NO_x) arise from a different source than do other air pollutants. Basically, nitrogen in the combustion air (plus, possibly, small quantities of nitrogen chemically contained in the fuel oil or coal) combines with the oxygen in the air during the combustion process for all fuels (coal, oil, or gas) in accordance with the reaction $N_2 + O_2 \rightarrow 2NO$. Later, most of the NO further oxidizes to form NO_2. This second reaction occurs to a limited extent within the boiler or gas turbine. The major amount of NO_2 formed from the further oxidation of NO is formed external to the boiler, often at a considerable distance downstream from the plant stack. A small amount of the NO formed within the boiler also reverts back to nitrogen and oxygen as the combustion gases are cooled.

Due to the basic nature of conventional boilers—designed to efficiently transfer the heat content of the fuel burned to the boiler water to produce steam—there are certain practical limitations on how the combustion operation can be conducted. There are practical engineering reasons for the nature of boiler designs. Obviously, the power plant boilers in use today were not designed primarily for minimal NO_x formation but, rather, to produce steam from water as efficiently as possible from the burning of available fuels; the NO_x formulation characteristics are a function of other design criteria.

In the case of gas turbines, the NO_x produced is also the result of the reaction between nitrogen and oxygen occurring in the combustion zone. Turbine design and method of operation both affect the extent of NO_x formation.

Carbon Monoxide

This air pollutant is formed in the combustion process because of incomplete reaction between the carbon in the fuel and oxygen in the air. In the case of burning oil or gas in conventional boilers, the quantity produced is considered negligible. Although measurable quantities of carbon monoxide are produced in the combustion of coal, the actual amount formed is considered insignificant in its impact on the environment in relation to the far greater quantities produced from other sources such as internal combustion engines used for transportation.

The magnitude of carbon monoxide emissions resulting from gas turbine operations *is* significantly high, however, and merits consideration.

Hydrocarbons

This air pollutant is formed during the combustion process as a result of incomplete combustion or the occurrence of other chemical reactions. The quantities produced by the combustion of natural gas in conventional steam boilers is considered negligible. The quantity of hydrocarbons produced by the combustion of coal and oil is of possible significance but can be minimized by good combustion control. Research is currently being conducted on the role of the various chemical compounds included in the broad category of hydrocarbons as possible precursors of particulate matter in the atmosphere and on their other possible environmental effects. Until further information is available on this matter, including technology for emission control, little can be said in regard to this type of air pollutant and its ramifications in respect to power plant site selection for conventional steam power plants.

Hydrocarbons are produced in greater quantities in gas turbine operation when burning gas, oil, or jet fuels. The extent of formation is a function of both turbine design and method of operation.

Carbon Dioxide

A major component of the products of any combustion process is carbon dioxide. Large amounts of this material are released into the atmosphere from many sources, such as home heating, transportation, industrial processes, open burning, etc., in quantities that total higher than the quantities attributable to electric power production. It is not considered toxic in the normal concentrations found in the atmosphere. Carbon dioxide emissions are viewed by some as a possible long-range problem since an increase of the carbon dioxide concentration in the earth's atmosphere could modify the heat balance of the atmosphere, resulting in possible climatic changes. There is no practical technology available to reduce the overall emission rate of CO_2 to the atmosphere and none available for reducing that much smaller portion due to electric power generation.

Hazardous Materials

The phrase "hazardous materials" has recently come into use to categorize a wide range of materials not generally considered elsewhere but which may have particular adverse environmental effects which warrant their consideration in efforts to control pollution of all types. Materials which could possibly be considered in this category are certain metals such as mercury, cadmium, and beryllium which occur in trace quantities in all coals and nondistillate fuel oils. As measurement techniques for determining the concentration of these metals in the fuel, combustion products, or the atmosphere are both difficult and subject to considerable error, there is a lack of sound knowledge in respect to their possible role as significant air pollutants. Much more research is needed in this area, including the development of control technology if it is deemed necessary. Until such time as this knowledge is available, little can be said in respect to these materials as factors in power plant design, operation, or site selection.

Radioactive Air Pollutants

When choosing between the alternatives of nuclear power and power obtained from the combustion of fossil fuels, one should be aware of the fact that both coal and oil contain trace quantities of naturally occurring radioactive isotopes.

Do power plants burning fossil fuels discharge significant quantities of radioactivity, and how do they compare in this respect with nuclear power plants?

In the case of coal containing 2.0 ppm of ^{232}Th and 1.1 ppm of ^{238}U (typical concentrations), the fly ash released annually from a hypothetical 1,000-MW(e) power plant burning 2.3×10^6 tons per year and having a 97.5% efficient fly ash removal system would contain 17.2 mCi of ^{226}Ra and 10.8 mCi of ^{228}Ra, which are the daughter products of ^{235}U and ^{232}Th. On the basis of AEC regulations covering exposure of airborne radioactive materials, Eisenbud (1964) states that this total of 28 mCi per year of mixed radium isotopes is approximately equivalent to 10^4 Ci of ^{85}Kr or 10 Ci of ^{131}I. Krypton-85 and iodine-131 were chosen for comparison since they represent two of the gaseous radionuclides of concern in nuclear power plant stack effluents (cf. Appendix A).

In the case of oil associated with the particulate emissions from a 1,000-MW(e) power plant burning 460×10^6 gallons per year, approximately 0.5 mCi of radium (^{226}Ra and ^{228}Ra) will be released annually— roughly equivalent to 200 Ci of ^{85}Kr or 200 mCi of ^{131}I.

Comparing these quantities of natural activity released from fossil-fueled power plants with that from nuclear power plants on the basis of hygienic equivalent quantities of typical radioisotopes associated with nuclear reactor operation, it is apparent that conventional fossil-fueled plants discharge relatively equal or greater quantities of radioactive material into the atmosphere than currently operating nuclear power plants of comparable size.

TRENDS IN POWER GENERATION

Power demand is increasing rapidly and may be ten times as great in the year 2000 as in 1968 [Council on Environmental Quality (CEQ), 1970]. The pattern of energy production is changing from a number of relatively small (under 50 MW) plants to a lesser number of large-capacity generating stations (over 200 MW) with long transmission lines. Increased fuel combustion at individual sites produces more intense emissions, further straining the capacity of the environment to dilute, disperse, and remove atmospheric contaminants. Current and anticipated demands for energy and trends with regard to power plant size lead to the conclusion that much greater consideration must be given to environmental problems associated with the expansion of existing and the location of future energy centers.

Fossil Fuel Consumption

Fossil fuels produced a little more than 80% of the electricity generated by utilities in 1968 [Federal Power Commission (FPC), 1970]. The utilization of fossil fuels for all purposes in the years 1920 to 1970 increased 3.5-fold in the United States (Land, 1970). Natural gas usage increased by 25-fold, petroleum by slightly greater than 9-fold while total coal consumption is approximately the same now as in 1920 (Land, 1970). A recent estimate indicates that coal consumption may rise by as much as 4-fold by the year 2000 in order to supply rapidly increasing power needs (Eisenbud, 1964). Thus, the greatest source of electrical energy is and will continue to be coal, notwithstanding the expanded use of oil, gas, hydroelectric, and nuclear power (Eisenbud, 1964). Obviously, pollutant emissions will increase significantly if the rate at which emissions are controlled is not greater than the increases in fuel utilization.

In recent years, gas turbines have been used with increasing frequency to provide extra generating power during high demand periods. These converted aircraft engines burn oil or gas and possess several advantages for utilities with limited reserve generating capacity. However, pollutants emitted from turbine engines appear to differ substantially from conventional boiler emissions in that particulate and sulfur oxide emissions are lower while nitrogen oxide, carbon monoxide, and hydrocarbon emissions may be considerably higher per unit of power generated. Unfortunately, specific information on emission factors is not yet available. In addition, turbine engines are not designed for continuous operation and may have higher noise and maintenance problems than conventional boilers.

In the past, many electrical generating stations have been constructed near the centers of greatest power demand. In addition to cost factors, it must be recognized that meteorological and topographic conditions often limit dispersal of pollutants and make locations such as river and mountain valleys less desirable sites.

Energy centers located close to potential biological receptors, such as areas of high population density or agricultural areas in which susceptible plant species are grown, increase the probability of harmful effects. The consequences of introducing gaseous and particulate pollutants into the environment (described later in this report) should be considered when energy centers are planned and sited. A brief summary of the nature of emissions from fuel combustion will follow prior to the review of the physical and biological effects of these air pollutants (Chapter 2).

Nature of Emissions

Electric power generation currently produces the following approximate percentages of the effluents emitted by all combustion processes in the United States: 50% of the sulfur oxides, 25% of the nitrogen oxides, 25% of the particulate matter, 5% of the gaseous hydrocarbons, and less than 2% of the carbon monoxide [Office of Science and Technology (OST), 1970].

Sulfur oxide emissions are dependent on the sulfur content of the fuel (usually 0.5% to 6% for coal) since almost all escapes during combustion. Concentration in the flue gas may be as much as 2,000 ppm by volume, or more if low-sulfur fuel is not used. Coal containing 1% sulfur emits approximately 1.5 pounds of sulfur oxides per million Btu heat input (Smith and Gruber, 1966). During the period 1920 to 1968, total sulfur oxide emissions increased by 30% but ground-level concentrations decreased in many large cities, presumably because of the shift from low to high elevation sources (Land, 1970). Recently, the use of low-sulfur fuel has aided this reduction in sulfur oxide concentrations. Total atmospheric emissions from power generation were 12 million tons of sulfur oxides in 1966 and may reach 36 million tons by 1980 [National Air Pollution Control Administration (NAPCA), 1969]. Controls on emissions will probably prevent increases from occurring thereafter despite rising fuel consumption.

Particulate matter emissions depend on the ash content of the fuel (usually 7% to 20% for coal), the efficiency of combustion, and the operating efficiency of fly ash collection devices (usually 80% to 97.5%) (NAPCA, 1969b). Particulate matter, both suspended and settleable, has decreased in many large cities because of the use of better grades of coal and particle collection devices. Current particulate emissions in ppm per million Btu heat input are approximately 0.02 for gas (no control), 0.07 for oil (no control), 0.67 for coal (90% fly ash removed), and 0.03 for coal (99.5% fly ash removed) (NAPCA, 1969b). Increased use of fly ash and other particle collection devices will probably prevent future increases in total emissions of solid matter from fossil-fuel combustion.

Nitrogen oxide concentrations in flue gases vary from 100 to 1,460 ppm (Smith and Gruber, 1966), depending on the type of boiler and conditions of combustion. Nitrogen oxide emissions have increased a total of 2.4-fold in the past 50 years (Land, 1966) because of increased use of petroleum and natural gas for internal combustion engines and electric power generation. Currently, natural gas-fueled power plants emit about 0.37 pounds of nitrogen oxides per million Btu (Duprey, 1968). Similar figures for oil- and coal-fired boilers are 0.69 and 0.84 pounds, respectively (Duprey, 1968). Total emission of nitrogen oxides to the atmosphere in 1968 by combustion of fossil fuels for electric power generation was 3.6 million tons. If current trends continue, this figure may be 2- to 3-fold higher by the year 2000 (Bartok et al., 1970). The contribution of power production to total nitrogen oxide emissions is expected to increase from 22% to between 35 and 40% by the year 2000. Domestic and commercial heating units currently make a total contribution of 7% to nitrogen oxide emissions (Bartok et al., 1970). It is anticipated that even if a moderate level of control were to be used, nitrogen oxide emissions would still increase between 1970 and 2000. Projections indicate that only the use of stringent controls will contain this problem (Bartok, et al., 1970). Nitrogen oxide emissions are of particular concern because they are "starting" materials for atmospheric reactions which lead to the production of photochemical oxidants. It has been suggested that a great reduction in the occurrence of photochemical reactions would take place if emissions for nitrogen oxides were held to less than 0.1 pound per million Btu for stationary sources and 50 ppm in the exhaust for mobile sources (Larsen, 1970).

A wide variety of other inorganic and organic substances are emitted, in smaller quantities, from power plants by fossil-fuel combustion. Since energy demand is increasing, it seems likely that these substances will be emitted to the atmosphere in greater amounts in the future. The list of inorganic pollutants includes hydrogen chloride, hydrogen fluoride, silicon tetrafluoride, iron, titanium, lead, copper, vanadium, barium, nickel, zinc, chromium, manganese, molybdenum, cobalt, beryllium, and cadmium [US Dept. of Health, Education and Welfare (HEW), 1970; Cuffe, 1964; Gerstle et al., 1965].

Coal contains between 0.1% and 0.7% chlorine (Meetham, 1964); consequently, hydrogen chloride has been measured in stack gases at a concentration of 49 ppm with an estimated emission factor range of 0.08 to 0.3 pound per million Btu (Smith and Gruber, 1966). The fluorine content of coal is usually less than 0.1% (Meetham, 1964), and about half is evolved during combustion, 90% in the particulate and 10% in the gaseous form. The gases are hydrogen fluoride and silicon tetrafluoride, and the concentration of fluorides in stack gases from coal combustion is usually less than 5 ppm. The U.S. Public Health Service has estimated that in 1968 about 16,000 tons of fluorine were emitted into the atmosphere of the United States from the combustion of coal (HEW, 1970). Specific data on amounts of the other inorganic substances emitted per unit of power produced do not seem to be available.

Polynuclear aromatic hydrocarbon compounds such as pyrene, benzo(e)pyrene, fluoranthene, and benzo-(e)pyrene are emitted in flue gases in varying amounts up to about 1.5 mg per million Btu (Cuffe, 1964; Gerstle et al., 1965; Hangebrauck, 1964). Other hydrocarbon compounds such as organic acids and formaldehyde are also produced in small quantities. Emission of hydrocarbon compounds is generally greater from combustion of coal than oil or natural gas and is affected greatly by combustion conditions (Hangebrauck, 1964). Emission rates are usually less than 0.1 pound (calculated as methane) per million Btu. Carbon monoxide may also be emitted in flue gases in amounts usually less than 0.5 pound per million Btu.

Noise pollution may be a problem for future consideration because of the construction of larger generating stations. It has recently been recommended that criteria for noise attenuation be established as part of the site selection process and considered in the location and design of power plants.

METEOROLOGICAL FACTORS IN POLLUTANT DISPERSAL

The obvious solution to the problem of maintaining or restoring air quality is to reduce the emissions of the offensive materials to the point at which acceptable air quality levels are reached. This alternative has the advantage of simplicity, but it is frequently very costly; it may create other waste (liquid or solid) problems that are more serious than the air quality situation; and, finally, it fails to take full advantage of the enormous capacity of the atmosphere to dilute, transform, and remove waste material at minute cost.

In this section, the attempt is made to segregate and discuss the meteorological and chemical factors that are involved in the atmospheric aspects of the problem. Frequently, the solutions or remedial techniques are so interrelated as to defy simple classification, but the principles should become clear as the treatment of the problem progresses. Similarly, the matters involving cost cannot be discussed intelligently without carrying the analysis to the point of describing all ramifications of the proposed solution. For example, if emission control involves solid waste removal or treatment, this must be included in the cost estimates.

Meteorological Factors

There are essentially four basic meteorological factors involved in atmospheric pollution from power plants: transport, dilution, modification and removal, and ambient background conditions, the last being a catch-all classification that includes what might be described as the meteorological milieu of temperature, humidity, solar radiation, and the like to which the pollutants are subjected.

Transport

Clearly, pollution from any source(s) must follow the trajectory of the air flow in which it is emitted, and this can be an overwhelmingly important factor in evaluating power plant siting. A classic example of the significance of this factor is the difference between locating a plant on the East Coast or the West Coast of the United States. In the former case, most of the flow would be directed offshore, towards the open ocean, whereas in the latter, the pollutant would normally be carried onshore. Whether the sources be single or multiple, the simple matter of the predominant direction of transport is critical to the effect of the pollutant.

Dilution

The atmosphere is normally turbulent and quickly dilutes waste gases and particles, although the actual rate varies considerably with meteorological conditions. It is a matter of everyday experience that visible plumes from continuous "point" sources of all kinds tend to be long and narrow, averaging perhaps ten times as long as they are wide. This is to say, the material in plumes is diluted, or spread laterally about a tenth as fast as it is transported downwind. Actual atmospheric dilution values applicable to plumes from nuclear plants (i.e., low-level, nonbuoyant plumes) have been detailed in the AEC publication, *Meteorology and Atomic Energy—1968* (Slade, 1968) and summarized in the International Atomic Energy Agency booklet, *Applications of Meteorology to Safety at Nuclear Plants* (IAEA, 1968). For tall stack, buoyant, conventional plant plumes, the appropriate dilution values can be found in the ASME publication, *Recommended Guide for the Prediction of the Dispersion of Airborne Effluents* (Smith, 1968), and the subject of plume buoyancy is dealt with extensively in the recent AEC publication, *Plume Rise* (Briggs, 1970).

The small-scale atmospheric turbulence that dilutes a plume, causing its characteristic shape, is three-dimensional. Plume elements spread out in all three dimensions as they are carried along by the mean wind. Large-scale atmospheric motions are, on the other hand, primarily two-dimensional and horizontal. The result is a large-scale, horizontal swinging of plumes, according to the prevailing wind direction. Thus, the average plume concentration experienced at a point on the ground is caused both by the instantaneous plume dilution by turbulence and the long-term effect of changes in the wind direction and speed.

The behavior and appearance of buoyant plumes from tall power plant stacks are strongly influenced by the kind and degree of turbulence present in the lowest few thousand feet of the atmosphere, the planetary boundary layer, sometimes called the mixing layer. At night, vertical atmospheric turbulence is strongly inhibited by the stabilizing effect of nocturnal ground cooling, and plumes typically spread slowly in the vertical direction and remain aloft. Vertical spreading of plumes can also be inhibited by atmospheric inversions, or reversals of the normal temperature decrease with height. An inversion is often present at the top of the planetary boundary layer. Where, as on the U.S. West Coast, this inversion is reinforced by large-scale subsiding motion of the atmosphere, the limitation to vertical mixing can be quite pronounced. Horizontal spreading is inhibited and the normal directional swinging of plumes is restricted by terrain.

During the daytime, vertical motion and dispersion of plumes is normally great, owing to the presence of more intense vertical turbulent mixing in the boundary layer. The plume can be brought down to the ground comparatively near the plant, sometimes at distances of only a few stack heights, causing brief periods of high concentration values. If the wind speed is high enough that turbulent air motions exceed the intensity of the buoyant plume rise, the plume can be carried downward to the ground over a wider area and for a longer period of time. Mixing downward of plumes can also occur in frontal zones and in the lee of terrain obstacles such as ridges or cliffs. Finally, the intensity of vertical mixing is affected by the roughness of the underlying terrain, being greater over cities and smaller over bodies of open water than over open country. Thus, the intensity and duration of ground-level pollutant concentrations from buoyant tall-stack plumes can vary widely, from short periods of quite high concentrations to longer, persistent episodes at lower values.

Ultimately, material emitted into the atmosphere from power plant sources becomes mixed through the lower atmosphere, the troposphere, if it is not removed by some natural process. Typically, the residence time for such materials in the lower atmosphere is measured in weeks.

Modification and Removal Processes

Airborne materials, the gases, aerosols, and small particles released from power plants, may be modified or removed by a number of processes. Chemical reactions can create new substances and are discussed more fully on page 42. Gravitational fallout, impaction on surfaces, washout and rainout by precipitation and condensation processes, and radioactive decay all can play a part. However, this brief list by no means does justice to the large number of complex microphysical and chemical effects that operate to modify the distribution of airborne gases and particles. Much research remains to be done in this area.

Ambient Temperature and Humidity

In addition to the foregoing, which represent meteorological processes that *affect* pollutants, there are two meteorological variables that are themselves *part* of the problem—temperature and humidity. Since power plant operation always produces waste heat, some of which usually appears as atmospheric moisture, one must consider the ambient temperature and humidity in evaluating the effects that such emissions will have.

The Cumulative Air Quality Effect of Multiple Sources

Because of the effect of tall stacks and plume buoyancy, the maximum ground-level concentration, which determines ambient air quality, occurs for conventional plants with tall stacks at some distance from the source. This distance is almost never less than about four times the stack height and probably averages ten or more times this figure. For nuclear plants, the source is lower, and the control point is the plant fence, or exclusion distance. Exclusion distances are typically ½ to 1 km for nuclear plants.

The downwind ground-level concentration X, at points more distant from a power plant source than the point at which the control concentration occurs, obeys the formula

$$X \propto Q(\sigma_y \sigma_z u)^{-1}, \qquad (1)$$

where Q is the source strength, u is the average wind speed and σ_y and σ_z are horizontal and vertical lengths characterizing atmospheric diffusion. Values of σ_y and σ_z for various meteorological conditions can be found in the references. Equation (1) can be justified in various ways (Slade, 1968) and is moreover obvious on dimensional grounds. The diffusion parameters σ_y and σ_z are often expressed as simple power laws (Slade, 1968; Smith, 1968) of the form $\sigma = ax^p$ where x is downwind distance from the source and a and p are constants determined by meteorological conditions.

When the long-term average concentration $\overline{X}_A$ is considered, the formula becomes

$$\overline{X}_A \propto Q(x\sigma_z u)^{-1}, \qquad (2)$$

because the influence of σ_y averages out. These formulas can conveniently be written in a form in which lengths are made proportional to the distance to the control point x_C and concentrations proportional to their values at that point $\overline{X}_{AC}$. When this is done for the average concentration value, it develops that

$$\overline{X}_A = \overline{X}_{AC}(x_C/x)^{p+1}. \qquad (3)$$

The air quality at any point near a power plant is governed by three contributions: (1) the direct contribution from the plant, as given by $\overline{X}_{AC}$, the control or ambient air quality level; (2) the contribution from other more distant plants; and (3) the global background contribution. The present global background values of air pollutant concentrations are small compared to their values near the sources. The buildup of these quantities must, of course, be watched very carefully, particularly those which could influence climate, like CO_2 and small particles. But it is unlikely that the particular sites chosen for power plants have any effect on the global buildup, which depends primarily on the total quantity of pollutants emitted. Present air pollution regulations have usually been interpreted as being concerned with the direct, local pollution contribution from a source, i.e., with contribution (1) above. Little attention has so far been paid to (2), the contribution from other sources, but its importance as more power plants are sited is obvious.

Consider a power economy in which generating plants are separated by some average distance s. The question can be asked, at what average spacing s should power plants be sited to hold the average contribution to the air pollution from all other plants to some fraction P of the direct contribution $\overline{X}_{AC}$?

The average contribution to the air pollution at a power plant from the surrounding plants at distance $x=s$ is just equal to $N\overline{X}_{AC}(x_C/s)^{p+1}$ where N is the number of surrounding plants. For a more or less even distribution of power plants, N should equal about 6. At the next ring of plants, $x=2s$. There are, however, twice as many plants at that distance, and the number increases linearly with distance. Thus, the average contribution to the air pollution near a power plant due to all other power plants is

$$\overline{X}_A = N\overline{X}_{AC}(x_C/s)^{p+1}\sum_1^n i^{-p}, \qquad (4)$$

where i designates each of the power plants, $i=1, 2, 3, \ldots n$. If $\overline{X}_A$ is to be equal to $P\overline{X}_{AC}$ then, solving for the spacing s, it is found that

$$s = x_C\left[\frac{N}{P}\sum_1^n i^{-p}\right]^{1/(p+1)}. \qquad (5)$$

It is interesting that contributions from fairly great distances $x=ns$ are significant. For $p=0.5$, corresponding to fairly poor atmospheric dilution, n must be taken to be equal to about ten.

If a uniform density of power plants corresponding to $N\simeq6$ is assumed and the value $P=0.1$ chosen, then $s\simeq40x_C$. If $P=0.01$, then $s\simeq180x_C$, the other quantities being the same. For conventional plants, the distance x_C to the point where pollution is controlled equals in some cases several kilometers; and for many nuclear plants, x_C is about a kilometer. Thus, for the average contribution to the air pollution around a plant from all other plants $\overline{X}_A$ to be limited to a few percent of the local air pollution control value $\overline{X}_{AC}$, the plant spacing must be of the order of 50 to 100 km. The effect of removal processes could modify this result, decreasing the required spacing considerably; and we are as yet, of course, very far from the density of large power plants that was assumed. However, we can conclude from the analysis that the influence of the air pollution imported from other sources is not negligible, and will have increasingly to be taken into account as the number and density of large power plants increases.

STACK GAS CONVERSION PROCESSES

The Effects of Chemical Conversion of Stack Gases

Atmospheric conversion processes are of major importance in determining the effects of power plant emissions on visibility, health, and climate. Visibility depends on the nature of the aerosol, particularly its size distribution and refractive index. Particle size distribution determines the site of deposition in the lung and, probably, particle transport to plant surfaces. Chemical composition in both the gas and particle phases determines the effect on biological tissue; there is good evidence that the degree of oxidation of sulfur oxides in emissions plays a key role so far as toxicity is concerned.

The effects on climate are of at least three types. First, there is the well-known "greenhouse effect" resulting from the accumulation of CO_2 in the atmosphere. It is interesting to note in this connection that the removal of SO_2 from power plant stack gases by the limestone injection process adds a few percent to CO_2 emissions since that gas is driven out of its carbonate form. Second, particulate matter in the atmosphere may be increasing as a result of increasing emissions and the conversion of sulfur oxides to particulate form as discussed above. This may lead to an increase in the earth's albedo and a reduction in the radiation reaching the earth's surface (McCormick and Ludwig, 1959). The importance of this effect has not been established yet. Third, particulate matter, either primary or secondary in nature, may serve as cloud condensation nuclei and so influence rainfall patterns. One possibility is that they may retard precipitation by producing an excessive number of droplets which are not sufficiently large to destabilize a cloud. On the other hand, the proper types of nuclei in sufficient numbers may seed clouds which would otherwise produce no precipitation. Thus the same cause can be invoked for two phenomena opposite in effect. This third effect may influence power plant siting the most. However, effects on climate are more likely to play a role in the choice of alternative technologies for power production and, possibly, in setting absolute limits on the amount of power which can be generated by fossil-fueled plants.

Chemical Processes

Stack gases and their mixtures with air are dynamic systems in which chemical reactions are taking place both in the gas and condensed (aerosol) phase, and the particle size distribution is changing as a result of coagulation and the transfer of materials between phases. These processes lead to the formation of new chemical species as well as changed physical form. Particulate material and gases formed in this way are termed secondary while the original fly ash and gases are regarded as primary.

An important process in such systems is the conversion of gaseous emissions to particulate matter. This can take place homogeneously (in the gas phase) or heterogeneously (in or on the surface of particles or drops) or by both mechanisms at the same time. The size distribution and chemical composition with respect to size play key roles in determining the effects of particulate pollutants on receptors.

Because of the complexity of the stack gas system, it is seldom possible to make more than semiquantitative estimates of the conversion rates under field conditions. If the conversion rates were known, it would, in principle, be possible to predict the concentrations of the products of the various chemical and physical processes downwind from a source, taking into account both mixing and advection. This is an active field of research at the present time (Urone and Schroeder, 1969). Although little progress has been made on the stack gas problem, it is clear that the point of maximum ground concentration for the products of the various conversion processes will not, in general, correspond to the point of maximum concentration for a nonreacting species.

It is important to have at least a rough idea of which conversion processes take place in order to predict the effect of the pollution products on visibility and health and, possibly, weather in the region downwind from the stack. In the sections which follow, the nature of these conversion processes is discussed for some of the major constituents of power plant emissions.

Sulfur Oxides

Sulfur oxides are highly reactive components in the atmosphere, even at low concentrations. Many laboratory studies simulating atmospheric conditions have been carried out and these have been reviewed by Urone and Schroeder (1969). Two field studies have been reported of conversion rates in stack gas plumes, and data are available on sulfate and sulfur oxide concentrations in urban areas.

The laboratory studies have shown that gaseous mixtures of olefins, NO_2, SO_2, and water vapor, when irradiated, produce aerosols containing sulfuric acid. Other experiments have shown that the absorption and oxidation of gaseous SO_2 are favored by the presence, in aqueous solution, of ammonia and the salts of manganese, copper, and iron. Similar substances are present in power plant emissions, although it has not been shown that they occur in the proper catalytic form or concentration.

Two field studies have been made of the rate of conversion of SO_2 in power plant plumes. Gartrell, Thomas, and Carpenter (1963) used a helicopter to collect the SO_2 and SO_3 components in progressive plume cross-sections of a large coal-burning TVA power plant. They found rapid oxidation rates—from about 0.1% up to 2% per minute—with a strong dependence upon the ambient relative humidity and the presence of water droplets in the plume. Weber (1970) found that SO_2 from electrical power stations was lost in the Frankfurt am Main atmosphere at rates between 0.8% and 2.5% per minute, depending on the meteorological conditions. The final state of the SO_2 was not identified. Weber stated that these rates are valid for the time period immediately after the plume leaves the stack, or for a highly polluted atmosphere under neutral or stable meteorological conditions in the winter period.

The relative concentrations of SO_2 and sulfate in urban atmospheres are another indication of the extent of the conversion processes. Values for the mass ratio of sulfate to SO_2 for different cities vary considerably, but most fall in the range 0.1 to 1.0 (HEW, 1966).

It is difficult to predict which of the competing processes studied in the laboratory control the loss of

SO_2 in power plant plumes. In the daytime and at low humidity, it seems likely that photochemical reactions involving SO_2, NO_2, water vapor, and perhaps hydrocarbons are of primary importance in the transformation of SO_2 into H_2SO_4 aerosol. At night and under conditions of fog or high humidity, or during rain, the absorption of SO_2 by aqueous solutions followed by reaction to form SO_4^2 within the drops is a well-documented process. Further investigations of conversion processes in actual plumes are needed.

Nitrogen Oxides

Nitric oxide (NO) is the principal oxide of nitrogen emitted from power plant stacks. In the presence of ozone, which may be generated photochemically or in the electrical precipitators used for fly ash removal, NO is rapidly converted to NO_2 by the thermal (non-photochemical) reaction

$$NO + O_3 \rightarrow NO_2 + O_2.$$

Nitrogen dioxide is relatively reactive and will form nitrates by such reactions as the displacement of chloride ion from aqueous solution

$$2NO_2 + NaCl \rightarrow NOCl + NaNO_3$$
$$NOCl + H_2O \rightarrow HCl + HNO_2$$
$$NOCl + h\nu \rightarrow NO + Cl$$

Little is known about the rates of such conversion processes in plumes, and there have been relatively few laboratory investigations simulating atmospheric conditions of the type discussed above for the sulfur oxides. Data are available for the mass ratio of nitrate ion to NO_2 for various cities in the United States (HEW, 1966), and these values tend to fall in the range of 2% to 4%. Hence the conversion rate for NO_2 under plume conditions is likely to be smaller than for SO_2.

Fly Ash

Physical processes involving particulate matter include coagulation, sedimentation, diffusion, and condensation or evaporation. Equations describing these processes individually and in combination can be written and have been solved for certain special cases (Pich, Friedlander, and Lai, 1970). However, the complexity of the stack gas system and the lack of information on many parameters have blocked realistic solutions for the complete problem. Qualitatively, however, the processes taking place are rather straightforward. As the hot moist gases mix with the ambient air and cool, water vapor condenses out either by homogeneous or heterogeneous nucleation (Hidy and Friedlander, 1964). Larger particles and/or droplets sediment out in the vicinity of the stack while smaller ones remain airborne. Eventually the droplets evaporate as mixing proceeds, with the moisture content of the particles reaching a value in equilibrium with the ambient conditions.

At the same time, the condensation and chemical

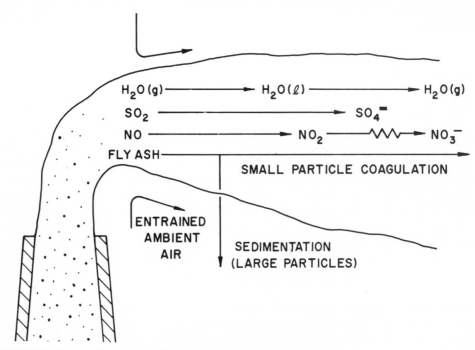

FIGURE 1 Schematic diagram of stack gas conversion processes.

reaction processes involving the sulfur and nitrogen oxides lead to the formation of sulfate- and nitrate-containing particles. These combine by coagulation with the fine fly ash particles and with background particulate matter in the ambient air entrained into the plume. The final result is a mixture of particles (containing fly ash, sulfate, nitrate, a background contribution, and water) suspended in a gas containing the residual sulfur and nitrogen oxides and water vapor. Figure 1 summarizes schematically the various conversion processes which occur in the plume.

Instrumentation

A major problem in the air pollution field is the difficulty in characterizing complicated reacting particle–gas systems such as power plant stack gases (Friedlander, 1970). In-stack measurements are particularly difficult because of extreme conditions of particle concentration, temperature, and water vapor concentration. Even ambient air measurements are not easy to make. Particle size distributions for the range between a few hundred angstroms and a few microns can be measured by hybrid devices (Whitby *et al.*, 1971) on a continuous, on-line basis. However, comparable methods do not exist for the measurement of the sulfate concentrations in particulate matter or, for that matter, any chemical component of pollution aerosols. Instruments are available for the continuous, on-line monitoring of major gaseous effluents including SO_2 and nitrogen oxides.

Using techniques such as neutron activation analysis (Guderian and van Haut, 1970), it is possible to determine the elemental composition of an aerosol sampled with a multistage impactor or filter. The analysis is carried out on a sample collected over a period of hours.

2

ENVIRONMENTAL EFFECTS OF AIR POLLUTANTS EMITTED FROM FOSSIL-FUELED POWER PLANTS

The purpose of this chapter is to review and summarize the effects of air pollutants emitted by electric power generating facilities on human health, vegetation, and the physical environment. Research needs and suggestions, both for baseline studies and for evaluation of environmental effects in the vicinity of fossil-fueled power stations, are given at the end of the report. The material brought together in this review should be useful for the planning of new power generation sites and for evaluating environmental hazards associated with the production of electricity.

HUMAN HEALTH

A few generalizations are in order at the outset. Except perhaps in the instance of CO, it is not possible to isolate or quantify the effects of the individual emissions. Health surveys customarily rely on measurements of only a few pollutants, but these choices are often dictated by technical or economic considerations. They do not necessarily represent the biologically active components of air pollution. Instead, in many instances, the selections may be considered primarily to be indicators of the general severity of the pollution. A combination of laboratory and epidemiological studies is probably needed to achieve a valid estimate of the importance of a particular emission.

To the extent that air pollutants affect health, they tend to do so in combination with each other and with other forms of stress, rather than independently. The other forms of stress may be quite diverse. Examples

are cold, damp weather which may act as a crucible for the production of irritant aerosols and may diminish the resistance of the upper airways to infection; socio-economic disadvantages that are attended by inadequate food, poor hygiene, overcrowding; and tobacco smoking, sometimes referred to as a personal form of air pollution. The latter appears to be the single most important factor causing chronic respiratory disease. It may be said that air pollution acts chiefly to aggravate rather than cause disease. As a corollary, it may also be said that there are segments of the population which, owing to disease (especially respiratory and cardio-vascular), age (very young and old), and genetic defects, appear especially vulnerable to pollution. These groups require particular consideration when air quality standards are formulated. The public's concern for their well-being will ultimately be reflected in the stringency of the standards.

Episodes of air pollution may be brief or prolonged, infrequent or commonplace. Severe episodes may be associated with striking increases in illness and death. Less susceptible to measurement in man are the biochemical changes in lungs and other tissues that may occur in response to chronic exposure to low levels of pollutant gases. Such slowly developing reactions have been shown in laboratory animals. They may have implications for the rate of aging of the lung and the onset of degenerative changes.

In many instances, low concentrations of pollutants that appear innocuous to man may cause damage to plants or animals. As we come to recognize the inter-relatedness of all living systems—as, for example, through food chains—we recognize that injury to other species of life may shortly impinge upon human welfare.

Some "nonhealth" effects of air pollutants, such as eye irritation and odors, may be highly objectionable; others, like reduced visibility, may interfere with skilled performance and increase the risk of accident.

Finally, biological systems have the potential to adapt to environmental stresses. Whether and to what extent adaptations occur to air pollutants is largely unknown.

Oxides of Sulfur

Sulfur dioxide (SO_2) is the most abundant and hazardous gas among the oxides of sulfur. In urban atmospheres, it is converted in part to either sulfuric acid (H_2SO_4) or sulfate salts, both of which can be more irritating to the lung than the parent gas. SO_2 is colorless; estimates of the odor threshold range from about 0.5–1 ppm. An important characteristic of SO_2 is its high solubility in tissue fluids at body temperature. As a consequence, virtually all of the gas in inspired air is taken up by the airways above the trachea, particularly during quiet, nasal breathing. Penetration to the peripheral airways increases when breathing is by mouth and rapid, as in exercise; consequently, at a fixed environmental concentration of SO_2, the exposure of the peripheral airways and alveoli increases with activity out of proportion to the simple increase in the volume of ventilation.

The results of laboratory investigations in man and animals indicate that SO_2 at realistic concentrations acts chiefly as an irritant that causes reflex constriction and narrowing of the tracheobronchial tree. The constriction increases the resistance to the flow of air so that the work of breathing is exaggerated. Furthermore, to the extent that the narrowing is unevenly distributed among the airways, there may also be an inequality in the distribution of ventilation. When inspired air is unevenly distributed in the lung, the exchange of oxygen and carbon dioxide may be impaired. In healthy volunteers, the reflex constriction reaches a maximum within minutes and then recedes. Whether the constriction becomes modified by intermittent exposure, whether repeated constriction predisposes the lung to disease, and, most importantly, whether the response of patients with respiratory disease differs in magnitude and type from that of healthy persons are not known.

SO_2 alone is not considered to be a hazard to health except in an occasional highly susceptible person. Its significance probably arises from its interactions with other airborne contaminants, particularly particles. Laboratory investigations have shown that the increase in flow resistance of the lung in animals due to SO_2 is potentiated if the gas is combined with an aerosol of sodium chloride. The aerosol alone is physiologically inert. Two explanations for this synergism have been offered. One is that the aerosol acts as a vector that takes up and transports the gas to more vulnerable portions of the lung. The second, more likely explanation, is that a chemical interaction(s) occurs between the two agents so that SO_2 is converted to a more irritating substance. Sulfuric acid and sulfate salts are potential products. In addition to the nature of the transformation, the size of the particle will also influence the magnitude of the synergism. For a specified concentration of the aerosol, synergism is greater if the particle diameter is under rather than over one micron. (Such evidence underlines the importance of reporting the size distribution of airborne particles in addition to their concentration and composition.) To date, the evidence for synergistic combinations of SO_2 and aerosols in man is controversial.

Epidemiologic surveys have shown that urban air pollution in which SO_2 and particulates are important constituents may be associated with elevated rates of respiratory infection (bronchitis and pneumonia) in childhood and adolescence, increased absenteeism from work, increased prevalence of chronic bronchitis, and increased mortality rates in patients with cardiorespiratory disease. There is evidence, requiring confirmation, for a "no-threshold" response among high-risk individuals: In a study recently conducted in Chicago, elderly patients with chronic bronchitis exhibited symptoms of increased illness at each elevated level of pollution (as measured by SO_2). The importance of considering SO_2 as acting in association with aerosols (and weather) rather than alone cannot be overstressed. Preliminary reports from London suggest that a reduction in dust (resulting from emission regulations) in the absence of a significant reduction in SO_2 levels has been associated with improvement in symptoms of patients with chronic pulmonary disease.

Particulates

Airborne particles may affect well-being or health by reducing visibility, causing disagreeable odors (some constituents may be aromatic), interacting with other pollutants to form new, possibly more toxic materials, and through direct effects on the receptor. Of the last two possibilities, reference has already been made to the consequences of combining droplet aerosols with SO_2. Dusts that are generated industrially may contain a variety of metals known to be associated with inflammatory and neoplastic diseases. Some of these metals, for example, nickel and cadmium, may be found in urban air; the nickel present in urban air arises chiefly from the combustion of coal and petroleum.

Particle Deposition and Clearance

The health effects of an airborne particle will be influenced in part by its rate and site of deposition in the respiratory system and thereafter by how efficiently and rapidly it is cleared from the body. Other significant modes of entry are the skin and gastrointestinal system. The upper limit of aerodynamic size for respirable particles is a diameter of about 15 microns. Larger particles settle too rapidly to be of consequence to health. Particles greater than 1 micron deposit chiefly through gravitational settling, inertial impaction and interception; those under 0.1 micron chiefly through diffusion (Brownian motion). The larger particles are deposited mostly in the upper airways and at the branching sites of the tracheobronchial system. The submicron particles settle more readily in the narrow peripheral airways and the alveoli. If the particle is relatively soluble, it may have a systemic effect that is independent of the site of deposition. If it is relatively insoluble, the site of deposition will influence the mechanism by which it is cleared from the respiratory system and, consequently, the time required for the clearance to occur. The particle may either be cleared mouthward or penetrate the alveolar lining. The time required by the mucociliary system to remove particles from the airways is measured in minutes or hours. Particles that pierce the alveolar wall may persist in the lung for years. The latter outcome offers considerable potential for disease.

Effect on Health

Associations have been shown between particulate concentrations and death rates from selected causes. In Buffalo, an increase in death rates was noted for both sexes, 50 to 69 years old, once the annual geometric mean particulate concentration equaled or exceeded 100 micrograms/m³. In Nashville, an increase in death rates was evident at levels equivalent to 1.1 or more coh units. (A coh unit is a measure of light absorption by particles, defined as the quantity of light-scattering solids that produces an optical density of 0.1.) SO₂ was also present as a pollutant in each of these studies. Additional surveys in Great Britain have shown that the severity of chronic bronchitis was worse when the daily average concentration of smoke exceeded 300 micrograms/m³ and that the absenteeism of workers due to illness rose when the concentrations of smoke rose above 200 micrograms/m³.

Oxides of Nitrogen

The oxides of nitrogen are prominent among the compounds that comprise photochemical air pollution, or smog. Two members of this family of compounds, nitric oxide (NO) and nitrogen dioxide (NO₂), reach concentrations in the atmosphere sufficient to represent possible hazards to health; of the two, NO₂ appears to be the more toxic. Nonetheless, the chief concern over these compounds arises from their capacity to combine in sunlight with a variety of hydrocarbons to form reactive, unstable oxidants, the most conspicuous of which are ozone and peroxyactyl nitrate. Brief reference will also be made to the latter oxidants.

Laboratory Studies, Mechanisms of Effect

The oxides of nitrogen may discolor the atmosphere, damage vegetation, and cause functional and structural changes in the lung. In particular, NO appears to play a role in causing eye irritation. Laboratory experiments have shown that concentrations of the gas above 1 ppm are associated with an increase in severity of eye irritation; this may reflect the contribution by NO to the production of acrolein, an aldehyde. NO₂ is a stable compound which appears to act chiefly on the smaller peripheral airways and alveolar capillary membrane. In animals, brief exposure to relatively high levels of the gas produces pulmonary edema. Recovery from the edema may later be attended by thickening and obliteration of smaller airways (bronchioles). Chronic exposure of animals to about 2 ppm of NO₂ has caused alveolar damage that is analogous in appearance to emphysema in man.

Ozone, like NO₂, appears to act principally on the more peripheral portions of the lung. It is particularly toxic to the capillary endothelial wall, causing an increase in capillary permeability that leads to pulmonary edema. The edema may follow exposure to concentrations of O₃ only slightly in excess of those encountered in Los Angeles.* O₃, like SO₂, may also induce reflex constriction of the larger airways.

NO₂ and O₃ oxidize fatty elements contained in cell membranes as well as in those membranes that encapsulate special subdivisions of the cytoplasm (such as the mitochondria and the lysosomal enzymes). This lipoxidative action may play a role in the increased susceptibility to specific respiratory infections that follows the exposure of animals to these gases. For example, prolonged exposure of mice and monkeys to 0.5 ppm of NO₂ increases their susceptibility to fatal infections with Klebsiella pneumonia while about 0.1 ppm of O₃ has been shown to reduce the phagocytic activity of the alveolar macrophages in mice. The macrophages are

---

* Experimentally, many species of animals develop tolerance to ozone with repeated exposure as measured by a reduction in the amount of edema that forms. Whether man develops similar tolerance is unknown.

responsible for clearing inhaled particles, both animate and inanimate, from the alveoli. They contribute importantly to the maintenance of sterility in the periphery of the lung.

NO_2 and O_3 can denature proteins, thereby altering the configuration and molecular weight of these complex molecules. There is evidence that the structural proteins of the lung, especially collagen, may be denatured in this way. Such changes in the structural proteins could influence the elastic behavior, rate of aging, and vulnerability of the lung to nonspecific degenerative diseases.

The precise chemistry of both the lipoxidative and denaturing actions of these gases in not known. Ozone is thought to act at least in part through its formation of free radicals. NO_2 produces free radicals less actively. There is virtually no information on the possible effects of peroxyacetyl nitrate on animals and man.

Studies in Man

Acute exposures of healthy volunteers to realistic concentrations of NO_2 have not produced pulmonary functional changes. By contrast, exposure to less than 1 ppm of O_3 has produced narrowing of the airways and reduction in the rate of diffusion of CO across the alveolar–capillary membrane; both changes appear to remit within hours of the end of exposure. Symptoms of pharyngeal irritation, cough, and substernal burning are commonly observed in such experiments. The results of surveys of large populations subjected to photochemical air pollution have been equivocal, although recent data suggest a significant association between oxidant levels and elevated morbidity and mortality rates in Los Angeles. The possible health effects of the oxides of nitrogen and ozone and its satellites on health is a field that is in urgent need of additional study.

Hydrocarbons

To date, over 50 hydrocarbons have been identified in urban air samples, the most common being the alkanes. The principal significance of the hydrocarbons, as the Air Quality Monograph of the Public Health Service * states, derives from their participation in photochemical reactions. The oxidants generated by these reactions have been discussed. Other important by-products are aldehydes, including formaldehyde and acrolein, which are thought to contribute to the annoying odor and eye irritation associated with photochemical pollution. Aldehydes may also act as irritants to the respiratory mucosa. For example, formaldehyde and acrolein have been shown to increase pulmonary flow resistance in animals at the same concentrations that evoke eye and

* See Bibliography.

nasal irritation in man. While the inhalation of vapors of the aromatic hydrocarbons may cause systemic injury, no such effects have been reported except at levels far in excess of those encountered in urban air. The polynuclear hydrocarbons are potentially carcinogenic. Of the group, benzopyrene is considered to be the most potent carcinogen. There is uncertainty over the role played by these polynuclear hydrocarbons in the higher rates of cancer of the lung found in urban environments as compared with rural environments.

Carbon Monoxide

Unlike the other pollutants under consideration, CO does not irritate or inflame the respiratory system. Instead, it is an asphyxiant that acts to deprive the peripheral tissues of oxygen. Although motor vehicle traffic is considered the chief source of CO in urban air, the greatest exposure to the gas ordinarily occurs during tobacco smoking. Estimates of the concentration of CO in inhaled cigarette smoke range from about 400 to 450 ppm. By comparison, several cities have established standards for CO ranging from 15 to 30 ppm, based on an 8-hour average.

Mechanism of Action

CO and O_2 are transported in the blood chiefly in combination with hemoglobin (Hb). CO has the greater affinity, being about 210 times more effective in combining with Hb than is O_2. The combination is termed carboxyhemoglobin (COHb). Based on several sources of data, the California State Department of Public Health has derived equations for estimating the fraction of circulating Hb that is present as COHb:

Before equilibration is reached, for concentrations between 100 and 2,000 ppm

$$COHb\% = CO \times KT, \qquad (6)$$

where CO is the inspired concentration in ppm, T is the time of exposure in hours, K is a constant that depends on the (alveolar) ventilatory rate which, in turn, increases with activity. At rest, K is 0.018; during light work it is 0.048.

At equilibration, for concentrations under 100 ppm

$$COHb\% = CO \times 0.16, \qquad (7)$$

where CO is the inspired concentration in ppm. The time required for equilibration depends chiefly on the (alveolar) ventilatory rate. The greater the individual's activity and ventilation, the shorter the time. For light activity, the half-time for equilibration is about 2 hours.

CO not only reduces the oxygen-carrying capacity of

the arterial blood but also affects the strength of the bond between O_2 and Hb so that the O_2 is less readily released by the blood to the peripheral tissues. Both of these actions of CO aggravate the tissue hypoxia. The two organs most sensitive to the deprivation of oxygen are the brain and the heart.

Effect on Performance and Behavior

Changes in visual and time perception and in various psychomotor tasks have been reported at COHb concentrations about as low as 2–5%. [According to NAPCA (1969b), continuous exposure to about 13 ppm of CO would produce a COHb concentration of 2%.] It should be noted that some of these studies have been repeated by different investigators without confirmation of the results. Additional research of this type is, therefore, needed. It is of obvious importance to determine whether temporary impairment of the highest cortical functions attends exposure to the CO levels encountered in urban atmospheres.

Effect on Health

The population at greatest risk is likely to include persons already hypoxic from anemia, reduced cardiac output, altitude, or chronic pulmonary disease; persons with increased O_2 needs as in pregnancy; and patients with coronary vascular disease. There is evidence that the fetus may be particularly vulnerable to the asphyxiant effects of CO; the aged may also fall into this category.

Epidemiologic studies have focused on the cause and aggravation of illness, on accidents, and on interference with performance. The well-established association between cigarette smoking and greater morbidity and mortality from coronary vascular disease may in part be related to the elevated COHb typical of smokers. In animals, it has been possible to show that continuous exposure to several hundred ppm of CO increases the likelihood of inducing atherosclerosis with a high-cholesterol diet; however, the levels of COHb achieved with such exposures are much higher than those ordinarily encountered in man. In man, COHb levels under 10%, less than those found in heavy smokers, can impair the metabolism of the heart, especially in the presence of coronary vascular disease. An increased level of COHb in drivers involved in auto accidents (relative to the levels found in control subjects) has been reported, but the evidence implicating CO in the accidents is still insubstantial. Whether acclimatization occurs to chronic CO remains unclear.

EFFECTS ON VEGETATION

Sulfur Dioxide

The toxicity of sulfur dioxide to vegetation has been known for about 100 years. Since then, numerous cases of plant injury due to exposure to sulfur dioxide have been documented from both field and experimental investigations. Absorption of sulfur dioxide by plant leaves during the normal gas exchange process is followed by oxidation to sulfite and sulfate. In many species, it has been shown that the absorbed sulfur is incorporated into organic compounds.

Under some circumstances, plants grown in soil deficient in sulfur may benefit from absorption of atmospheric sulfur dioxide as long as the concentrations are insufficient to cause toxic effects. However, rapid absorption leads to injuries at all levels of biological organization of plants. Poisoning of biochemical processes occurs and is followed by discoloration and death of leaf tissues, premature loss of leaves, reduced plant growth, decreased yield, deformations of leaves and branches, and increased mortality of plants (Guderian and van Haut, 1970). Sensitive species die out in an area repeatedly contaminated by sulfur dioxide, and resistant plants take over (Treshaw, 1968). As a result, the number of species is reduced and the plant density is decreased; the latter effect may result in climatic changes. Larger plants are usually affected first and are replaced by shrubs which may eventually give way to low-growing plants (Gordon and Gorham, 1963; Woodell, 1970). In cases where plant cover is totally eliminated, measurable changes have taken place in air and soil temperatures, wind speed at ground level, evaporation rates, humidity, and rainfall (Hursh, 1948).

In recent years in the United States, concentrations have seldom been sufficiently high to cause the devastation which occurred around smelters during the first half of the century (Scheffer and Hedgecock, 1955). However, acute and chronic effects have been demonstrated when continuous or intermittent exposure to relatively low concentrations of sulfur dioxide (Linzon, 1971; Dreisinger and McGovern, 1970; Wood, 1968) occur. Effects may not be observed for days, weeks, months, or, in some cases (conifers), even years after the initial period or periods of exposure.

Symptom expression and its severity depend upon the interaction of a complex group of factors. Predisposition to injury is determined by: (a) genetic susceptibility (species and variety), (b) age of plant, (c) stage of plant development, (d) vigor and growth rate, (e) concentration of pollutant and duration of exposure, (f) frequency of pollutant episodes and time between exposures, (g) time of day of exposures, (h) nutrition of the plant, (i) water supply, and (j) environmental conditions (temperature, humidity, light intensity, and photoperiod) before, during, and after periods of exposure. For these reasons, direct relationships do not usually exist among measurements of injury to field-grown plants by sulfur dioxide, growth or yield reduction, total emissions, atmospheric sulfur dioxide concentrations, or the amounts of sulfur accumulated by the plant from the air.

Experimental and field investigations have shown that some of the more sensitive plant species are alfalfa, buckwheat, barley, clover, pumpkin, squash, oats, rye, wheat, cotton, soybean, apple, pear, Eastern white pine, Ponderosa pine, larch, Jack pine, birch, elm, and aspen. The occurrence and severity of injury to sensitive species depend on several factors in addition to those listed in the previous paragraph. Continuous low concentrations of sulfur dioxide may be more injurious than intermittent exposures when total doses (product of concentration of pollutant and duration of exposure) are equivalent. Plants appear to have the capacity to recover from sublethal exposures between fumigation periods. However, peak concentrations may be more injurious than continuous low exposures when the intermittent concentrations are sufficiently high to cause severe injury. There is significant risk of injury to sensitive vegetation on exposure to long-term average sulfur dioxide concentrations of more than 0.02 ppm when there are also repeated short-term peak concentrations of more than 0.02 ppm (averaging time ten minutes or longer) (Guderian and van Haut, 1970). Injury to sensitive species may occur if these concentrations are exceeded. However, the protection afforded by keeping ambient SO_2 concentrations below these levels would reduce the risk of damage to vegetation by sulfur dioxide to a negligible level.

Nitrogen Dioxide

In general, plants seem to be less sensitive to injury by nitrogen dioxide than by sulfur dioxide. However, sensitive species such as bean and tomato, on continuous exposure to concentrations of about 0.5 ppm, have exhibited the following symptoms: leaf distortion, growth suppression, and discoloration of leaves (Taylor and Eaton, 1966). Of greater significance is the observation that growth suppression may occur without death of tissues (Taylor and Eaton, 1966). Premature loss of leaves of citrus and reduced fruit yield were produced by exposure to 0.25 ppm NO_2 for 35 days (Thompson *et al.,* 1970).

Photochemical Oxidants

Nitrogen dioxide and hydrocarbons undergo photochemical reactions to form ozone and peroxyacetyl nitrates which are very toxic to vegetation. In the United States, ozone, a major constituent of photochemical smog, probably causes more injury to garden crops, ornamental plants, and forest trees than any other air pollutant (Hill *et al.,* 1970). Concentrations of ozone sufficiently high to injure plants have been found in or near many cities of the United States and more than 70 miles from heavily populated areas (Dugger and Ting, 1970). Extensive damage to crops and reduced quality and yield occur in the Los Angeles area and in parts of eastern North America every year (Mukammal, 1965). Sensitive plant species may be injured by exposure to as little as 0.05 ppm ozone.

Peroxyacetyl initrate concentrations are relatively low in photochemically polluted atmospheres; however, their high phytotoxicity has also led to severe plant injury (Taylor and McLean, 1970). Parts per billion concentrations (by volume) are sufficient to cause easily recognizable symptoms on leaves of susceptible species (Taylor, 1968).

The complicated nature of atmospheric reactions makes future trends difficult to predict; however, it would appear that rapidly rising emissions of nitrogen oxides increase the potential for production of ozone and peroxy compounds. These pollutants are formed in and are carried by air masses which may cover wide areas (Mukammal, 1965). Furthermore, recent evidence indicates that combinations of pollutants may act synergistically on some plants (Menser and Heggestad, 1966), further increasing the possibility of harmful effects.

Hydrocarbon Compounds

Ethylene is toxic to sensitive plants in parts per billion concentrations and causes deformations and growth inhibition (Heck *et al.,* 1962). Aldehydes can also cause injury to vegetation, but, in general, other hydrocarbon compounds are not as phytotoxic as inorganic air pollutants (Heck and Dires, 1962).

Particulate Matter

Dust particles fall out and deposit on plant leaves. Dense stands of vegetation, strategically situated, are used in some parts of the world to slow air movement and aid in atmospheric cleansing (Bernatzky, 1965). If toxic elements are present in particles, it is possible that this approach may bring these components into the soil and the wildlife food chain. Deposition of small particles on leaves may directly injure plants by plugging stomata (pores for gas exchange) and interfering with essential physiologic processes. In addition, toxic water-soluble substances such as acids, in particulate matter, may be dissolved by rain or dew and cause injury to leaves.

Sulfuric Acid Mist

Acidic compounds are injurious to plants when they overcome the capacity of the plant to maintain its normal acid–base balance. Deposition of sulfuric acid aerosols can, in this manner, harm plant organs.

Carbon Monoxide

Vegetation appears to be very resistant to injury by carbon monoxide, contrary to this pollutant's effects on man. However, plants do seem to be involved in carbon monoxide production and utilization (Delwiche, 1970). Data are not yet sufficient to indicate whether carbon monoxide absorption by plants occurs rapidly enough to be of significance on a large scale.

EFFECTS ON THE PHYSICAL ENVIRONMENT

Effects of Heat and Water Vapor

Whereas one can visualize methods that would virtually eliminate most pollutant emissions, the generation of electric power inevitably implies the release of heat, and the very rapid increase in both the number and size of sources raises serious questions about the effects of this release. In view of the certainty of the continuing association between energy conversion and heat emissions, the potential problems should receive careful consideration on long, as well as short, time scales, for they will be with us indefinitely.

It seems clear also that the atmosphere must continue as the main "dumping ground" for excess heat, because its dispersive capacity is far greater than that of fresh water bodies available for such purposes, and also because air-breathing life forms are both more mobile and less vulnerable to environmental variations than aquatic life.

Unlike most of the other pollutants from power-generating facilities, heat and water vapor are not only natural properties of the atmosphere, but they are exceedingly variable properties as well. This variability, coupled with man's ability to tolerate a wide range of climatic conditions, makes the establishment of standards for heat and moisture a very different problem from that encountered in dealing with alien chemical compounds, where one need only accept a given concentration level as having significant effects. It is conceivable that one could transform the climate of an entire region from temperate to tropical without severely damaging the health of the human population, and limits must therefore be based on grounds of aesthetics, comfort, and cost rather than health.

Despite the current preoccupation with short-term, local problems, it is instructive to examine the possible global significance of heat liberation, if only to be certain that important, irreversible consequences are not overlooked. In comparing global and local heat problems, it is very important to distinguish between total heat and waste heat. Large-scale problems should be treated in terms of the total heat from all energy conversions, including the useful electric power as well as the waste, and the contribution from space heating, transportation, and manufacturing. Locally, it is appropriate to consider the waste heat alone, since it is a concentrated source producing its own immediate effects.

There are numerous means of comparing man's heat production with that of nature on a global scale. The latter is the solar heat available for driving the atmospheric heat engine ($\sim$150 watts/m^2). Currently, our mean global energy production is 0.01 watts/m^2, and it is unlikely that the increase will be greater than a factor of 5 by the year 2000. Man's contribution, therefore, will then still be much less than 0.1% of the natural heat, and it seems very doubtful that such a small change would seriously alter the global balance.

Acknowledging, therefore, the immediate priority of local problems, it is again useful to examine nature's heat budget, since some such yardstick must be used to assess the acceptability of new power installations. In terms of sensible heat, the mean value of absorbed solar energy at the earth's surface is 350 MW/km^2, and the 1,500–2,500 MW of waste heat from a 1,000-MW (electrical) power plant is, therefore, equivalent to the sun's influence over an area 2.5 km on a side. Translated in terms of water vapor, as is largely the case with an evaporative tower, the rate of moisture release is approximately equivalent to the natural evaporation from the soil and vegetation over an area 4.25 km on a side on a typical summer's day in the central United States. Thus, a 1,000-MW power plant represents a large but not enormous abberation of natural conditions in a local area.

The universal direct effect of the emission of waste heat to the atmosphere is to produce buoyant plumes of warm air whose volume depends upon the characteristics of both the generating facility and the cooling system. In a fossil-fuel plant, some of the waste heat issues from the stack as a compact hot plume (250–350°F) in which chemical pollutants are also contained. The remainder of the sensible heat is released in a much larger volume of warm (15–25°C above ambient temperature) air. With an evaporation tower, much of the waste heat is initially in the form of water vapor, but additional sensible heat is released as soon as the water vapor begins to condense into a cloud of droplets. This moisture returns to the latent form as these droplets evaporate near the periphery of the plume. In a nuclear plant, all of the waste heat issues from the tower since there is no exhaust stack for combustion, but the processes are otherwise similar.

The second direct effect of the typical heat emission from evaporative cooling towers is the formation of a visible plume of water proplets. These droplets are embedded in the rising column of air, and their extent and significance are directly related to the ambient meteorology and topography. The total capacity of the air to hold water vapor as a gas decreases very rapidly with temperature, and at 0°F it can maintain only 3% of the water vapor that it is capable of holding at 80°F. Thus, at 0°F temperatures, a visible cloud will almost always persist for long distances, since even with completely dry ambient air, the water vapor released in one minute from a 1,000-MW nuclear plant is enough to saturate 60 million m^3 of air. At high ambient temperatures, relative humidity does become important, and at 80°F and 50% RH, this same emission will saturate only 2 million m^3. To the general public, the most obvious effect of the tower release will be the creation of a visible plume in very cold or humid conditions which, together with the massive "gum drop" tower, creates an aesthetic problem.

It seems quite evident from the limited research and observations available that these plumes will usually remain above ground in flat or gently rolling terrain. They will cause small changes in the mean ambient temperature and humidity over a very broad area, but these would normally become significant only in conjunction with numerous other sources.

If, however, the terrain is rugged or man-made structures rise to great heights, serious environmental problems may be created. First of all, the intersection of such a plume with obstacles will produce local fog, with possible disruption of transportation and noticeable alteration of local temperature and humidity. When the ambient temperature is below freezing, such a cloud can produce icing on the vegetation, roads, and structures.

There may also be instances of very light precipitation falling directly from the cooling tower plumes, usually as drizzle or a dusting of snowflakes. There have been reports of the fall of large water droplets from cooling tower plumes, but these seem to have been associated with earlier tower designs in which the device known as a "drift eliminator" was not used. The drift eliminator is essentially a baffle designed to prevent the transport of large droplets directly from the tower structure, and all modern units are equipped with them.

In addition to these apparent and direct effects of cooling tower operation, there are indirect results to be considered. Natural processes frequently tend to develop cloud formations, and each locality has its individual peculiarities. For example, during the winter one finds a high incidence of low stratocumulus clouds and snow showers in the areas lying southeast of the Great Lakes; the central midwest receives much of its summer rainfall from thunderstorms; stratus clouds are common along the west coast of the continent. Natural sources of moisture and upward air motions favor certain cloud developments, and cooling towers may certainly do the same. The larger and more concentrated the heat and moisture release, the more likely the cloud generation becomes. Essentially, the plume may act as a trigger mechanism, initiating cloud development where it might not occur otherwise, or increasing the dimensions of the existing clouds. Instances have already been noted wherein a cooling tower plume has created cumulus clouds in otherwise clear air, and there is no reason to doubt that showers or thundershowers could be started in this way. One should not, however, overemphasize the importance of this triggering mechanism, since the cooling tower plume can initiate cloud formation only when nature itself permits the development. No cooling tower is likely to create thunderstorms in areas where nature does not, and generally clear regions are not suddenly going to become cloudy. At most, the tower plume could influence slightly the frequency and location of natural phenomena.

Another conceivable indirect effect of the tower plume might involve an interaction with the chemical pollutants from the exhaust stack of a fossil plant. Merging of the two may alter the rates and compounds derived from conversion of these pollutants, but it is by no means intuitively obvious that this effect would necessarily be detrimental or significant. Interaction between cooling tower plumes and radioactive by-products from a nuclear plant should produce no deleterious effects. In fact, deliberate release of radioactivity in the plume might reduce the local ground-level effects.

It should also be recognized that the problems associated with heat and water vapor release are not likely to include rude environmental surprises. The meteorological profession has long been dealing with clouds, fog, precipitation, and icing; and while the source of heat and moisture is new, the atmospheric problems are not. Obviously, as cooling towers grow in capacity and number, the effects will become more obvious and frequent, but nothing disastrous seems likely to develop suddenly.

Visibility

Emissions from power plants contain large masses, numbers, and types of particles. Small-diameter particles (0.1 to 1 micron in radius) contribute to haze and smog formation, lowering visibility and decreasing contrast between distant objects. Gaseous emissions from fossil-fuel combustion, namely, sulfur and nitrogen oxides and hydrocarbon compounds, also contribute to the atmospheric particulate load by indirect means. Both photochemical and thermal reactions occurring in the atmosphere generate large numbers of particles which reduce visibility. It has been estimated that more than half of the aerosols in the lower atmosphere come from chemical reactions involving volatile hydrocarbons, nitrogen oxides, ammonia, and sulfur compounds (Hidy, 1970). This man-made contribution is about 6% of the 10^7 tons per day emitted from all earthly sources. These man-made contributions may cause an 11% increase in aerosol concentration by the year 2000 (Hidy, 1970). Sulfur dioxide concentrations of 0.1 ppm reduce visibility significantly by aerosol formation when atmospheric relative humidity is greater than 50% (NAPCA, 1969c). Although it is very difficult to determine the exact consequences of power plant emissions to visibility reduction, it is clear that generation of electricity does make a significant contribution to the man-made production of visibility-reducing aerosols and particles.

Solar Radiation

Small particles are particularly effective in diffusing and reflecting, as well as absorbing, light. Recent estimates indicate that a reduction in the total amount of solar radiation reaching the earth may be occurring due to accumulation of particulate matter in the atmosphere (Bryson and Wendland, 1968).

Corrosion and Soiling

It has been demonstrated that materials of many kinds corrode and discolor more rapidly in atmospheres containing sulfur oxides and particulate matter. Buildings, surfaces, statues, bronze, steel, iron, clothes, dyestuffs, paints, and rubber are affected.

Acidification of Rain and Soil

Increased sulfate and sulfuric acid content of rainwater, lakes, and soil in areas near and far from pollution sources has been demonstrated in several parts of the world (Gorham and Gordon, 1960). The immediate consequences to the biota from current power plant emissions appear small. However, continued deposition coupled with rapidly increasing quantities of fuel consumed for electric power generation requires that more attention be paid to the possibility that soil microorganisms and plants might be harmed by fallout of pollutants in rain in areas downwind of large generating facilities.

Climate

Subtle changes in the environment, difficult to detect except after many years of observation and study, may have far-reaching consequences. Recent estimates indicate that we have already reached a point where man and his activities significantly affect climate (Bryson and Wendland, 1968). Increased particle loading of the troposphere may affect cloud formation and rainfall, as well as changing the heat balance of the earth. In addition, practically all combustion processes produce carbon dioxide from atmospheric oxygen and carbonaceous fuel constituents. Generation of power contributes to the demonstrated buildup of carbon dioxide in the earth's atmosphere. This accumulation reduces radiation of energy from the surface of the earth and may be responsible for large-scale climatic changes (Mitchell, 1968).

SUMMARY AND RECOMMENDATIONS

Clearly, combustion of fuel to produce electricity adds large quantities of pollutants to the environment, straining the capacity of natural processes to dispense with contaminants. Precise determination of the magnitude of energy production's contribution is difficult because of the complexity of the environment and the paucity of information. A clear definition of the character of the local environment prior to construction and during the operation of power plants would greatly assist the detection of environmental effects of power plant effluents. Four specific purposes would be served by making chemical, physical, and biological observations prior to choosing sites for energy production: (1) the susceptibility of the local environment to deterioration by the type, number, and mass of substances to be emitted would be determined; (2) the capacity of air, water, and land to absorb, dilute, and dispense with the by-products of energy production would be evaluated; (3) accumulation of baseline information would allow detection of changes in the environment due to effluents; (4) future decisions on type of fuel, size, and location of generating stations would be based on information rather than estimates in order to provide the least harmful environmental intrusion.

Baseline studies should be carried out by making observations and measurements of specific parameters of local climate, meteorology, soil, natural waters, vegetation, and wildlife (Treshow, 1963). The program should be described in detail when plans for construction of power generation facilities are designed and should be continued for the lifetime of the plant. Very sensitive vegetation and fish, either natural or introduced, could be used as indicators to give advance warning that emissions are approaching a level of wide concern. It is within the capacity of current knowledge to establish biological early-warning systems to protect human health and environmental values. This would place protection of public health and the environment on a level equal in importance with cost, engineering considerations, and the requirement to supply current and future demand for power.

Research support is appallingly low in relation to the potential magnitude of environmental problems (OST, 1970). Along with adequate public support, a much greater contribution of privately owned electric utilities to research should be made in order to obtain sufficient information on which to make rational decisions concerning energy production and power plant siting. Much more information is needed on (1) methods for reducing the emissions from fossil-fuel-powered generating plants; (2) the physical and chemical alterations and fate of pollutants after emission; (3) the role of vegetation in recycling elements and as temporary sinks for pollutants; (4) the basis for individual differences in biological susceptibility; (5) possible genetic influences on biological systems of repeated exposure to pollutants over many generations; (6) acclimation of biological systems to polluted environments; (7) ecological mapping of biota for judging the suitability of sites for power generation and the subsequent effects of these facilities on the environment; (8) climatic changes and global consequences of pollutants emitted by the generation of electricity.

Conflicting human wants and needs must be made compatible. Towards this end, information is needed in the social sciences as well as in the physical and biological sciences. Public understanding of the consequences of rising power needs is a responsibility which must be shouldered by scientists and engineers as well as public officials. Feedback, in terms of national attitudes and values, is an essential prerequisite to rational decision-making on energy sources and power plant siting. It is urgently necessary to obtain clarification of the social aspects of the broad problem of energy requirements. The technical questions can be solved only if the social issues are raised and reviewed publicly. If changes in values, attitudes, and economic policies are needed, the impetus must come from public review of the whole problem of rising energy needs (OST, 1970).

3 METEOROLOGICAL AND ENGINEERING FACTORS AND METHODS

METEOROLOGICAL FACTORS

Since no significant control of the meteorological processes of transport and diffusion of power plant effluents is possible, the choices of plant sites and permissible emission rates depend upon a reliable estimate of these factors for each site and region of the country. Both transport and diffusion may vary drastically from one site to another. For example, over relatively flat and windswept plains and coastal regions, transport will be in all directions (over some period of time), and diffusive mixing will be large. These factors lead to relatively low concentrations per unit pollutant emission, low chemical reaction products due to rapid dilution, and minimal exposure at any given site. At the other extreme, a narrow, steep-walled valley, subject to temperature inversions below the crest, provides the worst exposure conditions. Transport flows are limited to up- or down-valley (the likely location of people and farms) and the diffusive mixing is inhibited by minimal vertical and lateral mixing. The ratio of pollutant exposure can readily exceed 100 between the valley and the plain situations. In operational terms, this means that permissible emissions for a plains station may be more than 100 times that permitted for confined valley stations. More generally, it is well known that the meteorological potential for air pollution is significantly less in the plains areas than it is in the intermountain regions.

In more general siting problems where terrain effects are variable and multiple sources must be considered, meteorological factors must be evaluated from mathematical models which account for time and space variabilities of wind and turbulent diffusion. Significant progress has been made in the development of such models and, although they need further research and development, existing models can be usefully employed in evaluating air pollution problems associated with existing and planned power plants (Stern, 1970).

One of the simplest models can be constructed by including in the description of the model the distribution of pollution sources, their emission conditions, and the micrometeorological characteristics that directly affect the dispersion of pollutants. The mean concentration distribution of pollutants due to each source of pollution can be determined and summed to find the mean concentration patterns of pollution over the urban areas as a function of time. The relative contributions of each of the sources of pollution to the contamination at various points of the area can then be analyzed. This method of analysis can be extended to include the patterns of pollution originating from a multiplicity of sources of pollution. We may include individual point sources, such as a large power station or a steel mill, and area sources such as the automobile traffic distribution with appropriate hourly variation of the traffic concentration and variable geographic distribution.

The mathematical model can be used to determine (1) temporary measures to be taken when atmospheric pollution threatens to reach the allowable concentration levels; (2) efficacy of various plans to reduce pollution in an urban area; (3) effects of a new pollution source on mean concentration patterns; (4) pollution patterns for a city after future expansion; and (5) efficacy of various urban planning solutions as they affect pollution levels.

The use of the atmosphere is one of the most economical ways to dispose of pollutants. While the atmosphere has often been misused as a waste dump, its ability to disperse most contaminants without harmful effects can be increased. At present, the main problem in air pollution is related to the distribution of pollutants in the lower atmosphere as a function of time and space rather than to the overall quantity of each pollutant in the world atmosphere. The application of criteria permitting a selective use of the atmosphere for dispersal of pollutants seems to be desirable and could be very effective. At the same time, we should be concerned with the possibility of saturating the atmosphere, at least in the future, with some pollutants to a degree at which they become harmful and, indeed, could lead to some permanent damages.

The receptor of air contaminants usually comes in contact with the combined effects of the pollution originating from a multitude of sources. In some cases, two sources may produce harmless pollutants which will interact at a large distance from their origins to become an objectionable contamination. On the other hand, a pollution source close to the receptor may be responsible for a much larger contribution to that receptor's pollution exposure than other large but distant sources. The relative responsibility of all contributors could be evaluated by taking into account, through mathematical models, the quantities of pollutants emitted by each source, and the meteorological conditions which may produce areas and periods for which dangerous contaminations can be reached. Only from such analyses can land use planning be effectively related to air quality control.

The evaluation of the detriments and benefits associated with air pollution should be related to the basic needs of the population, including its need for clean air. Each community should be made conscious of the fact that most often those benefits must be paid for by the danger and nuisance of atmospheric contamination, the inconvenience and costs of its control, or a slower growth of the community.

THE ROLE OF TALL STACKS

So long as the standards and criteria for air quality are based upon pollutant concentrations and dosages in the "breathing zone" near the ground, tall stacks provide a potential means for meeting these standards at minimum costs. If, for any reason, future standards and criteria are based upon *total* pollutant loading of the atmosphere, the tall stack will, of course, become completely obsolete.

In the past fifteen years, the high stack has become almost symbolic of good industrial air pollution prac-

tice, and its use as an effective and economical device for reducing ground-level air pollution has become widespread. Today, almost every major power plant designed for fossil fuel includes a stack chosen to keep ground-level pollution within acceptable limits. Formal study of individual plants varies according to the complexity of the area, the size of the plant, and the sophistication of the organization, but air pollution abatement function of the stack is as much a part of the design consideration as boiler and turbine details.

There are several benefits to be obtained from the use of a tall stack (or any stack for that matter) about which no one will argue seriously. Experts are not in accord about the exact degree of benefit or the manner in which it may be achieved, but there can be no quarrel over the reality of the effects.

Under *any* meteorological conditions, a tall stack located in open uncomplicated terrain will produce a dramatic reduction in the local ground-level concentrations of gases or small particulates compared to the same emission released at low levels. The degree of reduction and the size of the area over which it occurs are dependent upon the stack height, the distance from the source, and the meteorological conditions, but there is always a region near the stack where the improvement is impressive. This reduction is accomplished simply by giving natural atmospheric turbulence an opportunity to dilute the pollutant before it reaches ground-level receptors.

A stack located in open terrain converts the least favorable meteorological condition into the most favorable. Within a well-developed surface temperature inversion, the vertical mixing process is extremely slow and ineffective. For all practical purposes, a plume emitted at any reasonable height above ground in an inversion remains aloft, producing no ground-level concentrations until the temperature inversion in which it is embedded is destroyed. Concentration within the plume changes little with distance, but it remains aloft.

It appears also that a stack is beneficial, at least with certain pollutants, in providing an opportunity for physicochemical processes to degrade the pollutant before it reaches ground level. This advantage is, of course, valid only if the end product of the conversion process is less objectionable than the original pollutant.

A stack can be a most effective tool for removing a plume from a specialized local wind circulation that tends to bring pollutants to ground level in higher-than-normal concentrations. Even an elevated source in the lee of a large building or terrain obstacle, as in Figure 2, may produce undesirable effects nearby, whereas an isolated stack would not.

Finally, a stack has the advantage of operating on all pollutants simultaneously, rather than reducing the con-

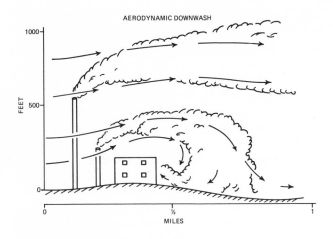

AERODYNAMIC DOWNWASH

FIGURE 2 Aerodynamic downwash. Standing eddies in the wind flow behind the building bring pollutants from a low stack quickly to the ground.

centration of only one, as is true of precipitators and SO_2 scrubbing systems. Thus, it is usually the most economical technique of reducing the general pollutant levels from a large source such as a fossil-fuel power plant.

Tall stacks possess no magic power to eliminate a pollutant. They do not reduce by one gram the total amount of pollutant released to the atmosphere. They distribute it in a different way than would be true of a low-level source, but a receptor at a great distance from a stack will receive substantially the same concentrations no matter what the source height. Thus, there is a most important distinction between local siting problems, where the tall stack is an excellent device, and the regional–global problem, where it accomplishes nothing.

The most important question that needs to be resolved is the role of the stack in regional pollution distributions. It is easy to see that the advantages of a stack disappear on a very large (continental) scale, but it is not clear how valuable the device may be on a scale of 50–100 kilometers. There is growing evidence that the theoretical equations that have proven effective in predicting the behavior of plumes from smaller stacks seriously overestimate the ground-level concentrations derived from installations of 500 feet or more. One can note qualitatively that these overestimates may result from improper specification of various atmospheric factors, such as the turbulence and wind shear at these higher elevations, or from the depletion of the pollutants by conversion processes, but it is most important to translate these notions into quantitative estimates if the stack is to be used intelligently as a control device in the next two or three decades.

AIR POLLUTION CONTROL: ENGINEERING ALTERNATIVES

In the generation of electric power, what are the engineering alternatives available to both reduce the level of air pollutant emissions from a power plant and minimize their effect on the environment? How are these alternative courses of action related to site selection and what are the ramifications involved?

On the next few pages, this complex subject will be discussed from several viewpoints. The treatment, by necessity, will be very general and will not cover each and every aspect in great detail. The detailed engineering technology involved is discussed fully in the reference material listed in the bibliography. Rather than to serve as a complete treatise on power plant air pollution control technology, the information and comments presented are offered to put the factors involving the sources and methods of control of air pollutants from electric power-generating facilities in proper perspective —to aid one in choosing the best possible combination of alternatives among those available in order to supply electric energy to those areas where needed with minimal environmental effects and at a practical cost.

To solve the environmental problems involved in power plant site selection in the near future, major decisions must soon be made. These decisions, in order to be as valid as possible, must be made on the basis of sound experience obtained from existing power generation facilities and on the basis of adequate theoretical and experimental data regarding new innovations. Decisions should not be made on the basis of predictions of nebulous validity concerning energy sources and pollution control technology still in the earliest formulative stages of development. We must, then, realistically confine our current efforts to working with the proven power sources available to us now—modifying and/or supplementing them by the employment of equipment and operational techniques currently being developed and proven effective in reducing the formation and emission of air pollutants to the atmosphere.

Hence, in accordance with these guidelines, the material which follows is concerned with steam boilers (coal-, oil-, or gas-fired) and gas turbines (gas-, oil-, or jet-fuel-fired). Nuclear power is not neglected, but due to some of the unique environmental problems involved with the radioactivity aspects of reactor operation, this topic is covered elsewhere in the COPPS report. However, in order that future options may be increased, it is strongly recommended that both basic and applied research be accelerated wherever feasible on such proposed power sources as those mentioned above. As work progresses in this field, the material presented in this report can be expanded to include the consideration of additional engineering alternatives.

Control of Air Pollution by Fuel Selection

From the previously presented material on the nature and source of various air pollutants, it is apparent that the selection of fuels provides a method for dealing with the emission of both particulate matter and oxides of sulfur in the case of combustion equipment burning oil, coal, or gas. The selection of fuel to achieve desirable end results for either pollutant cannot be considered independently, however, as the choice of a fuel on the basis of one property can create problems resulting from the other properties inherent in that particular fuel. For example, a type of coal selected on the basis of a low sulfur content may present difficulties due to a high ash content or other properties adverse to proper combustion in the boiler furnace. Similarly, a particular low-sulfur-content fuel oil may possess properties adverse to the operation of a gas turbine or be difficult to utilize as a boiler fuel because of its high viscosity, high pour point, etc.

Keeping this situation in mind, let us consider some of the feasible alternatives available.

In respect to coal, there is very limited availability of fuel having a sulfur content sufficiently low to meet current or proposed emission criteria for oxides of sulfur. Coals with sulfur contents in the order of 1.0% or less are in extremely short supply, are high in cost, and frequently have high ash contents or possess other adverse properties in respect to proper combustion. While research is in progress to develop methods for desulfurizing coal, little hope can be offered that this will be a viable solution to the sulfur oxide emission problem in the immediate future. Much more research in this area is required. Efforts to obtain low-ash-content coal are desirable, however, as this reduces the burden imposed on the fly ash removal systems employed. One ramification to be considered (and which is discussed more fully later) is the adverse effect of low-sulfur fuels on electrostatic precipitator performance.

In respect to fuel oil, the supply of crude oil having a naturally low sulfur content is extremely limited, and fuel oils derived from these sources are high in cost. This necessitates the employment of desulfurization processes at the refinery. Desulfurization of fuel oil, though adding to the cost of the fuel, seems a highly desirable route to be taken in the reduction of sulfur oxide emissions. Intensive research in this area, followed by the rapid installation of desulfurization facilities, is highly desirable.

Gas, as a fuel, alleviates the problems associated with both particulates and oxides of sulfur. The availability of this fuel is limited and is subject to governmental regulation. In general, the cost of gas on a Btu basis is higher than that of coal and oil.

A major step forward in the control of air pollutant emissions from both power plants and other combustion sources would be the establishment and implementation of a National Fuel Policy. Such action should be encouraged.

In the choice between the various types of fuel to be used—coal, oil, and gas—due consideration of the other factors involved is essential. These factors include the transportation and storage of coal and oil on site, the right-of-way of gas transmission lines, the reliability of supply, and the other costs related to the use of each type of fuel, e.g., fly ash removal equipment in the case of coal.

The selection of fuels for gas turbine operation involves many factors—the chemical properties of oil, for example, are critical in respect to both the combustion characteristics of the fuel and turbine maintenance.

There has been a considerable amount of discussion in technical journals and elsewhere of the use of such new types of fuel and combustion technology as the burning of coal in fluidized-bed boiler furnaces, gasification of coal and subsequent removal of sulfur oxides prior to combustion, etc. All of these procedures are of interest as possible alternatives to be employed in the future, but their current state of development makes serious consideration of their application in the near future unrealistic.

Control of Air Pollution by Combustion Equipment Design and Operation

Pollutants which can be controlled by combustion equipment design and operation include oxides of nitrogen, carbon monoxide, and hydrocarbons.

Nitrogen oxide formation can be affected by the design and method of operation of the combustion system. This essentially involves the reduction of flame temperature, as the formation of NO from the chemical reaction of O_2 and N_2 is accelerated by high temperature. Among the techniques available for NO_x emission control are the following:

a. Reducing the quantity of excess air in the combustion process.

b. Use of 2-stage combustion whereby part of the air required for complete combustion is provided at the burner, and additional air is provided above the burner to complete the combustion process

c. Modification of boiler heat transfer design to permit rapid heat removal from the burning gases in the flame, thereby reducing flame temperature

d. Recirculation of flue gas back into the combustion zone to reduce flame temperature by dilution

e. Use of increased amounts of atomizing steam when burning oil to reduce flame temperature

The extent to which any of these techniques for NO_x reduction can be applied to a given boiler depends greatly upon the basic boiler design and the practical extent to which boiler design and operation can be modified.

The reduction of both carbon monoxide and hydrocarbon formation can best be accomplished by good boiler furnace design and combustion control. Unfortunately, many of the procedures commonly used to insure complete combustion and, hence, less CO and hydrocarbon formation, involve techniques favorable to NO_x formation, e.g., use of higher quantities of excess air. Hence, to provide the optimum balance between complete combustion and NO_x reduction, refined techniques of combustion control are mandatory. These involve the use of complex combustion gas sampling systems, analytical instrumentation, etc.

Much work is currently being done in developing improved boiler furnace designs and combustion control technology to achieve these desirable goals. Intensive research and development in this area are essential to the design and operation of new power generation facilities which will emit minimal amounts of these pollutants to the atmosphere.

Control of Air Pollution by Combustion Gas Treatment

The removal of particulates from the combustion gases from a coal-fired boiler by electrostatic precipitators, mechanical collectors, etc., is currently being practiced. Further refinement of particulate removal technology is required, however, to meet present and future particulate emission criteria. Equipment capable of continuously removing better than 99.5% of all fly ash from coal-fired units may be considered standard equipment in the near future. Plants being built today quite frequently are designed to employ fly ash removal equipment which will meet these specifications.

The reduction of the sulfur oxide content of the combustion gases presents real difficulties for the design and operation of electrostatic precipitators for particulate removal. The benefit obtained by burning low-sulfur coal to achieve low sulfur oxide emission levels is somewhat offset by the change produced in the electrical resistivity of the fly ash, making it much more difficult to remove. Ramsdell and others (Ramsdell, 1968;

Cahill and Ramsdell, 1967) have conducted many studies of this problem, one of which culminated in the design, construction, and successful operation of a combination-type electrostatic precipitator for the 1,000-MW(e) coal- and/or oil-fired Unit 30 at the Ravenswood station of the Consolidated Edison Company in New York City. This precipitator (installed between the boiler economizer and the air preheaters) operates at elevated gas temperature to take advantage of the decrease in the electrical resistivity of low-sulfur fly ash with increase in temperature. This precipitator achieves a 99.5% removal efficiency for coal fly ash and removes most of the particulate matter contained in the combustion gases when the unit is oil-fired.

A vast number of processes of various types have been proposed and are in various stages of development for the removal of sulfur oxides from combustion gases. Among the types of technology involved are the following:

a. Catalytic conversion of the SO_2 in the combustion gases to form sulfuric acid which may possibly be suitable for sale

b. Wet scrubbing of the combustion gases by alkaline solutions (containing calcium hydroxide, sodium carbonate, etc.) to remove the SO_2 and SO_3. The scrubbing solution may be further processed to recover the sulfur compounds in some salable form or discarded.

c. Removal of SO_2 and SO_3 by the use of solid absorbent materials, followed by the regeneration of these absorbents to remove the sulfur compounds for either disposal or sale after further processing

d. Scrubbing of combustion products by a eutectic mixture of alkaline carbonates, the absorbed sulfur compounds being continuously removed from the eutectic mixture by means of a regeneration process. Further processing of these compounds produces a salable product.

The above is only a partial listing of processes potentially available in the near future for sulfur oxide removal. The development of these processes is still in an early stage (National Academy of Engineering, 1970). At the time this report is being written there are no full-scale plants utilizing any of these methods in successful operation. This is another area where much applied and practical development work is essential. However, the current state of the art in this field is such that future power plants could well be designed to incorporate such processes, if properly selected.

The ramifications involved with combustion gas treatment to remove particulates and/or sulfur oxides are many. For example, when the two types of treatment are to be used in conjunction, the removal of particulate matter from the gases prior to SO_2 removal may be essential to the operation of certain SO_2-removal processes, or the SO_2-removal process itself may be capable of particulate removal. The SO_2 removal process may involve the creation of liquid waste material that could create a water pollution problem. Process cost, both capital and operating, is large, and space requirements for the process equipment are great. The cooling of stack gases by scrubbing solutions may reduce stack plume temperatures to such a degree that wet, visible plumes can create an adverse environmental situation entailing both aesthetic considerations and the hindering of dispersion of other air pollutants still contained in the stack gases (such as NO_x) into the atmosphere. An additional factor to be considered is the limitations that the installation of certain types of SO_2 removal processes may impose upon the operation of the boiler plant—for example, the ability to start up or shut down rapidly or to readily follow varying electrical load demands.

Finally, the benefits to be derived from the use of a tall stack to disperse any residual air pollutants in the combustion gases into the upper atmosphere should be given adequate consideration.

Alternatives Involving Heat and Water Vapor

The alternatives having to do with heat and water vapor involve direct interactions between meteorology and engineering, and it is difficult to discuss them separately. They also suggest needed research, and its seems appropriate to introduce these ideas together with the alternatives rather than as separate topics.

There are certain generalizations that can be made regarding power plant cooling systems, even with our present limited knowledge. Clearly, the evaporative tower is most effective and least objectionable when located in the temperate and warmer portions of the country. Here the visible cloud will normally extend for short distances, and there is little chance of fog and precipitation. It is most likely to cause problems in the cold, northern areas where clouds may persist for miles and certainly can cause fog and icing on elevated landforms and structures.

The simplest alternative in northern areas may be a system which can utilize the tower during the warm months and substitute direct cooling by local waters during the cold season. A more expensive technique in terms of land use, capital costs, and aesthetics is the dry cooling tower. This device would probably be five times the size of an equivalent evaporative unit, but it would virtually eliminate the winter disadvantages. Despite extreme public concern over "thermal pollution" of water supplies, it is also sensible to use large lake and oceanic sites wherever possible in the northern areas. This type of heat disposal spreads the release over a larger area and permits more of it to be lost by direct radiation to space.

In the research and development sphere, we obviously need innovative thinking to make the most beneficial use of the excess heat. Attempts are being made to use waste heat to extend the agricultural growing season in experimental plots, and one can visualize other applications that could be helpful rather than harmful. There are no technical reasons to prevent our designing dry cooling systems that would tend to reduce the incidence of fog and icing at northern airports, and we might eliminate some of the fog pockets along critical portions of the nation's highways as well. We certainly have not fully explored the possibilities of utilizing excess heat as a source of residential heating; in the most sophisticated form, one might design an entire community around a power source.

Proper choice of sites should minimize the adverse effects of heat and water vapor release where one can find no useful and appropriate application. This is particularly true of oceanic sites, where careful attention to prevailing meteorological conditions should result in having most of the waste heat swept out to sea so that it has an enormous volume in which to dissipate.

There are several areas in the meteorological province requiring intensive research. First is an expanded effort to understand precisely how cooling tower plumes behave. This work will require modest extensions of our theoretical knowledge but major improvements in experimental data. The field studies required are intricate and demanding, but it is quite important to be able to predict accurately the trajectories and extent of plumes, as well as their interaction with airborne chemicals.

It is imperative also that meteorologists begin to concern themselves with the development of yardsticks that will assist planners in determining relationships between increased heat and water vapor release and climatology. We should try to predict accurately the climatic changes that may occur before we allow them to happen. As a corollary, one should point out that the knowledge of regional climatology in terms of heat and water vapor budgets is fragmentary and poorly organized. If one were ready to proceed with a projection of the influence of a new large heat source, one would probably find the data on the present characteristics of the region quite inadequate. It would be helpful for studies of both chemical and thermal problems if existing data were organized in proper form for these applications. An obvious step would be the accumulation and summarization, for general use, of the wind, stability, temperature, and humidity data obtained at the various research centers and reactor sites.

Also for the common goal of evaluating chemical and thermal problems. we need fundamental studies of regions, aimed at understanding the relationships among sources, dispersion, conversion, and sinks in the framework of an existing topography and climatology.

In a broader context, the problem of heat must be attacked from two viewpoints. The first is to increase the efficiency of generation as much as technology will allow, implying the minimum production of waste heat. This approach will probably focus attention on power generation systems quite unlike those in common use today. Second, we must reduce unnecessary heat generation wherever possible. For example, we are technically capable of producing systems that will use solar heat for residential heating and cooling. Development of these ideas has not been pursued vigorously because the systems are awkward in comparison with conventional electric power. Such systems are, however, possible within the framework of present technology and should require no more than determination for development.

SUMMARY AND RECOMMENDATIONS

In the determination of the best site for the installation of needed electric power generation facilities and of the type of facility to be installed, the subject of air pollution control is of major importance.

When choosing the general type of power plant to be built at a particular location, all facets of each alternative (the type and characteristics of the fuel to be used, the details of plant design and method of operation, and the type of air pollution control equipment to be installed) should be fully investigated and evaluated, not only separately but also as they interrelate in the overall impact on the environment.

In the consideration of the various options available to minimize air pollutant emissions and their effect on the environment, a strong word of caution seems appropriate.

Too often, the choice between various alternative courses of action has been made on the basis of insufficient or invalid data, with disastrous results. Numerous publications concerned with this subject have presented "predicted" or "generalized" design, performance, and cost data for major components of power plants and equipment for air pollution control. Similarly, a wealth of predictions have been made as to the cost and availability of certain fuels in the near and distant future. These "data," as interesting as they may be, should never be used as the basis for choosing among alternatives in any specific situation under consideration. The naive and perhaps hopeful acceptance of certain of these published "data," without qualifications, can lead one to conclusions which are far from realistic.

A first step, and probably one of the most important

in respect to site selection and the design of a power generating facility to be built on that site, is the following: the gathering of valid data regarding each of the items listed below that are applicable to the specific situation under consideration, and the dissemination of those data, properly explained and evaluated, to all parties concerned in the decision-making process. The data to be obtained, evaluated, and disseminated should cover, at the very minimum, the following items:

a. Fuel: cost and availability (both present and future) for coal, oil, gas, and nuclear fuels of various types and properties, e.g., sulfur content of coals and oils, pour point and viscosity of oils, ash content of coals and oils, etc.

b. The overall situation in respect to the utilization of these fuels, including transportation to the site, storage requirements, disposal of fly ash and bottom ash, etc. (covering costs, land use, esthetic factors, etc.)

c. The actual costs of the basic types of power-generating facilities under consideration (covering method of financing, taxation, local construction and operating labor wage rates, land values, etc.)

d. The true design and operating parameters of the basic types of power generating facilities under consideration (covering predicted nature and magnitude of pollutant emissions, operational flexibility to meet varying load requirements, etc.)

e. Full data on air pollution control systems that may be employed in respect to design and performance, cost of construction and operation at the specific sites being considered, the space requirements involved, sources and costs of materials required in the process (e.g., limestone, activated carbon) and their methods of handling and storage, the nature and magnitude of materials produced for sale or requiring disposal (e.g., sulfuric acid, calcium sulfate) and the facilities and costs involved in their disposal

f. The nature and magnitude of air pollutants already present in the atmosphere of the area and predicted for the future, properly correlated with the meteorological situation peculiar to the area

Only when valid information is available and put in a usable form can sound decisions be made in respect to the choice of alternatives. The Bibliography (p. 78), lists basic references on air pollution aspects of power generation; included in these references are such items as emission factors for NO_x in lbs/10^6 Btu for certain combustion processes, cost data for SO_2 removal processes in \$/kW of capacity, costs for desulfurizing fuel oil, etc. *These numbers should be ignored or, at most, used only as rough preliminary guidelines—nothing more!* Actual data, applicable to another given set of circumstances, may be not be even remotely similar to these published data.

4   REGULATIONS AND STANDARDS

Air pollution, while a recognized problem and the object of legal control for several centuries, has become a problem of *national* public concern in the United States only within the past two decades. Legislation designed to establish air quality standards and the regulatory mechanisms necessary to achieve these standards is even more recent and has undergone major revisions within the past four years. The adaptation of our political and legal processes to this facet of the common good is obviously in a state of flux and experimentation. At the heart of these essentially political activities, we find the same questions addressed by this study. "What are the equitable balances between environmental quality and waste disposal?" and "What are the alternative methods by which these balances can be achieved?" Electrical power generation is, of course, only one of many human activities which must be included in these considerations.

Because of the increasing tendency towards *national* standards for air quality, and regulatory measures to achieve these standards, we have concentrated on this level of governmental activities. This choice poses a distinct hazard—local and state governments may choose more restrictive standards and regulations than those imposed by federal law. We presume, however, that such choices reflect a considered evaluation of the balance between environmental quality and waste disposal in light of local or regional requirements, a process for which the present study of engineering considerations is designed.

The Clean Air Act of 1970 has moved the United States strongly in the direction of early restoration of air quality and the preservation of this environmental feature where it is presently satisfactory. Two features of this Act are particularly important in the context of the present study:

1. *The Act requires national air quality standards for various pollutants, standards which must be achieved or bettered in all areas of the country.*

While setting essentially *minimum* air quality standards for the nation, the methods for achievement of these standards is left to local and state government, subject only to satisfactory demonstration that these agencies' plans for achieving or maintaining these levels of air quality (the so-called implementation plan) are feasible and effective. Failing this demonstration, the federal government may impose its own implementation plan. Within this constraint, local and state governments may choose their own strategy of permissible activities and controls within their jurisdiction.

2. *On new or modified sources of air pollutants for which standards have been established, air pollution equipment which represents the "state of the art," i.e., the most effective available,* must *be installed.*

This requirement does not automatically permit operation of pollution sources if these devices are included. On the contrary, if a new, properly equipped source causes, in combination with existing sources, a violation of the air quality standards, it cannot be operated until such time as old sources are adequately diminished. This dual requirement of "state-of-the-art" *and* air quality standard observance is to be resolved in terms of the more restrictive of the two. That is, under no circumstance is the air quality standard to be violated (unless variances to do so have been approved) but, additionally, more restrictive controls than those necessary to achieve the air quality standards may be imposed on new or modified sources.

While the Act requires that economic factors be included in establishing what the "state of the art" is in air pollution control equipment, these requirements on new sources, *regardless of where they are to be located,* operates to resolve automatically in favor of environmental quality any consideration of trade-offs between environmental quality and atmospheric waste disposal. If one pursues this line to its obvious conclusion, it is clear that, ultimately, man-made sources of air pollution will be reduced to near-zero emissions and air quality standards will be set at "background" levels of naturally occurring pollutants. In the meantime, the constraints on new source siting will operate only to exclude them from presently saturated areas until such time as old sources are phased out.

If this path is indeed to be pursued, then the only enduring part of the present study is the discussion of alternative methods of emission control. Alternative choices of plant siting, so as to balance considerations of air pollution concentrations from multiple sources against their effects, will be helpful in setting and meeting interim air quality standards, of course, but these considerations will become less and less relevant to plant siting. From our limited perspective we suspect this "interim" requirement will exist for quite some time. In the first place, economically satisfactory technology for complete abatement of air pollutants, i.e., zero emissions, is not yet in sight. A second compelling reason for this judgment is that it is not at all clear that we can afford to forgo all uses of the environment for waste disposal. The present tenor of public and executive opinion with regard to any conflict between economic costs and environmental protection is that such conflicts must be resolved in favor of environmental quality. What is frequently overlooked, however, is that these opinions associate primarily with our urban and industrial locales, where past and present abuses of the environment by waste disposal do indeed call for drastic measures of pollution abatement and control to achieve even a minimal degree of environmental quality. The extension of these control measures and the costs they imply, to the much larger areas of the nation where no such environmental crises exist requires further justification.

Two avenues of justification appear to be at work now: (1) an honest conclusion that our environmental quality should be restored to or maintained at natural levels and an *a priori* assumption that the costs incurred in achieving this goal will not be socially and economically disruptive, and (2) that no competitive advantage should accrue to any one person or industry on the basis of differential waste control costs across the nation. The desirability of a natural environment, given that its cost is acceptable, is not to be argued. But, as noted previously, the technological feasibility and the costs of achieving this objective are not obvious to us. It would appear that an analysis of these costs is overdue.

The removal of environmental quality control costs from the considerations of alternative plant sitings carries economic implications and governmental philosophies far beyond our capability to judge. We would note, however, that the imposition of essentially fixed costs on new facilities, regardless of where they are located, automatically causes site selections to be based on other cost considerations, such as labor, logistics, fuel supplies, customer location, etc. *Since these considerations have been dominant in arriving at present sitings, their continued importance as the only manipulative variables in the plant siting equation should lead to minimal changes in site selection, there being no incentive to do anything more than comply with the letter of the law.*

ENGINEERING CONTRIBUTIONS TO STANDARDS AND REGULATIONS

Despite the present tendency to constrain technological solutions to the problem of power production and environmental quality by uniform emission control regulations, it is very clear that engineering considerations must be an integral part of any successful regulatory system. On the one hand, achievement of standards and regulations must be feasible. This can be determined only by sound engineering evaluation and design.

On the other hand, the *specification* of feasible alternative methods of achieving power production goals within acceptable environmental quality criteria and the evaluation of these alternatives also depends upon knowledgeable and broad-ranging engineering, scientific, economic, and social knowhow. In this arena, engineers must look beyond their traditional role of problem-solving within fixed and well-defined specifications.

Even more important, however, is the need for clear and unambiguous communication between the engineering community and the legislative and regulatory community. Each of these communities has its distinct and separate responsibilities; the professionals in engineering and the physical, medical, biological, and social sciences can be expected to define the risks inherent in environmental utilization; the political and regulatory professionals must decide the acceptable level of risk in terms of the benefits associated with taking that risk; and the engineering professionals must specify how that level of risk or a lower one can be achieved at minimum monetary and social costs. It seems clear that this process calls for a high level of integrated teamwork, not independent actions.

The early steps taken towards an effective program for technological assessment of alternative solutions here point to the larger problems still before us. Further steps in this direction should include even more consideration of the social aspects of power production and environmental quality than has been possible so far.

This process has just begun.

5 CONCLUSIONS AND RECOMMENDATIONS

CONCLUSIONS

This report presents a preliminary study, a partial contribution to a larger study, and only a first step towards the definition of the complex interactions of power generation and environmental quality. Firm and sweeping conclusions are, therefore, premature and we shall not offer any, except perhaps to note that failure to continue this line of inquiry is unthinkable.

On a more tentative basis, we may draw the following conclusions:

1. *The environmental effects of air pollutants associated with electrical power production are real and highly complex.* We suspect the true magnitude of these effects is only dimly perceived at present. That perception is sufficient, however, to conclude that the problem we have addressed is a real one and deserves close attention.

2. *The range of available alternative solutions of the plant-siting, plant-operation problem has not been explored to any significant degree.* Although the only variables which can be controlled in the power generation–environmental quality trade-offs are plant location and pollutants emitted, there is a much wider variety of combinations of choices in this regard than has been either evaluated or employed. This is true probably because there has been no demonstrated need for such an evaluation heretofore; successful combinations of engineering solutions for power generation problems have been perpetuated because they *were* successful. Only in the development of nuclear power has close attention been paid to environmental impacts and meth-

ods to control them. The environmental impacts of fossil-fuel power generation have only recently been recognized and are essentially a new dimension in conventional power generation.

3. *Available and developing technologies in plant siting and operation appear to be adequate to an initial assessment of alternative solutions of the power generation–environmental quality problem.* We say "appear to be" simply because their adequacy must be tested in the crucible of just such analyses. We have identified some weaknesses in our knowledge already, but it is from in-depth, integrated analyses that we will learn where error, ignorance, and uncertainty preclude adequate evaluation and solutions. This knowledge is the touchstone for rational specification of future research.

4. Finally, we must note that *an enduring and satisfactory solution of the power generation–environmental quality problem cannot be achieved on a piecemeal basis.* To be sure, there are many difficult scientific, technological, and political problems involved which should be addressed in depth by specialists within these fields. And each power plant site offers a peculiar and unique set of operational and environmental problems which must be solved. But, in solving the larger problem, the knowledge of specialists must be synthesized and the *combined* effects of multiple power plants, embedded in the other sources of environmental pollution, must be considered. Such coordinated and systematic analyses and considerations provide the only sure base for successful national and regional policies which are designed to assure adequate electrical power *and* a livable environment.

RECOMMENDATIONS

The following recommendations are offered in the order of priority and importance we judge to be appropriate to achieving an enduring solution to the problems of electrical power generation and the maintenance of equitable environmental conditions.

1. *The synthesis of engineering and technological know-how with the considerations of relevant scientific and social fields of knowledge should be sharply accelerated.*

Responsibility for the definition and evaluation of alternative solutions to the power generation–environmental quality problems should be vested in the President's Council on Environmental Quality, and the resources necessary for an expeditious accomplishment of these analyses should be provided. As noted in this report, decisions on power plant types and siting must be made soon if we are to avoid a serious disruption in electrical energy production. Therefore, we urge that the studies begun here be carried forward to a more certain base of evaluations as expeditiously as possible. The investment required is trivial in comparison with the costs which will be incurred in providing for electrical power generation and acceptable environmental quality.

2. *In support of and as an integral part of the analyses recommended above, analyses and research on the effects and indices of environmental quality variations should be significantly increased.*

The subtle and complex nature of the effects of air pollutants on animate and inanimate objects has been described in this report. It seems quite clear that no simple measure of environmental quality can describe these myriad effects. However, if we are to consciously protect ourselves and the things we hold valuable, we must develop an increasing understanding and quantification of these effects. In particular, we must avoid the all too common pitfall of inadvertently creating a new and more serious problem while single mindedly protecting ourselves against an already perceived effect. First emphasis in these studies should quite properly go to the effects of pollutants on human health. However, human welfare is also directly affected by the effects of pollutants on plant and animal life and on inanimate objects. These latter effects must also be defined more precisely.

3. *Research on the fate of airborne pollutants should be sharply increased.*

There is an urgent need for studies, experimental and theoretical, under field conditions and in the laboratory, of stack gas conversion processes. The challenge is to design sound basic studies which will shed light on the behavior of complicated systems composed of particles and gases in which chemical reactions and physical processes such as condensation and coagulation are taking place. At the same time, new instruments must be developed to monitor the particle size distribution and chemical composition of the particulate matter in plumes. New theoretical methods must be developed which take into account the dynamics of aerosol clouds as well as diffusion and advection. For example, the problem of simultaneous diffusion and chemical reaction of sulfur and nitrogen oxides in plumes remains unsolved.

Understanding of the fate of airborne pollutants is fundamental to rational control operations. Our present ignorance in these areas *must* be overcome as quickly as possible.

4. *The development of new and improved pollution-free methods of electrical power generation should be a matter of highest national priority.*

The possibility of providing for the nation's (and, eventually, the world's) electrical energy requirements without serious impact on the environment of earth, air, and water should be the ultimate objective of our power and energy policies. Intermediate steps of improved pollution control are clearly required, since this ultimate objective cannot be attained within the next two or three decades. As noted in this report, these intermediate methods of pollution control must be developed to the point of clear demonstration of feasibility before they can be incorporated in the present conventional power systems. These are the systems with which we must work during the next few decades. For the longer haul, both environmental quality and fuel supplies demand the development of fusion, solar, and, if possible, magneto-hydrodynamic power supplies.

APPENDIX A: NUCLEAR POWER PLANTS AND ENVIRONMENTAL RADIATION

The emissions to the atmosphere from a nuclear power plant consist of radioactive materials formed in the reactor either by fission or by the capture of a neutron in structural materials, the coolant, or impurities. These radioactive materials become airborne as a result of outgassing of the coolant, through leaks of high-temperature coolant and subsequent evaporation, or through the resuspension or evaporation of dusts or liquids from contaminated surfaces or bodies of water. The quantities and character of the actual emissions vary widely from one reactor type to another and with specific features of the design of each reactor. Thus, the major part of the emission from a boiling water reactor consists of the fission gases which have entered the coolant in the reactor, either by diffusion through the fuel cladding or through breaks in the cladding, and are sparged from the water at the exit of the turbine condenser. These gases are normally discharged to the atmosphere after a short delay time. In the pressurized-water reactor, the coolant for the reactor is recirculated in a closed loop, with the steam for the turbine generated in a separate loop isolated from the reactor coolant. In this type, the gases are removed in the coolant loop and are stored for periods up to months to eliminate the short-lived gases by radioactive decay. In each case, the quantity of fission gas eventually released to the atmosphere depends upon the condition of the reactor fuel. The emissions from the liquid-metal reactor, breeder reactor, and the high-temperature gas-cooled reactor will arise from the same general source but will vary both in nature and in quantity from those emanating from the light-water reactors. In the homogeneous reactor, which is still a viable concept for future power reactors in the form of the molten-salt reactor, the fuel is not contained in cladding, so that the fission gases and other volatile species of radioactive materials can enter the blanket gas without a barrier, and the emission will depend upon the efficiency of the gas cleanup system or the tightness of the containment for the blanket gas. However, in this type of reactor, the fission products can be continually removed from the fuel so that the long-term buildup of the fission products in the high-temperature, energy-producing portion of the reactor can be avoided.

The behavior in the environment of the materials emitted largely determines the radiation doses received by man. This behavior differs according to the chemical and physical nature of the radioactive material, with three main groupings normally considered: the noble gases, the particulates, and those vapors or gases which can deposit or enter into the ecological network of the environs. The noble gases are classified separately because they are chemically inert except in certain exotic laboratory conditions. They do not, therefore, enter into the metabolism of the organisms comprising the biosphere and are not of concern in food chains. The limiting radiation dose in relation to the standards is that from the external radiation from the material in the air surrounding the individual rather than the dose to the lung or any other organ of the body. The MPC's (maximum permissible concentrations) of the International Commission on Radiological Protection (ICRP) or the National Council on Radiation Protection (NCRP) are derived by calculating the radiation dose which would be received at the center of a cloud of such concentration. As a corollary, of course, the measurement of such a radiation dose in the environs provides a more direct measurement of the actual exposure to these materials than the measurement of the concentration with subsequent conversion to radiation dose through calculations.

The other two groups of emissions are distinguished primarily by the method of collection and measurement. The particulates consist of small particles of the normally solid radioactive materials or materials adsorbed on normal dust particles. They can be collected on filters either for determination or for removal from the gas stream. The vapors and active gases include materials such as iodine, tritium, and carbon dioxide, which cannot be collected on filters in many of their physical states, and special collection techniques are required either for determination or for removal from the stream. Both the particulates and the vapors of reactive gases require special consideration of more than external radiation effects because many of them are metabolized into the body and can be deposited and enter the ecological systems of the environment in such a way that the reconcentration can result in higher potential exposures than can the quantity in the ambient air. The MPC's of the ICRP or the NCRP for these materials are based upon both the possible uptake of the radioactive materials from the respiratory tract and movement to another organ of the body with consequent irradiation of this organ (soluble materials) and the retention of the particulates in the respiratory tract to produce a dose to these organs (insoluble materials). Since they were derived for control of breathing air in occupational situations with later extension to public use, they do not consider any deposition in the environs or other methods of possible reconcentration.

There are no standards presently available which take into account the full potential impact of such materials in the environment because of the lack of detailed data on many of the isotopes, or the variability in the behavior of the materials due to differences in their physical form as emitted from different plants, and on the differences in behavior of the same material in two different environments. The latter factor is well illustrated by the known increased uptake of ^{137}Cs in plants in certain regions of Florida where the sandy soils do not bind the material to the soil grain, thereby making it unavailable to the plant, or by the well-known lichen–caribou–Eskimo food chain, which has resulted in higher quantities of this material in the Eskimos than in other populations. Such reconcentration can occur by physical means as well as biological, e.g., sedimentary accumulations.

Early work at Hanford has shown that the limiting concentration for ^{131}I emitted as a vapor, based upon people eating vegetation from the area, is about 1/10,000 of that based upon people breathing the air. An important part of the siting of a plant which has emissions of these materials is, therefore, consideration of the environs which may be exposed to the effluents and the possible mechanisms of exposure over and beyond those considered in the formal standards. In some cases, these can be considered in detail only after the plant has started operation and small quantities are emitted. This can be done adequately through a well-designed monitoring program. Current regulatory standards on emission of individual materials are based upon the numerical values of the MPC's as given by the ICRP or the NCRP, although the regulations do permit limiting the emission in cases where the regulatory agency finds a necessity because of such factors. In practice, the limits for such plants are set on an individual basis, taking into account all information on the reconcentration. For example, ^{131}I concentrations are currently limited to 1/700 of the MPC's, presumably because of information available to the agency that this factor is appropriate for the iodine emitted from such plants. Although there is no indication that such reconcentration mechanisms are or have been important in the operation of any power reactor because of the low emissions from these devices, the above-noted factors must be considered by the designers and operators in setting emission limits for off-standard conditions.

Another source of particulates which cannot be measured directly but which must be inferred from the isotopic composition of radioactive wastes is the emission of short-lived noble gases which decay to other elements once they are discharged. These are usually fission gases and are listed in Table A-1 along with the quantity of the daughter product produced for a given emission rate of the parent.

The environmental persistence of radioactive emissions also depends upon the half-life of the radioactive materials. It is customary to express the output as the rate of emission, for example, μCi/sec, since this is the significant variable in most atmospheric dispersion formulations. However, if a short-lived isotope is involved, an additional correction for the decay during the period of travel from the point of release to the receptor must be included. For particulates, vapors, or gases which can concentrate in the ecology of the region, the half-life is even more significant because of the time required for these processes. For long-term buildup, the emission numbers can be somewhat misleading, particularly if converted to emissions in a significant period of time such as one year, because at a particular instant most of the material has disappeared or is not yet in existence. For consideration of such effects, it is possible to convert these numbers to quantity-in-existence, i.e., the amount which is or will be present at arbitrary points in the environment if the emission rate is continued for a time long with respect to the half-life. It is emphasized that these numbers indicate potential doses only from the possible long-term buildup, while the emission rate is more indicative of the dose rates from the immediate emission in the environs of the plant. In Table A-2, the emissions from a boiling-water reactor, as reported by the U.S. Public Health Service, are reported both in emission rate and as quantity in existence. It can be seen that the short-lived noble gases become comparatively unimportant for long-term buildup in comparison to some of the longer-lived isotopes even though the emission of the long-lived isotopes is much smaller.

At the low levels of radiation represented by the standards accepted for the public, and especially at the much lower levels actually encountered around nuclear power plants, the time of delivery of the radiation dose has little, if any, effect on the postulated harm. Thus, it makes little difference whether the annual dose is delivered uniformly over the full year or in a matter of hours or days. This situation is somewhat different from that with many chemical materials, since exceeding the chemical standards by a much smaller margin can, in some cases, lead to an acute effect from a relatively short exposure. However, with gamma emitters, the effective range of the radiation is sufficient so that the radiation originating in a plume somewhat above the ground can result in exposure at ground level even though the radioactive material has not reached the ground. The stack height for gamma emitters is, therefore, primarily of interest in indicating the amount of air available for absorption purposes between the

height of emission and the ground. Since the average dose over a period of time is the chief criterion, the more normal plume calculations (which show that the concentration decreases at a rate proportional to something between the distance to the 1.5 power and the distance squared) are no longer appropriate; instead, the average dose will decrease approximately as the inverse of the distance as the plume concentration is averaged over all conditions of wind direction, wind speed, and stability.

Experience with current plants indicates that the radiation dose to the populace from airborne materials is very low. Even with this low emission, recent evidence indicates that the nuclear plant emissions can be made ever lower with current technology and only a small percentage increase in cost as compared to the situation with the fossil-fueled plants. Such equipment should also help to ameliorate the effects of minor off-standard conditions which could otherwise result in increased emissions. There does not, therefore, seem to be any real limitations of the siting of such plants, based on routine emissions.

Before leaving the question of routine emissions, it is necessary to take notice of the allegations concerning radioactivity emitted from fossil-fueled plants, primarily as natural uranium, radium, and thorium, along with

TABLE A-1   Long-Lived Daughters of Noble Gases (Quantities for Emission of 1,000 μCi/sec of Noble Gas)

Parent	Daughter	Half-life	Equivalent Emission Rate of Daughter (μCi/sec)	Daughter in Existence (Ci)[a]
^{89}Kr	^{89}Sr	52.7 d	0.04	0.3
^{90}Kr	^{90}Sr	27.7 y	4×10^{-5}	0.05
^{91}Kr	^{91}Y	58.8 d	2×10^{-3}	0.01
^{135}Xe	^{135}Cs	3×10^{6} y	10^{-10}	—
^{137}Xe	^{137}Cs	30.0 y	4×10^{-6}	0.006
^{139}Xe	^{139}Ba	82.9 m	9.0	0.06
^{140}Xe	^{140}Ba	12.8 d	0.01	0.02

[a] Assuming operation time long with respect to the half-life of the daughter.

TABLE A-2   Typical Emissions from One Boiling Water Reactor

Isotope	Half-life	Emission Rate μCi/sec	Emission Rate Ci/y[a]	Quantity in Existence (Ci)[a,b]
Noble Gases				
^{83m}Kr	1.86 h	240	7,600	2
^{85}Kr	10.76 y	0.6	20	300
^{85m}Kr	4.4 h	400	13,000	9
^{87}Kr	76.0 m	1,300	40,000	8
^{88}Kr	2.80 h	1,400	44,000	20
^{89}Kr	3.18 m	150	4,700	0.04
^{133}Xe	5.27 d	400	13,000	300
^{135}Xe	9.14 h	1,400	44,000	70
^{135m}Xe	15.6 m	1,000	32,000	1
^{137}Xe	3.9 m	350	11,000	0.1
^{138}Xe	17.5 m	3,000	90,000	5
Others				
^{3}H	12.262 y	6×10^{-3}	0.2	3
^{58}Co	71.3 d	2.6×10^{-5}	8×10^{-4}	2×10^{-4}
^{60}Co	5.263 y	2.5×10^{-5}	8×10^{-4}	6×10^{-3}
^{89}Sr	52.7 d	9.7×10^{-4}	0.03	6×10^{-3}
^{90}Sr	27.7 y	5×10^{-6}	2×10^{-4}	6×10^{-3}
^{131}I	8.05 d	9.2×10^{-4}	0.03	9×10^{-4}
^{137}Cs	30.0 y	3.5×10^{-5}	1×10^{-3}	5×10^{-2}
^{140}Ba	12.80 d	4×10^{-4}	0.01	9×10^{-4}

[a] Assuming full-time operation through the year.
[b] After operation time long with respect to the half-life.

the associated daughter products in the fly ash. Studies of the activity in such fly ash indicate the specific activity to be very close to that of soils in various regions. Thus, addition of this fly ash to the soil should change the radiological conditions in the environs very little, if at all, at most locations. The inhalation of any significant quantity while the material is still airborne again seems highly unlikely in view of the large inert mass associated with the radioactivity. Thus, to attain the MPC for the public for uranium, radium, or thorium, the quantity of fly ash in the air must reach levels on the order of one gram per cubic meter, a completely intolerable condition for other reasons.

It appears that most of the siting considerations for nuclear power plants eventually consider the possibility of an accident which could release large quantities of fission products into the air, thereby exposing nearby populations to massive radiation doses. Such a possibility has been a concern of those responsible since the early days of the Manhattan Project and has affected their decisions and the entire course of the nuclear industry. A great deal of attention has been devoted to the question in research programs, design of plants, and in the construction and operation of plants, with the AEC taking the lead in all respects. As a result, the modern nuclear plant has many features of redundancy and of backup which are unparalleled in any other branch of industry. However, it is impossible to guarantee that the probability of an accident is zero, although most individuals familiar with the design and construction of such plants would make the value judgment that the probability is acceptable and is probably vanishingly small as compared to the risks of other types of accidents in other facilities.

The experience with these plants, while as yet limited, is favorable in that there has never been an accident which released or threatened to release radioactivity significantly above the normal emissions. Studies of accidents from other causes indicate that as the consequences of an accident become larger, the frequency or probability of an accident of this size decreases. In general, the two seem to be inversely related. While the precautions taken in nuclear power plants may reduce the probability of smaller accidents further than that of larger ones, the lack of any release in the few hundred reactor-years of operation to date is an indication that the probability of a major accident is, indeed, low as compared to many other industries. In this respect, it may be noted that the reactor is not necessarily the extreme type of device which can subject the public to the results of industrial accidents. It can be shown, for example, that the release of ten tons of chlorine is about equivalent in immediate lethality in the surrounding populace to the postulated maximum release of fission products from a 1,000-MW(t) reactor. This does not include consideration of late effects from radiation, but does assume similar meteorological conditions in the two accidents.

Until additional experience is gained, through either operation or more refined engineering studies, accident possibilities will remain an important factor in siting. Even with increased experience with present reactors, new reactor types or extrapolations of present types with different features will doubtless require some isolation until adequate assurance of safety is attained. Proposals to substitute underground locations for isolation seem to provide little real gain since the hazard results from the leakage of fission products from the containment. The major paths for such leakage are in the access openings for men, materials, steam, water, etc. It would appear, therefore, that the probability of leakage is only slightly reduced, if at all, by underground construction as compared to the conventional containment.

PLOWSHARE GAS STIMULATION

With the growing shortage of gas reserves, there has been considerable governmental and industrial interest in stimulation of flow in widespread but tight natural gas formations by the use of underground nuclear detonations. Several demonstration experiments have been made, with promising technical results, but the future usefulness of this technique is still uncertain because the economics are not clear, a number of different formations need investigation, and standards or regulations under which the product can be used have not been defined.

It can be expected that the gas from such stimulation will be slightly radioactive, with the major contaminants being ^{85}Kr and ^{3}H. If the technique is used, standards must consider the multiple source problems of localized additions from household use as well as the centralized additions from larger industrialized uses such as power plants.

Although results are available from the experimental detonations, it is too early to predict ultimate concentrations in gas reaching the consumer from possible widespread use of the technique. It is apparent that consideration of the total emissions will be needed in establishing standards, particularly in a metropolitan area with widespread gas usage.

FUEL RECOVERY

An important part of the economics of nuclear power, particularly in the breeder reactor, is the removal of fission products from the fuel and the recovery of the unburned uranium and plutonium. Even in the present light-water reactors, a significant quantity of plutonium is formed, and its recovery and sale or reuse is vital in the fuel cycle cost. To recover such materials, it is necessary to destroy the integrity of the fuel cladding and to dissolve or volatilize the spent fuel to permit chemical processing. This, of course, leads to a plant handling solutions of these materials, and of fission products, with consequent increased likelihood of some portion becoming airborne through off-gasing of solutions (^{85}Kr, ^{133}Xe), volatilization of high-vapor-pressure materials (^{131}I), or resuspension from liquids or surfaces of solid particulates. The plant can be, and is, designed to provide containment of many of these materials through treatment of the off-gases by filters, absorbers, etc., to keep the level of emissions as low as practicable. Further, a single plant can handle the fuel from a number of reactors and can be sited in an area of low population density so that the impact of any releases on people can be minimal.

The chief airborne material emitted, in terms of radioactivity, is ^{85}Kr. The environmental impact is low, however, since it is an inert gas and the radiations emitted are relatively low in energy. It is of possible future concern as a global contaminant due to its long half-life and lack of natural removal processes. Already it is impossible to use atmospheric krypton for some scientific uses, such as filling radiation counters, because of the accumulations of ^{85}Kr from past plutonium production programs plus a small addition from nuclear power. Techniques for removing this material are being investigated, and requirements for removal will undoubtedly be implemented in the future.

Tritium, formed as a fission product, will also occur as moisture in the stack, although the amount will be small, compared with the quantity in liquid wastes. Some schemes for controlling tritium revolve around complete recycle of water, since it is difficult to separate the tritium bound in water. In such a system, care must be taken to avoid increased atmospheric release due to evaporation of the recycled fluids in which the tritium concentration is increasing.

APPENDIX B: AIR POLLUTION FROM GAS TURBINE OPERATIONS

The use of gas turbine units for electric power production has risen sharply from a negligible amount 20 years ago until it now represents 20% of the total generating capacity of some utilities. For a variety of reasons, the relative importance of gas turbines will continue to increase in the decades ahead, making their air pollution performance, although still less important than that of steam plants, of much interest.

The first interest in gas turbines for electric power was for peaking service. The low efficiency of such units made their use for extended periods quite uneconomic, but their very fast start-up times (compared with those for steam units) and very low capital costs were ideal for peaking units. The short duration of most peak demand periods made high fuel consumption a minor disadvantage. More recently, a combination of factors has both increased the number of gas turbines in service and greatly extended the periods of operation of these devices, in some cases moving them into the base-load category. These factors include

a. The quick delivery time of gas turbines in the face of construction delays for both conventional and nuclear steam plants; delays have become common due to congestion of orders, manufacturing delays, and, in some cases, opposition to the construction of large conventional stations

b. The siting flexibility of gas turbines, which are free from cooling water requirements and, in many cases, from the organized opposition of environmental groups, since most gas turbines have no visible exhaust plume, as well as from large site requirements because of their compact size (in some cases, the units are even regarded as mobile and are moved from site to site as requirements demand, sometimes on barges)

The typical power-generating turbine installed today is a modified aircraft engine of a capacity in the 20- to 40-MW range. Units are often combined in pairs or fours to produce a power station in the 100- to 200-MW range.

It is possible that gas turbine usage will increase considerably if preliminary studies of combined gas turbine/steam plants are confirmed by demonstration plant experience. Such combined plants obtain about 75% of the total power produced from the gas turbine unit and about 25% from a steam plant employing the combustion products of the gas turbine as a heat source. Advanced design studies show overall thermal efficiencies of about 50%, an important advance over the 40% efficiencies of modern steam units. The efficiency increase may be enough to justify the extensive fuel treatments required (as discussed below) to adapt most fuels for gas turbine use, greatly extending the applicability of gas turbine units.

AIR POLLUTION PERFORMANCE OF GAS TURBINE POWER PLANTS AND CONTROL TECHNIQUES FOR IMPROVED PERFORMANCE

The emission performance of gas turbine power plants depends rather strongly on particular design features of the combustion unit and, of course, on the fuel com-

TABLE B-1   Approximate Annual Gas Turbine Emission Amounts for 1,000-MW(e) [a]

Pollutant	Annual Release (10^6 lb) [b]
Particulate carbon	2.5–6
Particulate ash	negligible
Oxides of sulfur	25
Oxides of nitrogen	25–160
Carbon monoxide	5–400
Hydrocarbon (other than carbon)	1

[a] Smith *et al.*, 1968; Sawyer *et al.*, 1969; Cornelius *et al.*, 1968, 1970; Fenimore *et al.*, 1970; Starkman *et al.*, 1970; Lozano *et al.*, 1968; Durrant *et al.*, 1969; Sawyer, 1970; Heywood *et al.*, 1970; Korth and Rose, 1968; Sawyer and Starkman, 1968.
[b] For #2 distillate oil, sulfur content 0.2%, full load conditions, heat rate 15,000 Btu/kw.

position. This variability makes emission estimates difficult. Furthermore, there are no emissions data on modern large gas turbines in use as electric power units. Emissions estimates must, therefore, be made from data from aircraft engines and from small industrial turbines. Table B-1 presents some estimated emission quantity amounts which may be compared with the generally larger amounts cited for steam plants.

The following paragraphs will discuss each pollutant in detail.

Particulate Carbon

The mass rate of soot emission of gas turbine engines is rather small even for engines not designed specifically to minimize this emission. However, the soot particles almost always have dimensions less than 1 μ, ranging from 0.01 μ to 1 μ (Taylor *et al.*, 1968). The upper part of this range maximizes visible light absorption, resulting in a dense black plume. Consequently, engine designers have devoted considerable research to the problem of minimizing or eliminating soot emission. Addition of calcium and manganese compounds to the fuel has proven helpful in some cases, although the soot suppression mechanism of these activities is not understood (Taylor *et al.*, 1968). Recent study has pointed the way to a generally preferred solution—modifications in the fuel/air mixing patterns in the cumbustion chamber. Soot generation has been connected with certain local air/fuel ratio and temperature values. Appropriate redesign of the fuel introduction mechanism and the air introduction guidance through the burner can into the combustion chamber has effectively eliminated visible smoke (Durrant, 1968; Durrant *et al.*, 1969; Taylor *et al.*, 1968; DeCorso *et al.*, 1967; Ambrose and Bott, 1970; Toone, 1968; Faitani, 1963; Fiorello, 1968).

The necessary new equipment represents only small modifications of older burner cans and fuel nozzles, so smoke reduction has not required costly equipment. The amortized cost of smoke reduction research is estimated at about \$1/kW, compared with total capital costs of about \$80/kW for gas turbine equipment. The problem of particulate carbon and visible black smoke from gas turbines can be regarded as solved.

Particulate Ash

Particulate ash emissions derive from the ash content of the particular fuel used. In contrast to the case of steam boilers, where substantial amounts of ash can be tolerated by the equipment, most ash constituents can harm the blading of a gas turbine (Tipler, 1968). Attempts to burn residual oils in conventional gas turbines, for example, usually result in unacceptable maintenance and overhaul requirements. Even distillate oils often require careful pretreatments to remove harmful ash constituents before they can be used as a practical gas turbine fuel (Carpenter). Natural gas has no ash content.

The amount of ash emission from conventional gas turbines is, therefore, small, a result of the very clean fuels used. This situation will probably not change appreciably with the coming decade. Progress in attempts to minimize ash damage to turbine blading has been quite slow. An alternative approach would employ a closed cycle gas turbine in which the combustion products do not pass through the turbine but instead heat (in a suitable heat exchanger) the actual turbine working gas. Technical difficulties with the large, high-temperature heat exchangers required have made this approach uneconomical. If a closed-cycle gas turbine were perfected for use with "dirty" fuels, the ash problem would be rather similar to that for conventional plants for the fuel used (coal or oil), except that metallurgical problems require substantial dilution of the combustion product stream, resulting in correspondingly higher gas flow rates and, thus, higher particle collection equipment costs.

Oxides of Sulfur

The general situation for sulfur in gas turbine fuels resembles that for the ash constituents, although (except for special circumstances) sulfur is not so harmful to turbine blades. Gas turbines commonly burn distillate oils with 0.1% to 0.5% sulfur content. As gas turbines become more used and as fuel usage patterns shift, it might become desirable to burn fuels with higher sulfur percentages. Air pollution emission standards will require desulfurization of such fuel, which will cost about $0.50–$1/barrel, compared with a nominal initial cost of $2/barrel.

The possibility for the removal of sulfur oxides in gas turbine stack gases is similar to that described for steam plants above. Potentially important engineering differences, however, are the greater air flow per kW of the gas turbine plants (nominally four times greater than for steam plants), which implies correspondingly larger and costlier flue gas treatment units. In addition, gas turbine exhaust temperatures are high (nominally 600°F) and may require still further dilution to lower temperatures before treatment.

Oxides of Nitrogen

Nitrogen oxides are a major pollution problem for gas turbines. The high-temperature combustion process and the subsequent dilution of the combustion products with air containing available oxygen result in substantial NO_x emissions. In some cases, the turbine even produces a visible brown NO_x plume. Intensive research work is in progress to minimize gas turbine NO_x production, since the amount of NO_x from unmodified engines exceeds many contemplated emission standards.

It seems likely that combustion process changes of the same sort (but differing in detail) as those used for soot reduction will be successful in reducing NO_x emissions to acceptable levels. The research problem is difficult since the remedial measures conflict to some extent with those used to minimize smoke, but turbine manufacturers are generally confident that soot and NO_x emissions targets can be met simultaneously.

Carbon Monoxide

Full-load carbon monoxide data from gas turbines show levels higher than those for steam plants, but these levels are still quite low. However, at reduced loads, CO emissions can rise by two orders of magnitude (Sawyer, 1970; Korth and Rose, 1968; Sawyer and Starkman, 1968). Control of gas turbine facilities may have to take this partial-load CO production into account by reducing capacity by total shutdown of some units rather than by reduced load operation of all units.

Unburned Hydrocarbons

Under full-load operation, the total amount of hydrocarbon is negligibly small. Under partial-load operation, the emission amounts can rise by two orders of magnitude, reaching several hundred ppm (Sawyer, 1970, Korth and Rose, 1968; Sawyer and Starkman, 1968). This level is still fairly low. As with CO, proper plant control can minimize hydrocarbon emissions.

Hydrocarbon emissions can become very large during faulty turbine operation, resulting in visible clouds of unburned fuel. A change in fuel without corresponding changes in fuel treatment and injection equipment can cause such excess emissions. Temporary excess emissions are sometimes caused by faulty starts, which in turn are often due to improper fuel-injection equipment. Use of heavy fuels often requires air-atomizing fuel injectors, which may represent additional capital costs of about $1/kW. There is some evidence (Sawyer, 1970; Cornelius et al., 1968) that the HCO (aldehyde) content of the hydrocarbon emissions may be objectionably high.

REFERENCES
AND
BIBLIOGRAPHY

REFERENCES

Ambrose, M. J. and Bott, J. Zero Exhaust Visibility—A Goal Attained for Peaking Gas Turbines. ASME Paper No. 70-PWR-15, 1970.

Bartok, W., et al. Stationary Sources and Control of Nitrogen Oxide Emissions. Preprint, 2nd International Clean Air Congress, Washington, D.C., U.S. Government Printing Office, December 1970.

Bernatzky, A. Climatic Influences of the Greens and City Planning, Anthos 11, 29–36, 1965.

Brar, S. S., et al. Thermal Neutron Activation Analysis of Particulate Matter in Surface Air of the Chicago Metropolitan Area: One Minute Irradiations, Environ. Sci. Technol. 4, 50, 1970.

Briggs, G. Plume Rise, USAEC, TID-25075, 81 pp. Oak Ridge National Laboratory, Oak Ridge, Tennessee, 1970.

Bryson, R. A., and Wendland, W. M. Climatic Effects of Atmospheric Pollution. In: Global Effects of Environmental Pollution (S. F. Singer, ed.), AAAS Symposium, Springer-Verlag, New York, December 1968.

Cahill, W. J., and Ramsdell, R. G. Low Sulfur Coal Cuts Precipitator Efficiency, Electrical World, November 13, 1967.

Carpenter, R. J. Fuel Systems for Gas Turbines. Power Engineering 74, 4, 34–35.

Cornelius, W., Stivender, D. L., and Sullivan, R. E. A Combustion System for a Vehicular Regenerative Gas Turbine Featuring Low Air Pollutant Emissions. SAE Trans. 76, 4, 3140–3159, 1968.

Cornelius, W., and Wade, W. R. The Formation and Control of Nitric Oxide in a Regenerative Gas Turbine Burner. SAE Paper No. 700708, 1970.

Council on Environmental Quality. Environmental Quality; The First Annual Report. Washington, D.C., U.S. Government Printing Office, August 1970.

Cuffe, S. T., et al. Air Pollutant Emissions from Coal-fired Power Plants. Report No. 1, J. Air Pollution Control Association 14, 353–362, 1964.

DeCorso, S. M., Hussey, C. E., and Ambrose, M. J. Smokeless Combustion in Oil-Burning Gas Turbines. ASME Paper No. 67-PWR-5, 1967.

Delwiche, C. C. Carbon Monoxide Production and Utilization by Higher Plants, Ann. New York Acad. Sci. 174, 116–121, 1970.

Dreisinger, B. R., and McGovern, P. C. Monitoring Atmospheric Sulfur Dioxide and Correlating Its Effects on Crops and Forests in the Sudbury Area. Impact of Air Pollution on Vegetation Conference, Toronto, Can., April 1970.

Dugger, W. M., and Ting, I. P. Air Pollution Oxidants: Their Effects on Metabolic Processes in Plants, Ann. Rev. Plant Physiol. 21, 215–234, 1970.

Duprey, R. L. Compilation of Air Pollution Emission Factors. U.S. Public Health Service Publication No. 999-AP-42, Washington, D.C. 1968.

Durrant, T., et al. Combuster Design Changes Reduce Smoke from Gas Turbine Engines. SAE Journal 77, 6, 59–64, 1969.

Durrant, T. Control of Atmospheric Pollution from Gas Turbine Engines. SAE Paper No. 680437, 1968.

Edison Electric Institute Committee on Environment, Plant Siting Task Force, Major Electric Power Facilities and the Environment. New York, N.Y., Edison Electric Institute, February 1970.

Eisenbud, M., and Petrow, H. G. Radioactivity in the Atmospheric Effluents of Power Plants That Use Fossil Fuels, Science 144, 288–289, 1964.

Faitani, J. J. Smoke Reduction in Jet Engines Through Burner Design. SAE Trans. 77, 1080–1090, 1968.

Federal Power Commission. 1969 Annual Report. Washington, D.C., January 1970.

Fenimore, C. P., Hilt, M. B., and Johnson, R. H. Formation and Measurements of Nitrogen Oxides in Gas Turbines. ASME Paper No. 70-WA/GT-3 (ASME Winter Meeting, New York, November 29–December 3, 1970).

Fiorello, S. C. The Navy's Smoke Abatement Program. SAE Trans. 77, 1051–1058, 1968.

Friedlander, S. K. The Characterization of Aerosols Distributed with Respect to Size and Chemical Composition, J. Aerosol Sci. *1*, 295, 1970.

Friedlander, S. K., and Seinfeld, J. H. A Dynamic Model of Photochemical Smog, Environ. Sci. Technol. *3*, 1175, 1969.

Gartrell, F. E., Thomas, F. W., and Carpenter, S. B. Atmospheric Oxidation of SO_2 in Coal-Burning Power Plant Plumes, Am. Ind. Hygiene Assoc. J. *24*, 113, 1963.

Gerstle, R. W., *et al.* Air Pollutant Emissions from Coal-fired Power Plants. Report No. 2, J. Air Pollution Control Association *15*, 59–64, 1965.

Gordon, A. G., and Gorham, E. Ecological Aspects of Air Pollution from an Iron-Sintering Plant at Wawa, Ontario, Can. J. Bot. *41*, 1063–1078, 1963.

Gorham, E., and Gordon, A. G. The Influence of Smelter Fumes upon the Chemical Composition of Lake Waters near Sudbury Ontario and upon the Surrounding Vegetation, Can. J. Bot. *38*, 477–487, 1960.

Guderian, R., and van Haut, H. Detection of SO_2-Effects upon Plants, Staub (Engl. Transl.) *30*, 22–35, 1970.

Hangebrauck, R. P. Emissions of Polynuclear Hydrocarbons and other Pollutants from Heat-Generation and Incineration Processes. J. Air Pollution Control Association *14*, 267–278, 1964.

Heck, W. W., *et al.* The Effects of a Low Ethylene Concentration on the Growth of Cotton, J. Air Pollution Control Association *11*, 549–556, 1961.

Heck, W. W., and Dires, E. G. Growth of Plants Fumigated with Saturated and Unsaturated Hydrocarbon Gases and Their Derivatives, Texas Agriculture Experimental Station Publ. No. MP-603. College Station, Texas (associated with Texas A.M.U.), August, 1968.

Heywood, J. B., Fay, J. A., and Linden, L. H. Jet Aircraft Air Pollutant Production and Dispersion. AIAA Paper No. 70–115 (AIAA 8th Aerospace Sciences Meeting, New York, January 1970).

Hidy, G. M. An Assessment of the Global Sources of Tropospheric Aerosols. Preprint, 2nd International Clean Air Congress Paper No. ME-26A. Washington, D.C., U.S. Government Printing Office, December 1970.

Hidy, G. M., and Friedlander, S. K. Vapor Condensation in the Mixing Zone of a Jet, A.I.Ch.E. J. *10*, 115, 1964.

Hill, A. C., *et al.* Ozone. In: *Recognition of Air Pollution Injury to Vegetation, A Pictorial Atlas* (J. S. Jacobsen and A. C. Hill, eds.), Air Pollution Control Association, Pittsburgh, Pa., 1970.

Hursh, C. R. Local Climate in the Copper Basin of Tennessee as Modified by the Removal of Vegetation, U.S. Dep. Agr., Circ. No. 774. Washington, D.C. January 1948.

International Atomic Energy Agency. Application of meteorology to safety at nuclear plants, IAEA Safety Series, No. 29, 27 pp. IAEA, Vienna, Austria, 1968.

Korth, M. W., and Rose, A. H., Jr. Emissions from a Gas Turbine Automobile. SAE Trans. *77*, 1327–1338, 1968.

Land, G. W. The Changing Patterns of Fossil Fuel Emissions in the United States. Preprint, 2nd International Clean Air Congress, Washington, D.C., U.S. Government Printing Office, December 1970.

Larsen, R. I. Relating Air Pollutant Effects to Concentration and Control. J. Air Pollution Control Association *20*, 214–225, 1970.

Linzon, S. N. Economic Effects of Sulfur Dioxide on Forest Growth, J. Air Pollution Control Association *21*, 81–86, 1971.

Lozano, E. R., Melvin, W. W., Jr., and Hochheiser, S. Air Pollution Emissions from Jet Engines. J. Air Pollution Control Association *18*, 6, 392–394, 1968.

McCormick, R. A., and Ludwig, J. H. Climate Modification by Atmospheric Aerosols, Science *156*, 1358, 1957.

Meetham, A. R. *Atmospheric Pollution: Its Origins and Prevention*. 3rd Ed. The MacMillan Co., New York, 1964.

Menser, H. A., and Heggestad, H. E. Ozone and Sulfur Dioxide Synergism: Injury to Tobacco Plants, Science *153*, 424–425, 1966.

Mitchell, J. M., Jr. A Preliminary Evaluation of Atmospheric Pollution as a Cause of the Global Temperature Fluctuation of the Past Century. In: *Global Effects of Environmental Pollution* (S. F. Singer, ed.), AAAS Symposium, Springer-Verlag, New York, December 1968.

Mukammal, E. I. Ozone as a Cause of Tobacco Injury, Agr. Meteorology *2*, 145–165, 1965.

National Academy of Engineering. Abatement of Sulfur Oxides from Stationary Combustion Sources. A report to the National Air Pollution Control Administration. Available from the Clearing House for Scientific and Technical Information, Document No. PB 192887, May 1970.

National Air Pollution Control Administration. Control Techniques for Sulfur Oxide Air Pollutants. Publ. No. AP-52. Washington, D.C., U.S. Government Printing Office, January 1969.

National Air Pollution Control Administration. Control Techniques for Particulate Air Pollutants. Publ. No. AP-51. Washington, D.C., U.S. Government Printing Office, January 1969(b).

National Air Pollution Control Administration. Air Quality Criteria for Sulfur Oxides. Publ. No. AP-50. Washington, D.C., U.S. Government Printing Office, 1969(c).

Office of Science and Technology, Energy Policy Staff. Electric Power and the Environment. U.S. Government Printing Office, Washington, D.C., August 1970.

Pich, J., Friedlander, S. K., and Lai, F. S. The Self-Preserving Particle Size Distribution for Coagulation by Brownian Motion—III. Smoluchowski Coagulation and Simultaneous Maxwellian Condensation, J. Aerosol Sci. *1*, 115, 1970.

Ramsdell, R. G. Design Criteria for Precipitators for Modern Central Station Power Plants. A paper presented at the American Power Conference, Chicago, April 23, 1968.

Sawyer, R. F. Reducing Jet Pollution Before It Becomes Serious. Astronautics and Aeronautics *8*, 4, 62–67, 1970.

Sawyer, R. F., Teixeira, D. P., and Starkman, E. S. Air Pollution Characteristics of Gas Turbine Engines. Trans. ASME, Ser. A, J. Eng. Power *91*, 4, 290–296, 1969.

Sawyer, R. F., and Starkman, E. S. Gas Turbine Exhaust Emissions. SAE Trans. *77*, 1773–1779, 1968.

Scheffer, T. C., and Hedgecock, G. G. Injury to Northwestern Forest Trees by Sulfur Dioxide from Smelters. U.S. Dept. Agr., Tech. Bull. No. 117. Washington, D.C. June 1955.

Slade, D. (editor). Meteorology and Atomic Energy 1968, 455 pp, USAEC, TID-24190, Washington, D.C., 1968.

Smith, D. S., Sawyer, R. F., and Starkman, E. S. Oxides of Nitrogen from Gas Turbines. J. Air Pollution Control Association *18*, 1, 30–35, 1968.

Smith, M. E. (ed.). Recommended guide for the prediction of the dispersion of airborne effluents. New York, N.Y., Am. Soc. of Mech. Eng., 85 pp. 1968.

Smith, W. S. and Gruber, G. W. Atmospheric Emissions from Coal Combustion—An Inventory Guide. U.S. Public Health Service Publication No. 999-AP-24. Washington, D.C., 1966.

Starkman, E. S., Mizutani, Y., Sawyer, R. F., and Teixeira,

D. P. The Role of Chemistry in Gas Turbine Emissions. ASME Paper Nȯ. 70-GT-81, 1970.

Stern, A. C. (ed.). Proceedings of Symposium on Multiple Source Urban Diffusion Models, APCO Pub. No. Place Pub. AP-86, Washington, D.C., U.S. Government Printing Office, 1970.

Taylor, O. C. Importance of Peroxyacetyl Nitrate (PAN) as a Phytotoxic Air Pollutant. Preprint, 61st Annual Meeting of the Air Pollution Control Association, June 1968.

Taylor, O. C., and Eaton, F. M. Suppression of Plant Growth by Nitrogen Dioxide, Plant Physiol. *41*, 132–135, 1966.

Taylor, O. C., and MacLean, D. C. Nitrogen Oxides and the Peroxyacetyl Nitrates. In: *Recognition of Air Pollution Injury to Vegetation, A Pictorial Atlas* (J. S. Jacobsen and A. C. Hill, eds.), Air Pollution Control Association, Pittsburgh, Pa., 1970.

Taylor, W. G., *et al*. Reducing Smoke from Gas Turbines. Mech. Eng. *90*, 7, 29–35, 1968.

Thompson, C. R., *et al*. Effects of Continuous Exposure of Navel Oranges to Nitrogen Dioxide, Atmos. Environ. *4*, 349–355, 1970.

Tipler, W. Combustion in Industrial Gas Turbines. In: *Combustion in Advanced Gas Turbine Systems* (I. E. Smith, ed.) Pergamon Press, New York, N.Y. 1968, p. 21–44.

Toone, B. A Review of Aero Engine Smoke Emission. In: *Combustion in Advanced Gas Turbine Systems* (I. E. Smith, ed.) Pergamon Press, New York, N.Y. 1968, p. 271–296.

Treshow, M. The Impact of Air Pollutants on Plant Populations, Phytopathology *58*, 1108–1113, 1968.

Urone, P., and Schroeder, W. H. SO₂ in the Atmosphere: A Wealth of Monitoring Data, But Few Reaction Rate Studies, Environ. Sci. Technol. *3*, 436, 1969.

U.S. Department of Health, Education and Welfare, Air Quality Data, 1966 Edition. Washington, D.C., U.S. Government Printing Office, 1966.

U.S. Department of Health, Education and Welfare, Environmental Health Service. Control Techniques for Fluoride Emissions, Washington, D.C., U.S. Government Printing Office, 1970.

Weber, E. Contribution to the Residence Time of SO₂ in a Polluted Atmosphere, J. Geophys. Res. *75*, 15, 2909, 1970.

Whitby, K. T., *et al*. The Minnesota Aerosol Analyzing System Used in the Los Angeles Smog Project, Paper presented at the Los Angeles ACS Meeting, 1971.

Wood, F. A. Sources of Plant-Pathogenic Air Pollutants, Phytopathology *58*, 1075–1084, 1968.

Woodwell, G. M. Effects of Pollution on the Structure and Physiology of Ecosystems, Science *168*, 429–433, 1970.

BIBLIOGRAPHY

General

American Chemical Society, Committee on Chemistry and Public Affairs. Cleaning our Environment. Report by the Subcommittee on Environmental Improvement, Washington, D.C., 1963.

Edison Electric Institute. Major Electric Power Facilities and the Environment. Prepared by the Plant Siting Task Force of the Committee on Environment, New York, December 1969.

Hangebrauck, R. P., and Spaite, P. W. Pollution from Power Production. Presented at the National Limestone Institute 25th Annual Convention, Washington, D.C., 1970.

Magill, P. L., Holden, F. R., and Ackley, C. *Air Pollution Handbook*. McGraw-Hill, New York, 1956.

National Air Pollution Control Administration. Process Control Engineering R&D for Air Pollution Control. Washington, D.C., U.S. Public Health Service, November 1969.

Office of Science and Technology. Energy Policy Staff. Consideration Affecting Steam Power Plant Site Selection. Washington, D.C., U.S. Government Printing Office. (81.25).

Stern, A. C. *Air Pollution*, Vol. III. Academic Press, New York, 1968.

U.S. Congress. Senate. Joint Committee on Atomic Energy. Environmental Effects of Producing Electric Power. Hearings before Joint Committee on Atomic Energy (part 1). Washington, U.S. Government Printing Office, October and November 1969.

U.S. Congress. Senate. National Emission Standards Study (Public Law 90-148). 91st Congress, 2nd sess. Washington, D.C., U.S. Government Printing Office, March 1970.

U.S. Congress. Senate. Joint Committee on Atomic Energy. Joint Committee Print. Selected Materials on Environmental Effects of Producing Electric Power. 91st Congress, 1st sess. Washington, D.C., U.S. Government Printing Office, August 1969.

U.S. Congress. Senate. The Cost of Clean Air. Report of Secretary of Health, Education and Welfare (Public Law 90-148). 91st Congress, 2nd sess. Washington, D.C., U.S. Government Printing Office, March 1970.

U.S. Public Health Service. Washington, D.C. U.S. Government Printing Office:

Emissions from Coal-Fired Power Plants	PB 174708, 1967
Atmospheric Emissions from Coal Combustion—An Inventory Guide	PB 170351, 1966
Atmospheric Emissions from Fuel Oil Combustion—An Inventory Guide	PB 168874, 1962
Rapid Survey Technique for Estimating Community Air Pollution Emissions	PB 190240, 1966
A Compilation of Air Pollutant Emission Factors	PB 190245, 1968

National Air Pollution Control Administration. Washington, D.C., U.S. Government Printing Office:

Air Quality Criteria for Particulate Matter	AP-49, 1969
Air Quality Criteria for Sulfur Oxides	AP-50, 1969
Control Techniques for Particulate Air Pollutants	AP-51, 1969
Control Techniques for Sulfur Oxide Air Pollutants	AP-52, 1969
Air Quality Criteria for Carbon Monoxide	AP-62, 1970
Air Quality Criteria for Hydrocarbons	AP-64, 1970
Control Techniques for Carbon Monoxide Emissions from Stationary Sources	AP-65, 1970

Wetch, J. R. Air Pollution and Power Generation. Presented at American Power Conference, Chicago, April 1970.

Wright, J. H. Power and the Environment. Presented at American Power Conference, Chicago, April 1970.

Wright, J. H., Champlin, J. B. F., and Davis, O. H. The Impact of Environmental Radiation and Discharge Heat from Nuclear Power Plants. Presented at the International Atomic Energy Agency—United States Atomic Energy Commission Symposium on Environmental Effects of Nuclear Power Stations, August 1970.

Air Quality Criteria

National Air Pollution Control Administration. Washington, D.C., U.S. Government Printing Office.

Sulfur Oxides	AP-50 (January, 1969)
Particulate Matter	AP-49 (January, 1969)
Nitrogen Oxides	AP-84 (January, 1971)
Photochemical Oxidants	AP-63 (March, 1970)
Hydrocarbons	AP-64 (March, 1970)
Carbon Monoxide	AP-62 (March, 1970)

Health

Anderson, D. O. The Effects of Air Contamination on Health. Can. Med. Assoc. J. *97,* Part I, 528–536; Part II, 585–593; Part III, 802–806, 1967.

Ferris, B. G., Jr., and Whittenberger, J. L. Environmental Hazards. Effects of Community Air Pollution on Prevalence of Respiratory Disease. New Eng. J. Med. *275,* 1413–1419, 1966.

Higgins, I. T. T., and McCarroll, J. R. Types, Ranges, and Methods for Classifying Human Pathophysiologic Changes and Responses to Air Pollution. In: *Development of Air Quality Standards* (A. Atkisson and R. S. Gaines, eds.), Charles E. Merrill, Columbus, Ohio, 1970.

Lave, L. B., and Seskin, E. P. Air Pollution and Human Health. Science, *169,* 723–733, 1970.

National Academy of Sciences. Report of Committee on Effects of Atmospheric Contaminants on Human Health and Welfare, National Research Council. Effects of Chronic Exposure to Low Levels of Carbon Monoxide on Human Health, Behavior and Performance, Washington, D.C., 1969.

Working Group I(b)

Environmental Protection: Water

EDITOR'S NOTE: Many of the citations that appear throughout the report of the Working
Group on Water have been left incomplete by the authors and thus are not included in
the References at the end of the report. We regret any inconvenience that this may cause.

1 INTRODUCTION: AQUATIC ENVIRONMENTAL ZONES AND SITING CONSIDERATIONS

The Water Working Group is making the practical assumption that for the near term, the water resource will be used in some manner in the production of electrical energy. We recognize that if alternative methods of electric generation or cooling are employed that do not require water in the process, then there does not appear to be any near-term problem-solving to be accomplished with regard to the water environment. We also recognize, however, that release of emission products to the atmosphere can, from the long-term viewpoint, have important climatological effects and, therefore, impact, on regional water resources.

Thus, our task is to determine what alternatives are available to blend electricity production in the most compatible way with the aquatic environment. Obviously we must first assess the possible ecological consequences of electric generation upon aquatic environments. It is clear that because of the dynamic nature and the seasonality of events in various aquatic systems, definitive answers are not easy to obtain, especially in light of the still relatively primitive nature of the field of aquatic ecology. In spite of these limitations, it is possible to develop some environmental framework that indicates when and where these systems are vulnerable to the impacts of electric generation processes and when and where they are flexible.

The next task is to take maximum advantage of planning and engineering capabilities to situate and design installations in a manner that avoids encroaching on the utility of aquatic systems and, conversely, to take advantage of the flexibilities these environments provide.

We have recognized six ecologically different eco-

systems that can be used in electric generation: (1) oceanic; (2) estuarine; (3) riverine; (4) lake; (5) reservoir, and (6) cooling pond. All have certain diverse physical, chemical, and biological features which may require slightly different siting and design considerations from a national, regional, and local siting basis. A considerable number of publications have covered the physical, chemical, and biological nature of aquatic systems. Many of these are directly applicable to our present problem while others are less so. Still other publications have dealt with various phases of the electric generation–environment interaction. We felt it unnecessary to rewrite again that which has already been said, but rather to condense as much as possible and perhaps list key publications in the reference section. Thus, our report attempts to accomplish the following:

1. Establish a procedural pattern for siting considerations
2. Give high visibility to areas of possible incompatibility or of problem interactions
3. Suggest solutions or alternatives to minimize conflicts
4. Provide some suggestions for research priorities in order to assist in providing useful information needed in problem solving

Figure 1 is an oversimplified flow diagram depicting the kind of step by step considerations that must be made in order to develop workable solutions to power plant siting. The first two boxes deal with policies that must evolve from other multidisciplinary efforts outside the scope of the working group. Tables 1–6 capsulize the characteristics of aquatic system components as they relate to power plant siting. The subsequent sections attempt to follow the logic of Figure 1.

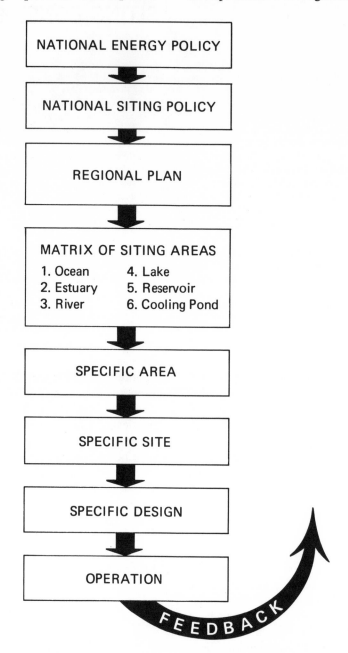

FIGURE 1   Flow diagram: Power plant siting considerations.

TABLE 1 Aquatic Systems—General Considerations for Siting and Design

Characteristics	Why Important *re* Environment Interaction with Electric Generation [a]	Alternatives	
		Short term	Long term
A. PHYSICAL			
I. Temperature	a. Biological "master factor" in that it can be a lethal, controlling, and directive factor to biological systems (Gunter, 1957; Brett, 1960; Kinne, 1963, 1964; Naylor, 1965; Mihursky and Kennedy 1967; Krendel and Parker, 1969). SES thermal discharges have been reported to exceed tolerances of aquatic species (Trembley, 1965; Coutant, 1962; Churchill and Wojtalik, 1969; Mihursky, 1969).	1. Release heat to atmosphere by system other than "one pass cooling" *re* closed cooling system. 2. Obtain rapid mixing of thermal discharge to reduce temperature change to tolerable levels. 3. Reduce temperature differential across condensers. 4. When feasible and appropriate, employ accessory cooling methods such as ponding, sprays, cooling towers.	a. Regional distribution to atmosphere to prevent local overloading and indirect effects on water resource.
	b. Southern latitude species in warmer seasons are already near their upper temperature tolerance limit (Naylor, 1965). There is more biological temperature flexibility to the north.	1. Revert to seasonal closed cycle cooling systems. 2. Site SES in northerly latitudes.	a. Move to deep oceanic sites. b. Closed loop systems using cooling reservoirs; no encroachment on natural public resource. c. Operate southern SES during cooler seasons only.
	c. Low temperatures can limit metabolic activity during winter seasons in northerly latitudes.	1. Proper use of waste heat in a constructive manner during cooler seasons to stimulate biological production.	
	d. Low temperatures in winter can cause cold kills in southerly latitudes.	1. Proper use of waste heat in winter can prevent cold kills.	
	e. Faunistic displacement options exist in heating northern waters; options are lost in heating southern waters.		
II. Pressure	a. Entrained organisms may be subjected to transient pressure due to bubble collapse in cooling water system. This may be significant but applicable data is not available.	1. Design to avoid cavitation, if necessary. 2. Decrease volume of H_2O used.	
	b. Increased pressure due to pumping.	1. Design to avoid pressure change. 2. Decrease volume of H_2O used.	
	c. Entrained organisms are subjected to decreased pressure with deep intake and shallow discharge.	1. Minimize intake–discharge depth differentials if necessary.	
	d. Entrained organisms are subjected to increased pressure with shallow intake and deep discharge.	1. Minimize intake–discharge depth differentials if necessary.	

TABLE 1—Continued

Characteristics	Why Important *re* Environment Interaction with Electric Generation [a]	Alternatives	
		Short term	Long term
III. Turbulence	a. Turbulence in jet discharges may cause disorientation in organisms and affect predator-prey relationships.	1. Nonjet discharge, if necessary.	
IV. Light intensity changes	a. Deep intake and shallow discharge will expose entrained phytoplankton to increased light; depending on species complex and adaptation, light may stimulate or inhibit photosynthesis.	1. If significant, minimize intake-outfall differential. 2. Use small percentage of new available water 10-20% to minimize effect to system.	a. Use inhibition to reduce over-enrichment *re* tendency toward eutrophication. b. Use stimulation to improve biological production.
	b. Shallow intake and deep release may take out of production the light-adapted phytoplankton.	1. If significant, minimize intake-outfall differential. 2. Use small percentage of new available water.	
	c. Upwelling etc. may cause increased turbidity and reduce light availability and photosynthetic activity.	1. Decrease velocities to minimize upwelling and scanning effects near soft bottom sediments and detrimental areas.	a. Use effect to reduce over enrichment *re* local tendency toward eutrophication.
	d. If abrasive material such as diatomaceous earth is used for condenser cleaning, turbidity will increase, and light availability will decrease.	1. Eliminate use of such materials if they have significant environmental impact.	a. Use nonfouling condenser materials.
V. Mechanical	a. Impingement of invertebrates and fish on intake screening.	1. Locate screens flush to shore line—not recessed. Also, locate 2nd set of screens at intake itself if intake is recessed. 2. Use improved traveling screens and water flushing techniques on screens with system to return organisms to receiving water body. 3. Hydraulic avoidance mechanisms.	
	b. Abrasion and damage of specimens passing through intake structures and cooling system.		
B. CHEMICAL			
I. Dissolved oxygen	a. Deep water intake in summer below thermocline in low or no DO water with surface release may discharge low DO water into high DO environment.	1. Add aeration system or surface turbulence to discharge system. 2. Seasonal use of closed cooling system. 3. Use rapid mixing.	a. Improve treatment of organic wastes before release into system.
	b. If low DO due to organic overloading surface waters, temperature addition can increase rate of organic digestion and aggravate low DO condition in summer and possibly convert the system from a marginal to poor water quality system.	1. Add aeration system or surface turbulence to discharge system. 2. Seasonal use of closed cooling system. 3. Avoid site.	a. Stimulate digestion as management tool. b. Improve treatment of organic wastes before release into system.

TABLE 1—Continued

Characteristics	Why Important *re* Environment Interaction with Electric Generation [a]	Alternatives Short term	Long term
II. Nutrients	a. Intake below thermocline in summer with surface discharge can release warmed water with higher nutrients in surface waters and possibly aggravate summer eutrophic problem if one already exists.	1. Alter intake–discharge design if necessary. 2. Seasonal use of closed cooling system.	a. Nutrient removal at organic treatment plant.
III. Chlorine	a. Used as biocide to control fouling and can also kill organisms entrained in cooling water supply (Mihursky, 1969; Morgan and Stross, 1969; Heinle, 1969; Hamilton *et. al.,* 1970; Hirayama and Hirano, 1970; Lewis, 1971; Marshall, 1971).	1. Maintain chronic low-level concentration that will not kill but will prevent fouling organisms from "setting". 2. Use Amertap system of flushing sponge balls through condenser to obtain physical cleaning instead of chemical. 3. Intake structures may need special biocide treatment; therefore, keep dosage to minimum concentration. 4. At intake may backflush with heated water to control fouling.	a. Fouling-resistant surfaces.
IV. Trace metals; heavy metals	a. Corrosion and/or erosion of metal surfaces (or other factors?) can release heavy metals in discharge water. These metals can be accumulated in the benthos and in biological "concentrators" such as oysters with detrimental biological and economic effects. (Roosenburg, 1969; Rothwell, 1971).	1. Improve metallurgy. 2. Alter design to reduce cavitation. 3. Avoid overcholorination.	a. R&D in development of resistant materials.
	b. Recycling water from cooling towers can be chemically treated with compounds containing chromium, zinc, and boron. Discharge of these materials may permit accumulation in tissues of biological "concentrators" (Marshall, 1971).	1. Recover heavy metals and retain. 2. Use chemical compounds not containing heavy metals. 3. Avoid using cooling towers.	
V. Miscellaneous chemicals	a. Discharge from cooling tower blowdown and other wastes from various water treatment processes, cooling towers, etc., may contain a variety of chemical compounds (Marshall, 1971). The biological effects of these are presently not documented.	1. Neutralize when necessary. 2. Release gradually for maximum dilution.	

[a] The authors of this report did not provide complete bibliographic references for this section. We regret any inconvenience this may cause.

TABLE 2 Oceanic Zone

Characteristics	Why Important *re* Environmental Interaction with Electric Generation	Alternatives	
		Short-term	Long-term
A. Physical			
I. Area—361.059 × 10⁶ km² (Sverdrup *et al.*, 1942)	a. Large surface area available for cooling		
II. Volume—1370.323 × 10⁶ km³ (Sverdrup *et al.*, 1942)	a. Large mixing and dilution capacity		
III. Temperature *Stream* East Coast influenced by warm Florida Current & Gulf Stream; Gulf Coast partially influenced by intrusion of warm Caribbean waters; Pacific Coast influenced by cold California Current plus upwelling of cold deep water	a. Warmer waters seasonally bring organisms near upper level of temperature tolerance while cooler waters apparently more easily permit modest addition of heat	1. Colder regions potentially able to accept more waste heat than warmer regions	
IV. Temperature–depth relationship	a. Thermocline is a permanent feature in tropical regions, while in temperate and boreal regions the summer thermocline is broken up in fall by natural cooling and wind b. Lose deep intake–shallow discharge design advantage when water is not thermally stratified c. Deep cold waters of Arctic and Antarctic form the deep and bottom waters and create the major deep water currents which assist in slowly mixing the world ocean. Must avoid any large-scale detrimental altering of ocean currents	1. Deeper waters may be used for cooling water supply with discharge into surface waters causing less change in receiving waters	a. Colder waters useful as cooling water supply
V. Density	a. Density differences are a function of temperature and salinity. These differences cause many of the circulation and flushing patterns and rates of many water bodies b. Density gradient in thermocline may prevent mixing and upwelling of heated water if released below the thermocline	1. Pump to surface if necessary and avoid deep release	

$I. Area$ placeholder

TABLE 2 Oceanic Zone—Continued

Characteristics	Why Important *re* Environmental Interaction with Electric Generation	Alternatives Short-term	Long-term
VI. Circulation	a. Caused by density gradients; unequal incident radiation; prevailing wind; and tides. Knowledge of circulation patterns important for siting and heated discharge patterns		b. Avoid excessive unequal atmospheric heating that might influence circulation patterns and climate
VII. Tides and currents	a. Makes more new water available for mixing and dilution b. Used by larval organisms and larger species as a transport and distribution mechanism	1. Use seasonal closed cooling system when necessary 2. Employ improved intake screening or avoidance techniques	
	c. Increase erosion and scouring	1. Use construction precautions	
VIII. Surface waves	a. Increase mixing capabilities b. Breakers from storms can destroy structures	1. Design to withstand stress	
	c. Can erode beaches	1. Proper bulkheading	
	d. Can cut new inlets and fill old inlets and thus alter circulation patterns	1. Proper bulkheading	
IX. Internal waves and seiches	a. Can exert force on submerged installations	1. Design to withstand stress	
X. Sediments	a. Determine load-bearing capabilities for structures	1. Locate structures on sediments that will bear calculated loads	
XI. Submarine topography	a. Can give indications of fault zones and probability of submarine landslides triggering turbidity currents. Areas near submarine canyons such as found off California and in Central Atlantic are probably most subject to turbidity current problems (Kuenen, 1950)	1. Avoid siting in areas that would present construction problems	
B. Chemical			
I. Overall chemical composition (see Harvey, 1963; Sverdrup *et al.*, 1942)	a. Coastal areas can have considerable variation in chemical constituents due to land runoff influence. May have bearing on metallurgical decisions		
II. Dissolved oxygen	a. Generally saturated in upper layers and falls to minimum at 500–900 m with increased values below these depths. b. In areas of upwelling (Gulf of California) oxygen minimum layer may be shallow as 100 m	1. Low-oxygen water drawn from deep should be aerated or quickly mixed if released to high-oxygen area	
III. Other gases, *viz.*, nitrogen and carbon dioxide	a. If cooling water pumped from deep layers and released in shallow layer, these gases can come out of solution, cause gas	1. Improve intake and screening device to prevent entrainment of fish species	

TABLE 2   Oceanic Zone—Continued

Characteristics	Why Important *re* Environmental Interaction with Electric Generation	Alternatives	
		Short-term	Long-term
	bubble mortality in organisms. Fish pumped up can have swim bladder gases expand and result in mortality		
IV. Salinity	a. Levels high, about 35 ppt & may present metallurgical problems, SES coupled with a desalinization plant can release hot brine which may tend to sink and perhaps have detrimental effects on stenothermal and stenohaline benthic species	1. Incorporate discharge design that will prevent heated brine from reaching benthic areas	
V. Nutrients, *viz.,* phosphates and nitrates	a. Tend to increase with depth. If deep water pumped to surface, it may stimulate primary productivity and subsequent food web. Problems may occur if overenrichment possibilities exist	1. Site in areas where overenrichment possibilities do not exist	1. Nutrient removal from inland sources
VI. Trace elements, *viz.,* copper, iron, etc.	a. Tend to decrease in concentration as one progresses offshore. This tendency may be an advantage for dilution of any heavy metal releases (Roosenburg, 1969; Rothwell, 1971)		
VII. Phytoplankton	a. Phytoplankton abundant from the surface to 100–150 m	1. Pump from below 100–150 m if design harmful to phytoplankton, or ecologically significant	
VIII. Zooplankton and larger pelagic organisms	a. Tend to concentrate at 800–1,650 m, but undergo vertical migration—up at night and down in daylight (Leavitt, 1938)	1. Establish variable intake options to pump from levels that will avoid concentrations of vertical migratory organisms	
IX. Meroplankton	a. Squid, pelagic fishes, and many deep-water organisms may produce larvae that are present in upper water layers (Marshall, 1954; Idyll, 1964)	1. Establish variable intake options or closed-circuit cooling	
X. Benthos	a. Presently, exploitable species are found in less than 500-m depth. Potential exists for exploitation of certain deep-water species. Generally, deep-water forms are not able to adapt physiologically to rapid environmental changes	1. Avoid discharging to benthic regions 2. Avoid siting in commercial harvesting areas, as harvesting gear may interfere with undersea transmission lines	

TABLE 3 Estuarine Zones

Characteristics	Why Important *re* Environmental Interaction with Electric Generation	Alternatives Short-term	Long-term
A. Physical			
I. Area	a. Many estuarine systems provide larger surface area for heat dissipation than many inland water bodies		
II. Volume	a. A few systems, such as the Chesapeake, have considerable volume ($1,852 \times 10^{13}$ gallons) (Cronin, 1967) available for mixing and dilution		
	b. Volume of new water passing by a single point tends to increase as one progresses towards full oceanic condition	1. Dilution advantage accrued as siting progresses towards full oceanic conditions	a. Locate in offshore oceanic areas for maximum dilution capacity
III. Temperature	a. Thermocline often seasonally present, offering lower temperatures for cooling water supply during warmer seasons if pumped from deeper waters (Mihursky *et al.*, 1970)	1. Deep water is coolest water available in warmer seasons, but shallow shelf water coolest supply during winter seasons	
IV. Density	a. Density differences, i.e., due to deep intake or down-estuary intake in higher salinities, and release in surface waters or up-estuary in lower salinities can cause heated discharge to sink and reach resident benthic organisms	1. Appropriate location of intake and discharge to minimize this effect if important to site and benthos 2. Proper use of momentum mixing to decrease densities	
	b. Combined ses and desalinization plants may result in discharge with higher densities	1. Proper use of momentum mixing to decrease densities 2. Avoid locating in estuarine environments	a. Proper regional planning b. Oceanic discharge with high dilution
	c. If heated water sinks, it will retain heat longer in the water mass due to insulation. Water that is still retaining its heat recirculates more easily	1. Appropriate location of intake and discharge to minimize this effect if important to site and benthos · 2. Proper use of momentum mixing to decrease densities	
V. Circulation & hydrology	a. A one-pass design can alter normal local estuarine circulation patterns which may be important to migratory behavior of organisms and critical if in site of restricted water supply	1. Use available water so as to provide zone of passage and not interfere with migration 2. Do not locate in areas of restricted water supply or circulation or in "passes" and bottleneck areas important to species 3. Release heat into embayment where no migration occurs	a. Locate in oceanic reaches
	b. High velocity near the bottom to obtain rapid mixing may cause scouring and redeposition of bottom sediments	1. Discharge into surface waters to minimize benthic impact, if significant 2. Use diffuser system on bottom	
	c. Heated water can return to intake area before cooling back to ambient temperature, which	1. Proper location and design of intake and outfall to minimize recirculation if critical to site	

TABLE 3 Estuarine Zones—Continued

Characteristics	Why Important *re* Environmental Interaction with Electric Generation	Alternatives Short-term	Long-term
	can cause uneconomical operation and result in increased discharge temperatures above those planned for in original design		
	d. Can redistribute larval stages of benthic organisms and alter normal "setting" patterns	1. Seasonal use of closed cooling system when appropriate 2. Avoid siting in areas important to setting of key species if necessary	a. Use redistribution as management tool
	e. Redistribute sediments		
VI. Turbulence	a. Excessive turbulence may be damaging to fragile life history stages such as fish larvae and to certain species such as jellyfish and combjellies. Turbulence effect on biota not properly documented with applicable experiments and observations	1. Design changes in condenser box, etc., if necessary 2. Nonjet discharge	a. Use turbulence to control certain noxious species such as jellyfish
B. Chemical			
I. Salinity	a. If intake located in shallower depth or down-estuary from discharge area, may have water of higher salinity being discharged into low salinity area. Could alter biotic distribution and community structure of receiving area as certain species are salinity-dependent	1. Minimize differential by intake–outfall relocation if local situation warrents 2. Proper momentum mixing	
C. Biological			
I. Spawning migration	a. If unseasonal heated water areas attract species, behavior may be altered to prevent fish from reaching spawning grounds at appropriate time	1. Determine migratory location in water column and locate discharge away from this area 2. Seasonal closed auxiliary cooling system	
	b. Congregation in intake area due to currents, etc.	1. Keep intake screens flush with shoreline, etc. 2. Seasonal closed cooling system	
II. Horizontal seasonal migration of juveniles and adults	a. (same as above)	1. [same as above—a(1) & (2), b(1) & (2)]	
III. Horizontal migration of postlarvae from spawning to nursery grounds	a. Weak swimming stages may be subjected to entrainment in cooling water supply	1. Determine tolerances to proposed design and alter to maximize survival, *re* ΔT and time of exposure, etc. 2. Seasonal closed auxiliary cooling system	
IV. Vertical migration (seasonal)	a. In systems such as Chesapeake Bay, organisms tend to move inshore or stay in upper layers of water column during warmer seasons. During winter seasons many species move to	1. Locate intake in deep water during summer and in shallow water during winter 2. Storage reservoirs to take in water at appropriate tidal condition and close intake gate to prevent	

TABLE 3 Estuarine Zones—Continued

Characteristics	Why Important *re* Environmental Interaction with Electric Generation	Alternatives	
		Short-term	Long-term
	deep waters for overwintering, i.e., fish and crabs. Certain other species (plankton) may be found throughout the water column only during colder seasons	organisms from entering when abundant in water column due to behavioral migratory response to ebb–flood tide condition. In general, use ebb water for cooling during spring and early summer and flood water during late summer and fall	
V. Vertical migration (daily)	a. Certain planktonic species or life history stages apparently depend on daily vertical migration patterns that are tied into halocline in order to maintain position in tidal estuary. This behavior apparently is tied into feeding and predator relationships. Disruption of this pattern and effect on species not documented	1. Attempt to maintain design, ΔT, and time exposure to tolerance of key planktonic organisms 2. Use only 10–20% of available net water	
VI. Meroplankton stages	a. Location of SES at key nursery or spawning areas, *viz.*, striped bass spawning grounds at fresh–salt interface at head of estuary will result in seasonal occurrence of passive drifting early life history stages which can be subjected to entrainment in cooling water supply	1. Avoid siting at key spawning areas 2. Use seasonal closed cooling system when appropriate 3. Design ΔT exposure time, etc., to tolerance of species	

TABLE 4 Riverine Zones

Characteristics	Why Important *re* Environmental Interaction with Electric Generation	Alternatives	
		Short-term	Long-term
A. Physical			
I. Location	a. In midcontinent and in many land centers, it is the only available source of cooling water. Few rivers, however, are large enough for 1,000 and larger MW(e) plants if only one-pass cooling is used	1. Employ supplementary cooling systems or a closed-loop system	a. Utilize specially designed regional canal system for dispersal of heat from load centers
II. Hydrology	a. Unidirectional flow, consequently emissions from electric generation are reasonably predictable (Jaske and Synoground, 1970)		
	b. Great seasonal variability in flow; low flow often coincides with highest annual temperatures	1. Cooling capacities need to be calculated on the basis of statistical low flows, rather than annual flows, average flows, or other groupings	
	c. Generally well mixed vertically	1. Use turbulence to enhance vertical mixing and provide rapid	

TABLE 4 Riverine Zones—Continued

Characteristics	Why Important *re* Environmental Interaction with Electric Generation	Alternatives	
		Short-term	Long-term
		drop in temperature (Jaske and Synoground, 1970)	
	d. Generally well mixed horizontally. A discrete plume (mixing zone) gradually blends into a mixed river, all of which will be at the elevated temperature	1. Two predictive models are thus necessary, one for the mixing zone, the other for temperature decay proceeding downstream. Temperatures may drop rapidly (Jaske and Synoground, 1970)	
	e. Seasonal variation in rainfall may cause flooding of floodplain	1. Power plant may have to be located some distance from river-bank, entailing long discharge canals or piping. However, length is undesirable as it prolongs thermal exposures of organisms pumped through the condensers (Coutant, 1970)	
	f. Low flows or other factors may create extensive shallow areas. In shallow water, a heated discharge impinges quickly on the bottom, thus affecting benthic forms with warmest water	1. It is preferable to have some stratifications of the warmest portions of the plume during the phase of rapid dilution in order to avoid benthic organisms until temperatures are reduced by nearly 80% (Coutant, 1962)	
III. Temperature	a. Thermal input from solar radiation is high since there is a high surface-to-volume ratio. Megawatt equivalents from solar radiation may dwarf power plant inputs and be a large part of total energy budget (Jaske and Synoground, 1970)	1. May require seasonal use of supplementary cooling systems	
	b. Thermal patterns closely follow air temperature patterns, particularly when in shallow areas. Large seasonal (sometimes daily) variations in temperature often affect engineering designs (e.g., cooling efficiencies) and thus have ecological effects. Latitudinal variations in thermal patterns are great. Need statistical representation of "normal" temperatures since there may be wide variations within a river channel at any one time	1. Must develop regional thermal predictive models for various river systems and employ supplementary cooling systems when needed	
IV. Sediments	a. Water flow and sediment flow are dynamic characters of importance in tributary mouths. Removal of cooling water can effect sandbars and siltation	1. Proper designs if potential damaging environmental impact	
	b. Dredging channels, inducing siltation, scouring by intake and outfall flows may greatly influence the habitat and thus community structure of benthic organisms	1. Proper designs if potential damaging environmental impact	
B. Chemical			
I. Oxygen	a. In absence of impoundment or high organic loads, dissolved		

TABLE 4   Riverine Zones—Continued

Characteristics	Why Important *re* Environmental Interaction with Electric Generation	Alternatives Short-term	Long-term
	oxygen levels for aquatic life are usually high. Rapid reoxygenation rates are possible		
II. Pollutants	a. Rivers are natural sewers for the landscape, particularly man's. Siting thermal or chemical discharges from a power plant may aggravate already polluted situations. Polluted water (e.g., high oxygen demand) may negate some of the advantages of rivers for cooling water supplies	1. Zones that are already biologically degraded and appear to be unrecoverable in near future should be considered for electric generation siting	
III. Nutrients, i.e., nitrates, phosphates, etc.	a. Rivers are often high in nutrients from mineralized sewage and land runoff. Thermal additions may stimulate unwanted algal production and accelerate eutrophication problems	1. Avoid siting in such marginal areas 2. Use heat as management tool to increase algal production where useful and reduce production if ecologically sound	a. Nutrient removal at source
C. Biological			
I. Behavior	a. Most riverine organisms are highly flow-oriented, and water circulation patterns may be critical for some species. Induced flows at intake and discharge of a power plant may sharply affect fish behavior (e.g., attract them). This effect may be greater than the thermal effect, or may enhance thermal effect by attracting organisms	1. Utilize proper intake–outfall designs and devices to protect mobile species	
	b. Invertebrate species often periodically undergo downstream movement (invertebrate drift). Possibilities exist for entrainment and mortality in heated water, both passing through the power plant and inclusion in the mixing plume	1. Utilize proper intake screening devices to avoid pulling in drifting invertebrates 2. Design system to give time–temperature condition that permits survival of species in question	
II. Benthos	a. Riverine systems are noted for high benthic biological productivity in areas of high water flow. If temperature rise is too high, selective exclusion of invertebrate herbivores may occur with consequent nuisance accumulation of vegetation (Coutant, 1970)	1. When necessary, use supplementary cooling systems to avoid damaging impact	
III. Fish	a. Migratory movements of fish populations are often quite extensive and important for life history completions. Thermal or flow disruptions at tributary mouths by cooling water flows can disrupt fish migration	1. Proper location of sites to avoid interfering with important migratory behavioral patterns	
	b. Migratory fishes are often present seasonally (even "resident" fishes have seasonal migrations)	1. Power plant effects can be highly seasonal, since important species may be in the river for	

TABLE 4   Riverine Zones—Continued

Characteristics	Why Important *re* Environmental Interaction with Electric Generation	Alternatives	
		Short-term	Long-term
		only short periods of time. Off-stream cooling may be needed for only brief periods. Critical ecological periods may *not* coincide with maximum annual temperatures	
	c. Migrations are often localized to specific routes in the river channel, e.g., to certain shorelines and to particular depths	1. Discharges must not disrupt migration behavior, both of upstream migrants (usually adults) and downstream migrants (usually embryos, larvae, or juveniles). Care must be taken in selecting locations of intakes and outfalls in order not to interfere with migrants. Areas not used for migrations can have heat-absorbing capacities greater than might be expected if migration routes were not considered	

TABLE 5   Lake Zones

Characteristics	Why Important *re* Environmental Interaction with Electric Generation	Alternatives	
		Short-term	Long-term
A. Physical			
I. Volume	a. Volume in lake quite constant, unlike many other aquatic systems; however, level control is important to other shoreline uses. Southern lakes generally shallow		
II. Circulation	a. Northern lakes circulation generally predictable. Central and southern lakes generally have circulation driven by wind	1. Proper spacing of intake–outfall to avoid recirculation conditions	
III. Temperature	a. Exhibit strong to weak thermal stratification. Northern lakes relatively colder than central and southern lakes		
	b. In northern lakes the volume of cold hypolimnion water is important for maintenance of coldwater species	1. Avoid pumping from hypolimnion and reducing volume 2. Thermal discharges either to surface or subsurface through momentum mixing requires individual lake considerations	
	c. Temperature influences rate of evaporation, and such water loss is important to certain water-poor central and southern areas		
B. Chemical			
I. General chemistry	a. Quite variable, with buffering capacity having considerable influence on water quality		
II. Dissolved oxygen	a. Many northern lakes tend to be oligotrophic and oxygen is	1. Use aerating system to replenish oxygen supply	

TABLE 5 Lake Zones—Continued

Characteristics	Why Important *re* Environmental Interaction with Electric Generation	Alternatives	
		Short-term	Long-term
	seldom limiting. Certain northern lakes that are (or are tending towards) eutrophy tend to have deeper waters containing little oxygen. Rate of eutrophication is more rapid in central and southern lakes and can result in low-oxygen conditions in hypolimnion. Low-dissolved-oxygen water pumped from hypolimnion can have detrimental influence on all biological population in discharge and mixing plume	2. Obtain rapid mixing	
III. Nutrients, *viz.,* nitrates and phospates	a. Nutrient inputs from surface drainage are quite variable; however, man's influence from sewage releases is very important locally and regionally. Many northern lakes are lower in nutrients than central and southern lakes. Nutrients influence primary production, which in turn influences light penetration and consequent water quality conditions at various depths	1. Careful considerations must be made of intake–outfall designs for each lake situation.	
C. Biological			
I. General	a. In the coldest northern lakes, the hypolimnion is extremely important for maintenance of cold-water species and food web, thus the cold-water fauna is chief asset. Certain northern lakes have a two-layered fauna: deeper–cold and shallow–warm. Many lakes, because of water quality conditions (especially central and southern lakes) are able to support only warm-water fauna that tend to depend on the shallow-water zone	1. Heating of recreationally important cold-fauna lakes is not desirable if tolerance thresholds are exceeded 2. Heating of two-layered-fauna lakes will probably result in greater energy flow to warm water fauna, perhaps at loss of cold-water fauna 3. Amount of additional heating tolerated by southern lakes must not exceed tolerance of useful warm-water fauna	
II. Migration	a. Littoral zone or tributary use for spawning	1. Natural reproduction requires avoidance of thermal influences in spawning areas, *viz.,* proper location of intake–discharge or seasonal use of supplementary cooling 2. Margin fluctuations during spawning activities must be controlled for natural reproduction success	
	b. Horizontal migration to and from center of lake	1. Avoid entrapment of schooling fish or attraction through intake system hydraulics	
	c. Seasonal vertical migration to deeper waters during warm season (northern lakes)	1. Maintain deep, cool waters (northern lakes)	
	d. Daily vertical migration of both fish and invertebrates, including zooplankton	1. Avoid entrapment of schooling fish or attraction through intake system hydraulics	

TABLE 5   Lake Zones—Continued

Characteristics	Why Important re Environmental Interaction with Electric Generation	Alternatives	
		Short-term	Long-term
III. Fish	a. Northern lakes—cold water species of fish—trout, walleye, northern pike, smallmouth bass, "Muskies," etc.	1. These fish require chilling in cold months as well as relatively cold water in the summer. Caution is required where hypolimnion intake sources may reduce the living space of cold-water fishes	
	b. Warm-water species of fish —crappie, white and largemouth bass, catfish, etc.	1. Upper temperature range must not be exceeded in large areas of the lake	

TABLE 6   Reservoir Zones

Characteristics	Why Important re Environmental Interaction with Electric Generation	Alternatives	
		Short-term	Long-term
A. Physical			
I. Geographic	a. Can be constructed from edge of sea to midcontinent and designed in the size needed to serve cooling requirements		
	b. Certain regions presently have reservoirs built for other purposes	1. Avoid using reservoirs for thermal discharges if additional evaporation losses cannot be tolerated in water-short areas	
	c. Can be formed by damming either a flowing or dry stream or constructed on high ground. The damming of a flowing stream destroys an old natural habitat for which organisms have evolved and replaces it with a habitat that did not exist in the area. Many species are locally exterminated. Anadromous fish activity is usually eliminated	1. Use existing reservoirs whenever possible for siting plants 2. Build new cooling ponds and reservoirs on dry streambeds or on upland areas	
II. Temperature	a. (In general, the same considerations prevail as described in lake section)		
	b. Downstream releases from reservoirs can be cold if released from deeper levels. If downstream release temperatures fluctuate because of alternating effects of shallow runoff over dam in combination with deep releases, this may have detrimental effects on downstream biota	1. Maintain consistent source of release or combine SES at base of dam where cold, deep release water can be used as cooling supply and thus by proper flow control maintain consistent and desirable downstream temperatures for biota 2. Design should permit option for depth of water release from reservoir	
III. Circulation	a. Circulation of water in a reservoir is dependent on wind and the passage of river water through the reservoir. A major effect of a steam-electric plant is circulation of water. Some species	1. Locate intake where it is least likely to entrain organisms 2. Either repell organisms from the intake structure or remove entrained organisms before passage through plant and return	

TABLE 6   Reservoir Zones—Continued

Characteristics	Why Important *re* Environment Interaction with Electric Generation	Alternatives	
		Short-term	Long-term
	avoid flowing water, planktonic organisms swim too weakly to escape a current, and many species make seasonal migrations either with or against the current. Fish such as rainbow trout, walleye, and white bass go upstream to spawn. A power plant may act as a barrier to large migrating organisms, and smaller organisms may be entrained in the cooling water supply and be subjected to chlorine, turbulence, a sudden rise of temperature, etc.	them safely to the environment 3. Put a weir on the lower end of the discharge canal to prevent fish from concentrating in the discharge canal in such numbers that they exhaust their food supply 4. Design intake and discharge for circulation of nutrients, melting of ice cover, and transport of oxygen to oxygen-deficient layers of reservoir	
B. Chemical	(Lake systems section applicable here)		
I. Dissolved oxygen (see lake and aquatic system section)	a. Nutrient levels are quite variable in reservoirs. Rich land, sewage, and fertilization are the major sources of nutrients in a reservoir. Nutrients tend to accumulate in deeper waters	1. In reservoirs deficient in nutrients in the surface water, fish production can be improved by designing plants to transport nutrients to the surface where phytoplankton can utilize them 2. Build reservoir where it receives drainage from rich land or a source of domestic sewage. If clear water is desired more than fish production, avoid sites where much nutrient will enter the reservoir 3. Use plants to aerate water and melt ice cover if organic production results in low oxygen levels	
II. Dissolved solids	a. Some reservoirs contain unusually high levels of dissolved solids which may reduce utility of water for human consumption or for agricultural use	1. Consider using such reservoirs having high dissolved solid levels as cooling water supply for SES	
C. Biological	(See Lake systems section, as well as section on hydroelectric considerations)		

2   HYDROELECTRIC CONSIDERATIONS

The bulk of the Water Working Group report deals with electric generation by steam electric stations; however, there is some electric production by hydroelectric installations. These latter facilities also interact with the environment. It is recognized that most dam sites in the United States have been developed, and future electricity contribution from hydroelectric installations will become less significant than at present. Nevertheless, it is felt that a brief section on environmental problems caused by hydroelectric installations and suggested methods for reducing environmental damage would be useful.

Hydroelectric plants vary in their character, their method of operation, and their effects on the environment. Small plants, large plants, plants operated as a single unit, or a succession of plants operated as a system have their special effects and problems. The method of operation of a plant or a hydroelectric system is of prime importance in determining environmental effects. The location and type or character of the plants are also important. When used to meet peak power needs, flow is intermittent and may be irregular. The streams below plants so operated may experience a flood at one time and very low flow or practically no flow at other times. Plants operated continually to provide a constant supply of power, with peaking from other sources, produce much more uniform flows and reservoir levels, as do a series of plants operated as a system.

SOME ADVERSE EFFECTS OF THE OPERATION OF HYDROELECTRIC PLANTS AND THEIR STORAGE FACILITIES

1. The prevention or serious curtailment of the migration of anadromous and resident species.
2. The production of intermittent stream flows that
    a. Periodically dewater large areas of streambed, rendering them nonproductive.
    b. Destroy benthic organisms.
    c. Destroy spawning beds, fish nests, eggs, and fry.
    d. Bring about the destruction of small fish and other organisms caught in depressions.
    e. Severely limit through drawdowns the growth of valuable aquatic plants and encourage the growth of semiterrestrial plants which may cause problems during periods of high water.
    f. Destroy a large portion of the benthic biota important in the natural purification process and thus seriously limit the natural purification capacity of the stream.
    g. Destroy or greatly reduce the production of aquatic life resources; reduce or prevent the use of boats; reduce sport uses and destroy aesthetic values.
    h. Decrease the dilution capacity of the stream for organic wastes and severely increase the toxic effects of wastes.
    i. Severely limit the capacity of the stream to

100

receive sewage and other organic materials, both treated and untreated, because the amounts which can be safely added are governed by low flows.

3. The production of serious outbreaks of gas bubble disease in aquatic organisms due to supersaturation with nitrogen when water passes over the dams.

4. Serious increases in water temperature due to the increase in water surface area and the creation of extensive shallow areas.

5. The production of dissolved-oxygen problems because of oxygen depletion in the deeper layers of reservoirs.

6. The creation of conditions which bring about anaerobic decomposition in the profundal bottom areas with the production of H_2S and other harmful metabolic products.

7. Increases in the incidence of parasites and diseases due to the crowding of fish in the vicinity of and in fish ladders.

8. Increased production of algae and actinomycetes which produce tastes and odors in domestic water supplies.

9. Increased production of algae which cause filter clogging problems in water treatment plants.

10. The creation of color problems in water used for domestic supplies through the concentration in and periodic release from bottom deposit of manganese compounds and other materials.

11. The increase in certain organic and metabolic products which interfere with certain uses of the water by such users as the soft drink industry.

12. Fluctuations in water levels in reservoirs which prevent or curtail the spawning of nest-building fishes, destroy food plants of waterfowl, and alter or destroy wildlife habitat.

13. The creation and operation of impoundments for power production can have a variety of other adverse environmental effects, among which are the following:

a. The production of health problems through the creation of environmental conditions conducive to the production of large numbers of disease vectors, such as Anopheline mosquitoes.

b. The production of pest problems by the creation of environmental conditions favorable for the production of large numbers of Culicine mosquitoes and other biting Diptera.

c. Periodic drawdowns which expose extensive bottom areas and destroy aesthetic values.

d. Fluctuating water levels and drawdowns resulting in the exposure of large areas of bottom and the reduction or elimination of important benthic food organisms; the growth of undesirable semiaquatic plants and coppice; the destruction of waterfowl food plants and the destruction of waterfowl and wildlife habitat.

e. Serious alteration or destruction of game animal habitat through the flooding of bottom lands.

14. The use of pumped storage in connection with the operation of hydroelectric plants can have undesirable effects such as

a. The flooding of prime recreational, camping, hunting, or game-producing areas.

b. The use and enlargement of existing lakes whose aquatic life and aesthetic values are reduced or destroyed by periodic severe drawdowns.

c. The destruction of organisms pumped up to the storage site and placed in a different and unstable environment.

d. Possible adverse effects on the aesthetics of an area due to pumphouses, transmission lines, and pipes.

METHODS FOR REDUCING DAMAGES DUE TO HYDROELECTRIC PLANTS

1. Use hydroelectric plants for firm power load and use other plants to meet peaks.

2. Place outlets which cannot be closed near the base of all hydroelectric dams so there is a constant flow below the dam of sufficient volume to keep the stream bottom covered.

3. Maintain uniform stream flows to the greatest extent possible.

4. Develop a series of dams on a river system so that flows can be more uniform and low water flows augmented while still maintaining constant water levels in the run of the river reservoirs.

5. Enhance low water flows to prevent toxic concentrations of waste and provide more capacity for receiving discharges of waste treatment plants.

6. Maintain constant water level in reservoirs during spawning periods.

7. Minimize or prevent conditions which cause supersaturation with nitrogen.

8. Provide better facilities for the movement of anadromous fishes.

9. Maintain water levels in reservoirs as constant as possible; especially prevent the growth of undesirable vegetation on flats which may result in health and pest problems as well as limiting wildlife values.

10. Dike and dewater shallow feather-edge problem areas.

11. Dredge and fill shallow shore areas and provide steeper shore lines.

12. Provide multiple-level intake towers so that

a. Cooler water can be used during periods of critical water temperature.

b. Well-oxygenated waters can be used during critical low DO period.

c. Strata containing taste and odor organisms can be avoided.

d. Strata containing blooms and filter-clogging algae can be avoided.

e. Strata containing undesirable colors or odors can be avoided.

13. Make provisions for the dewatering and flooding of backwater areas so they can be managed for the production of waterfowl foods and habitat.

14. Provide an early spring surcharge for the stranding of floatage.

3 ENGINEERING ALTERNATIVES FOR HEAT DISSIPATION SCHEMES

SOURCE OF WASTE HEAT

The efficiency of electric power production by either fossil or nuclear fuels is governed by the thermodynamics of the heat cycle. The ideal or Carnot efficiency is determined by the temperature of the heat source and by the temperature of the surrounding air or water which acts as a heat sink. The ideal efficiency is given by

$$E_i = \left[1 - \frac{T_{sink}}{T_{source}} \right] 100,$$

where the temperatures are measured on an absolute scale. In a modern fossil fuel station, maximum temperatures are of the order of 1200°F (1660°R), and the average annual temperature of the heat sink may be 60°F (520°R); thus, the ideal efficiency is approximately 68%. The temperature of the heat sink would have to approach absolute zero (−460°F) in order for the ideal efficiency to approach 100%. As in all mechanical and thermodynamic processes, the working efficiency is less than the ideal. New fossil fuel stations achieve about 60% of the Carnot cycle efficiency for an overall efficiency of about 40%. In a nuclear unit, maximum temperatures are limited by protective requirements for the fuel rods to about 650°F (1110°R), the ideal efficiency is reduced to 53% and the overall efficiency (at 60% of the ideal) to 32%. Thus in a nuclear plant, for every kilowatt of electrical energy produced, the equivalent of two kilowatts of energy is rejected to the surrounding air or water in the form of heat.

Taking account of differences in plant and stack losses between fossil and nuclear fuel units, the following heat rejection rates to condenser cooling water may be used to quantify the heat disposal problem for a unit producing 1,000 megawatts (MW) of electrical energy: Fossil —4.2×10⁹ Btu/hr; Nuclear—6.6×10⁹ Btu/hr. Thus nuclear units reject approximately 1½ times as much waste heat as a corresponding fossil unit. The heat sink consists of a condenser through which water is circulated. A typical condenser water flow rate for a 1,000-MW unit is about 1,500 cubic feet per second (675,000 gallons per minute) or 3.4×10⁸ pounds per hour. The temperature rise for the water passing through the condenser is obtained by dividing the heat rejection rate in Btu per hour by the water flow rate in pounds per hour. On the basis of the above numbers, the temperature increase through the condenser is 12°F for the fossil unit and 20°F for the nuclear unit. These figures are based on the current state of technology; however, most authorities see little likelihood of a significant increase in power cycle efficiencies within the next decade or two. Even if a major change in efficiency were to occur, it would be quickly overcome by the growth curve for electrical energy.

HEAT DISSIPATION TECHNIQUES

The ultimate heat sink for all forms of energy dissipation is through the earth's atmosphere to outer space. The SCEP * conference considered the climatic effects of all

* Man's Impact on the Global Environment, Report of the Study of Critical Environmental Prolems (SCEP), MIT Press, 1970.

forms of heat release on a global and regional basis. The estimated thermal waste of the world in 1970 was 5.5×10^6 MW, and this was not considered to have any significant global climatic effects. In the northeastern section of the United States, where 40 percent of the national energy use occurs, the thermal waste is currently equal to approximately 1 percent of the absorbed solar energy, and it is projected to reach 5 percent by the year 2000. Within the 4,000 square miles of the Los Angeles basin, it is estimated that the current waste has already reached the 5 percent level.

When waste heat carried by condenser cooling water is discharged into a water body, the transfer to atmosphere occurs over relatively large areas by evaporation, radiation, convection, and conduction. When cooling towers are used, heat is rejected directly to the atmosphere, primarily by evaporation in wet cooling towers and by convection in dry cooling towers.

Condenser cooling water systems are termed "once-through" when the cooling water flow is circulated only once through the system and waste heat is discharged into natural water bodies such as rivers, lakes, or coastal waters. They are termed "closed-loop systems" when the cooling water is continuously recirculated and waste heat is rejected to the atmosphere by cooling ponds, spray ponds, and wet or dry cooling towers. In certain cases supplemental cooling systems are used, where a cooling pond or cooling tower is combined with a once-through system to discharge to the atmosphere a portion of total waste heat before the rest is discharged to a natural water body.

Once-Through Cooling

The design of discharge structures for once-through systems is the most important factor in determining the magnitude, extent, and distribution of thermal effects in the receiving water bodies. The increased size of both nuclear and fossil-fueled units and growing concern with the environmental effects of water temperature changes have combined to limit the number of sites where once-through cooling can be utilized. However, by combining good engineering design with proper assessment of biological effects, once-through cooling should remain a viable mechanism for heat dissipation in major rivers, reservoirs, large lakes, and coastal waters.

Heat dissipation from the receiving water surface will ultimately return the water to its natural temperature state within a certain distance from the point of heat addition. This distance depends on the amount of mixing or dilution between the heated condenser water discharge and the receiving water. Heat dissipation from a water surface occurs through the combined mecha-

nisms of evaporation, radiation, conduction, and convection. The percentage of the total heat dissipation by evaporation increases as the temperature of the water surface increases above its natural or equilibrium state. The smaller the heat loss by evaporation, the lower the consumptive use of water. For a water surface at 5°F above equilibrium, the heat loss by evaporation is about one third of the total.

Through design of the heated water discharge structure there is a high degree of flexibility in tailoring the temperature distribution in the receiving water to minimize the biological impact. At opposite ends of design capability are complete stratification and complete mixing of the heated effluent. In the former case, mixing is avoided and the heated water is "floated" onto the receiving water in a relatively thin surface layer. Heat dissipation to the atmosphere is at a maximum rate, and there are no temperature changes at or near the bottom of the receiving water. Because of the ability of the heated layer to spread, precautions must be taken to prevent recirculation at the condenser water intake. This can be accomplished by means of an intake with a bottom opening known as a skimmer wall.

Thermal discharge regulations which prescribe maximum temperature differentials in the receiving water usually prohibit this extreme type of nonmixing discharge. However, the velocity at the exit of the surface discharge channel can be increased to achieve a desired surface temperature through entrainment of the surrounding water.

At the other extreme is the complete mixing of the heated discharge with the available flow past the site. The condenser cooling water is conducted through a diffuser pipe or tunnel and discharged through nozzles or ports near the bottom of the waterway. Entrainment of surrounding water into the high-velocity jets produces a rapid dilution. This type of discharge device provides the most rapid temperature reduction within the smallest area. On the other hand, a maximum amount of heat is stored in the water since surface heat dissipation is relatively slow at the very low temperature rises permitted.

An important area for once-through cooling is in coastal water where auxiliary cooling alternatives are most limited primarily due to salt-water drift from cooling towers. Except in shallow embayments, tidal and wind driven currents are generally available for rapid dispersion and dissipation of the added heat.

In designing the discharge structure for heated water, consideration must also be given to effects on overall water quality. For example, in dead-ended channels or embayments, promoting a stratified flow condition can substantially increase flushing rates. In naturally stratified large reservoirs, including some originally constructed for hydroelectric development, the addition of waste heat from steam-electric plants may, under certain circumstances, be beneficial for water quality by promoting re-aeration of hypolimnetic waters through new circulation patterns.

The primary advantages of once-through cooling are the low consumptive use of water, the ability to tailor the temperature distribution field in the receiving water to meet biological objectives, and heat dissipation to the atmosphere over a large area.

Factors which must be carefully investigated in a judicious design of discharge structures include entrainment of aquatic organisms, discharge jet effects on benthic organisms, bottom erosion, and effects on navigation.

The overall design of intake and discharge facilities for once-through systems should include consideration of changes in circulation patterns in the natural water body when water is withdrawn and then returned. Potential causes of such changes include warm water recirculation, if any; creation of a two-layer system when a density flow can be sustained in the receiving water body; entrainment of colder ambient water into discharge jet; drawing water from greater depths and discharging near the surface. Depending on the physical, chemical, and biological characteristics of the water body and on the existing level of water pollution, some of the changes in circulation patterns may be detrimental or beneficial.

When heated water is discharged into a water body, the resulting temperature field can be divided into two distinct zones: (a) an initial, or *near-field*, region in which temperature changes are governed primarily by the geometry and hydrodynamics of the discharge. The mechanisms which affect the temperature reduction in the near-field region are the dilution and entrainment due to the momentum of the discharge jet and the buoyancy effects due to the temperature difference between the discharge and the receiving water. (b) A *far-field* region in which the temperature distribution is governed by conditions in the receiving water. The important properties of the receiving water body are natural temperature stratifications, advection, diffusion and dispersion due to tidal currents, wind-driven currents and wave action, and heat dissipation from the water surface.

Temperature Distributions in the Near-Field Region

From an engineering viewpoint, the near-field region is the most important one, since the temperature in this region can be controlled by proper design of the discharge structure. The basic design parameter is the densimetric Froude number of the discharge, defined as

$$IF_o = \frac{V_o}{\sqrt{g \frac{\Delta \rho}{\rho} l}}$$

where

$V_o =$ mean velocity at the exit section of the condenser water discharge structure
$g =$ acceleration of gravity
$\Delta \rho =$ density difference due to temperature difference between condenser water at exit and receiving water
$\rho =$ density of receiving water
$l =$ length dimension characteristic of discharge structure

Surface Discharge The basic features of a surface discharge structure are shown in Figure 2. The width of the rectangular discharge channel is $2b_o$ and the depth is h_o. The densimetric Froude number is calculated with $l = h_o$. Stolzenbach and Harleman (1971) have made analytical and experimental investigations of surface discharges of heated water. Analytical procedures are given for the prediction of the three-dimensional temperature distribution in the near-field region. The parameters considered in the theory are the initial densimetric Froude number; the aspect ratio, $A = h_o/b_o$, of the discharge channel; the bottom slope in the receiving water; the current in the receiving water parallel to the shoreline and the dissipation of heat from the water surface. If

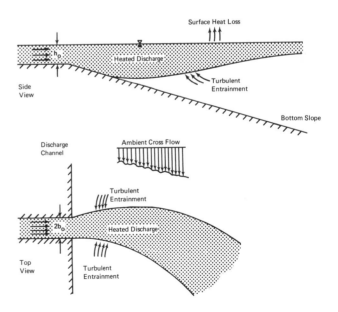

FIGURE 2 Schematic of heated surface discharge.

the discharge channel is designed with an initial densimetric Froude number nearly equal to or less than unity, the warm water tends to spread at the surface of the receiving water in the outlet area essentially at its initial temperature. While a small amount of mixing takes place, the buoyancy effect causes a large lateral spreading in a relatively thin, well-defined surface layer. Surface heat dissipation, under this condition, is at a maximum rate and is primarily responsible for temperature reductions in the region away from the outlet. A comparison of analytical and experimental results for this type of discharge channel ($IF_o = 1.03$) is shown in Figure 3. The ordinate for the upper plot is $\Delta T_c / \Delta T_o$ where

$$\Delta T_c = T_c - T_a \text{ and } \Delta T_o = T_o - T_a,$$

where

$T_o =$ temperature at exit of discharge channel
$T_c =$ centerline temperature in plume at the surface
$T_a =$ ambient temperature in receiving water

The abscissa is $x/\sqrt{h_o b_o}$, where x is the horizontal distance from the exit of the discharge channel along the centerline of the plume.

The bottom plot is a vertical profile through the centerline of the plume showing the depth at which the temperature in the plume is one-half the local centerline temperature at the surface.

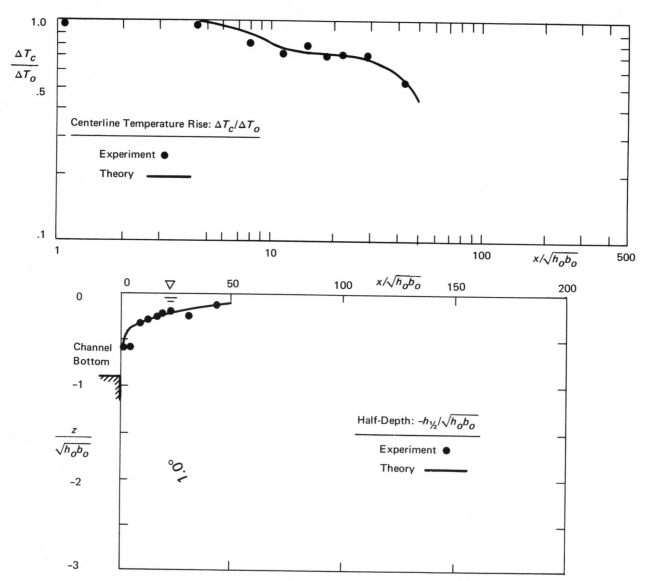

FIGURE 3 Theoretical calculations and experimental data for surface discharge: $IF_o = 1.03$; $K/u_o = 10.5 \times 10^{-5}$; $V/u_o = 0$; $A = 0.82$; $S_x = \infty$.

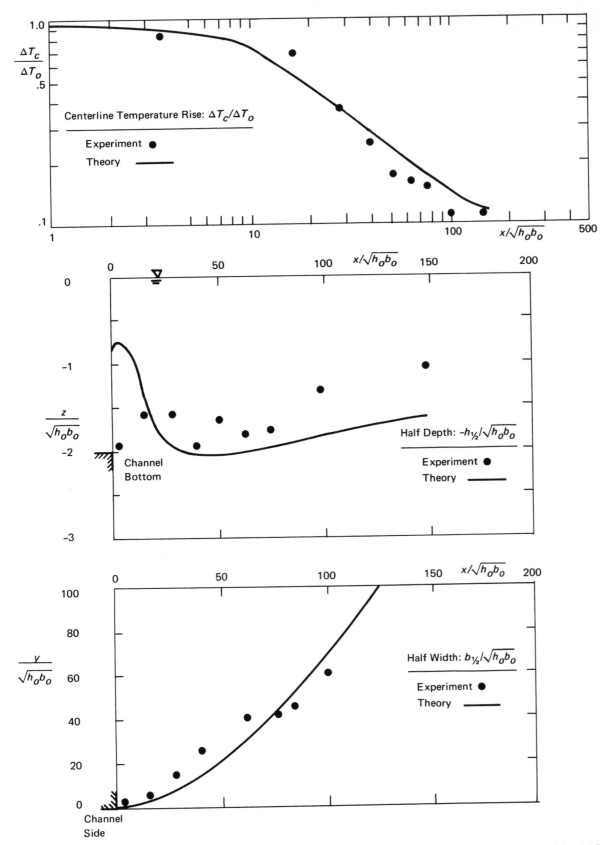

FIGURE 4 Theoretical calculations and experimental data for surface discharge: $IF_o = 6.60$; $K/u_o = 2.1 \times 10^{-5}$; $V/u_o = 0$; $A = 3.95$; $S_x = \infty$.

Figure 4 shows similar results for a surface discharge channel designed such that the initial densimetric Froude number $IF_o = 6.6$.

The bottom plot shows the lateral distance y, in the plane of the water surface, to the points at which the temperature is one-half the local centerline temperature at the surface.

A comparison of Figures 3 and 4 indicates the importance of the initial densimetric Froude number in determining the characteristics of the near-field temperature distribution. The larger centerline temperature reduction for the discharge channel with the higher initial Froude number is primarily due to the greater entrainment of the receiving water. Since the surface temperatures are reduced, the rate of dissipation of heat from the water surface is also reduced inasmuch as more heat is stored in the receiving water. In addition to the dilution process, the shape and direction of the plume is changed by the momentum of the current in the receiving water until the centerline of the warm water plume is essentially parallel to the direction of that current. When the receiving water is limited in width, the mixing zone may involve the entire width, as in Figure 5(A). In a large stream or in coastal waters, only a portion of the width is involved in mixing, as shown in Figure 5(B).

Submerged Discharge The basic features of both single and multiport submerged discharges are shown in Figure 6. When sufficient water depths are available, submergence of the outlet is an effective means of increasing the path of a buoyant jet. This has the effect of increasing the total entrainment of the cooler ambient water. By contrast with the surface discharge, lower water surface temperatures are produced, hence the rate of surface heat dissipation is relatively small. Single jets require a greater path length, hence greater sub-

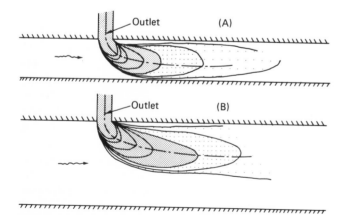

FIGURE 5 **Deflection of surface discharge by current in receiving water.**

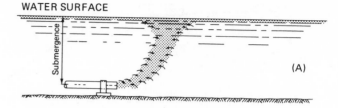

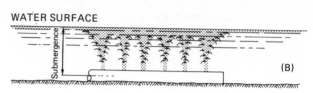

FIGURE 6 **Schematic of single and multiport submerged diffusers.**

mergence and water depths than multiport diffusers in order to achieve the same temperature reduction at the surface of the receiving water. Thus, if the available depth is limited, it may be necessary to use a multiport diffuser in order to prevent surface temperatures from exceeding a specified rise above the ambient temperature.

Multiport diffusers are large pipes, placed on the bottom of the receiving water body, from which heated condenser water is discharged through a number of ports. The net dilution can be large provided that the diluted water is removed from the discharge area by river, tidal, or wind-driven currents in the receiving water. If this is not the case, a buildup of temperature in the vicinity of the diffuser will occur as a result of re-entrainment of the heated mixture.

For a single, round, buoyant jet discharging horizontally near the bottom of a stagnant body of receiving water, the basic parameters which determine the temperature rise at the water surface are the relative submergence, y/D, and the densimetric Froude number of the jet at the point of discharge where $y =$ depth of receiving water above centerline of discharge port; $D =$ diameter of jet at discharge; and $l = D =$ characteristic length in densimetric Froude number IF_o.

Curves of maximum dilution S_m as a function of these parameters are shown in Figure 7 from both experimental and analytical sources (Abraham, Cederwall, Fan, and Rawn). The temperature rise at the surface of the receiving water, ΔT_c, at the centerline of the trajectory, is related to the dilution and the initial temperature difference ΔT_o by

$$\frac{\Delta T_c}{\Delta T_o} = \frac{1}{S_m}.$$

The practical range of initial densimetric Froude numbers is between 20 and 50. The single jet is not a very

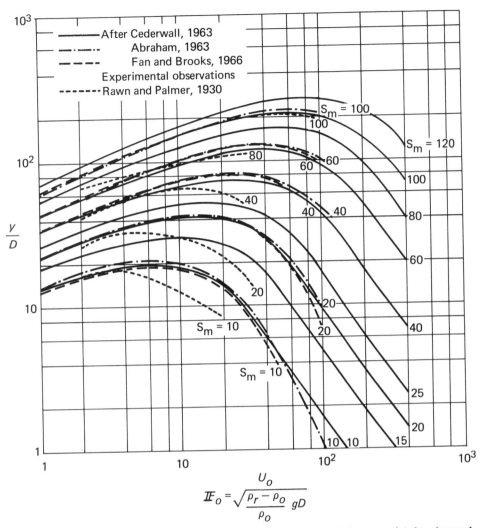

After Cederwall, 1963
Abraham, 1963
Fan and Brooks, 1966
Experimental observations
Rawn and Palmer, 1930

$$\mathbb{F}_o = \dfrac{U_o}{\sqrt{\dfrac{\rho_r - \rho_o}{\rho_o} gD}}$$

FIGURE 7 Theoretical solutions of dilution S_m for horizontal buoyant jet in stagnant receiving water of uniform density.

efficient dilution device at relative submergences below $y/D = 15$.

Liseth (1970) has presented experimental results for a submerged multiport diffuser with jets spaced a distance L apart along the diffuser. The jets discharge horizontally in alternate directions from the diffuser pipe in a stagnant body of receiving water. The experimental results are represented by the curves shown in Figure 8, when the parameters are as previously defined.

The results shown in Figure 7 are not valid in relatively shallow water, where $y/D < 15$, and for conditions in which the initial densimetric Froude number $IF_o > 10$. Recent studies at the Massachusetts Institute of Technology (Harleman, *et al.*, 1971) on shallow water diffusers have considered the combined effect of the jet-induced entrainment and the dilution effect of a current in the receiving water body. The maximum temperature rise at the water surface ΔT_s can be estimated from the relation

$$\frac{\Delta T_s}{\Delta T_o} = \frac{1}{\sqrt{\left[\dfrac{UyL}{V_o \pi D^2/4}\right]^2 \pm \dfrac{yL}{\pi D^2/4}}},$$

where U is the current velocity in the receiving water and the other variables are as previously defined. It is assumed that the axis of the diffuser pipe is at $90°$ to the direction of the current velocity vector. The jets discharge horizontally and when all jets discharge in the direction of the ambient current, the positive sign is used in the radical. If all jets discharge in the direction opposed to the current, the negative sign is used. For jets discharging in alternating directions, the second term in the radical is zero and the temperature reduction is given by the ratio of the jet discharge to the discharge due to the ambient current in a lateral distance L equal to the port spacing. In this type of shallow water diffuser, there is almost no temperature variation from surface to bottom beyond a relatively small mixing zone.

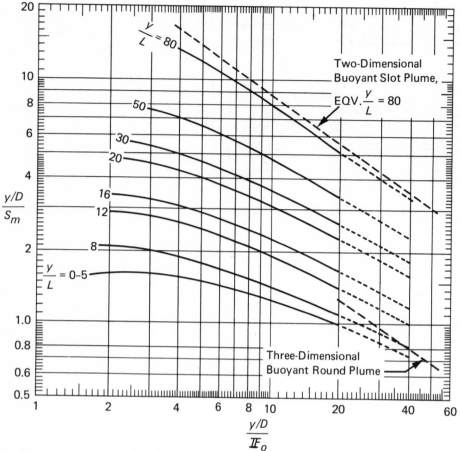

FIGURE 8 Minimum dilution along the center of merging buoyant jets from a manifold in stagnant receiving water of uniform density.

The above relations can be used in the preliminary design of submerged diffusers to meet specified temperature criteria in the receiving water. In all cases the basic design variables are the spacing of the jet ports and the total number of ports (i.e., length of diffuser). Design procedures to insure uniformity of lateral flow distribution from multiport diffusers are discussed by Vigander and Camp.

The following are some examples of diffusers designed for special purposes. The proposed diffuser for the 3,400-MW(e) Tennessee Valley Authority Browns Ferry nuclear station on a Tennessee River reservoir is shown schematically in Figure 9. The diffuser is designed to achieve complete and rapid mixing of the condenser water with the portion of the river flow passing over the diffuser. The diffuser consists of three pipes ranging from 17 to 20 feet in diameter extending 1,800 feet across the 30-foot channel of the river. The condenser water is discharged through more than 20,000 2-in. holes in the diffuser pipes. Temperature distributions, measured in a 1 to 15 scale laboratory model (Harleman, 1969) of a portion of the diffuser shown

in Figure 10, indicate that complete mixing is obtained about 50 feet downstream of the diffuser.

A diffuser suitable for shallow coastal waters is shown in Figure 11. The condenser flow rate of 1,000 cfs and $\Delta T_o = 15°F$ would represent the heat dissipation from a 500-MW(e) nuclear unit. Water surface isotherms (temperature rise above ambient) for a submerged diffuser located 2,000 feet from the shore and extending 3,000 feet parallel to the shoreline are indicated. The diffuser consists of 100 ports, 1 foot in diameter, discharging condenser water horizontally near the bottom at a velocity of 13 ft/sec in the offshore direction. A prevailing ocean current of 0.6 ft/sec moves parallel to the shoreline. The water surface area within the 1.5°F isotherm is approximately 6 acres.

Comparison of Surface and Submerged Discharge Schemes An alternative to the submerged diffuser shown in Figure 11 is to discharge the condenser water from a single surface discharge channel 10 feet deep and 20 feet wide located near the shoreline. The discharge velocity is 5 feet per second and the resulting surface

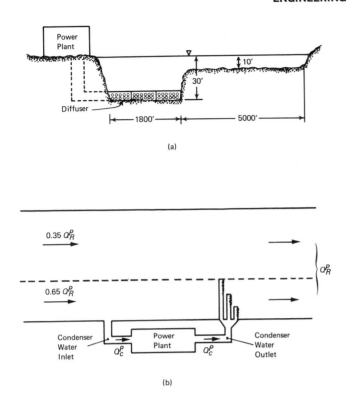

(a)

(b)

FIGURE 9   River section at power plant location.

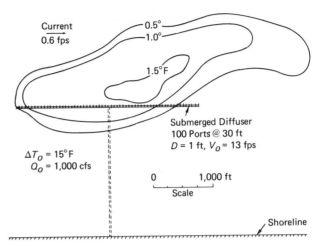

FIGURE 11   Isotherms of surface temperature rise for a submerged diffuser.

temperature isotherms are shown in Figure 12 (plotted to the same scale as Figure 11). The surface area enclosed by the 1.5°F isotherm is larger than for the diffuser scheme. In both schemes the temperature rise through the condenser is 15°F and the cooling water flow rate is 1,000 cubic feet per second.

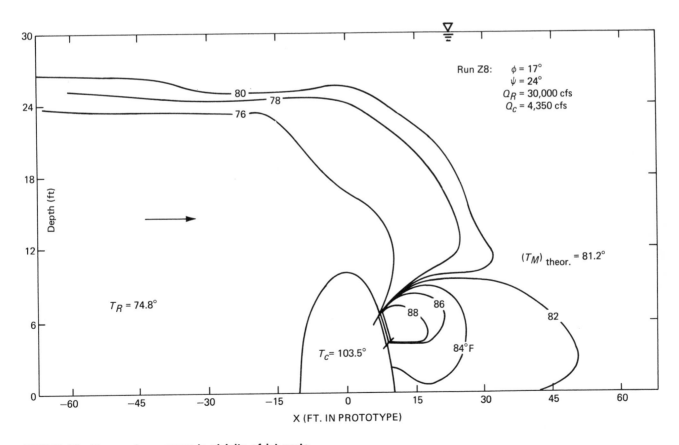

FIGURE 10   Temperature survey in vicinity of jet ports.

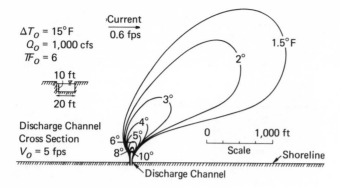

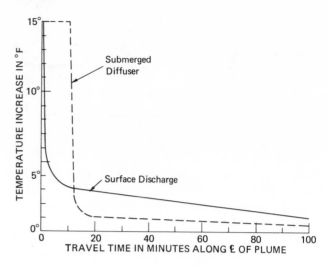

FIGURE 12 Isotherms of surface temperature rise for a surface discharge.

FIGURE 14 Comparison of temperature versus travel time for surface and submerged discharge.

A comparison is shown in Figure 13 in which the vertical isotherms along the plume centerlines for the two schemes are shown for a water depth of 30 feet. The surface discharge channel produces no change in temperature or velocity along the bottom, whereas the submerged diffuser results in significant temperature increases and higher velocities near the bottom. More heat is stored in the water with the submerged diffuser scheme because the smaller surface temperature rise results in a lower rate of surface heat dissipation. A comparison of the relative time of exposure of marine organisms, which may be entrained in the condenser water, is shown in Figure 14 for both the surface discharge and the submerged diffuser schemes. The preced-

ing discussion of surface and submerged discharges illustrates the wide range of design alternatives available to the engineer in controlling the thermal effects of heated effluents.

Temperature Distributions in the Far-Field Region

Beyond the initial mixing region in the vicinity of the condenser discharge, the far field temperature distribution depends primarily on flow conditions in the

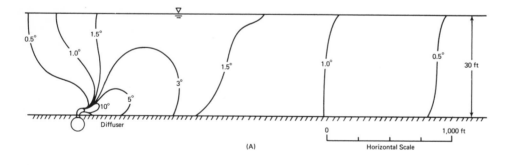

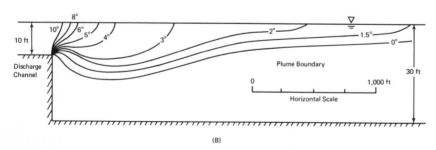

FIGURE 13 (A) Submerged diffuser—vertical section—plume centerline. (B) Surface discharge—vertical section—plume centerline.

receiving water body. In stagnant receiving water or when flow velocities are very low, stratification of warm water over cooler water can be developed and sustained. The basic heat exchange process is heat transfer to the atmosphere, although some dilution may take place due to mixing at the interface between warm and cooler layers. Temperature stratification in the far field region can develop even when there has been intensive mixing in the initial zone near the outlet. An example is the layer of water extending upstream from the mixed zone shown in Figure 10. When stratification has been developed in the near field region, as from a low Froude number surface discharge, the stratification in the far field region will be more stable, since the temperature difference between the upper and lower layers is greater. The advantage of a well-developed stratified condition is that a large proportion of the waste heat is transferred to the atmosphere by evaporation, radiation, and conduction, and only a small amount of heat is stored in the receiving water body. The disadvantage is the prevalence of relatively high temperatures over fairly large areas in the upper layer.

When flow velocities and turbulence in the receiving water are relatively high, stratification cannot be sustained even if it has been generated initially in the near field region. In addition to surface heat dissipation, heat will be transferred laterally and downward into the receiving water by turbulent mixing.

An important problem in both the near and far field temperature distribution is the possibility of recirculation of heated water between the condenser water discharge and intake. If not prevented, this condition can lead to a rapid heat buildup. Recirculation can be avoided in certain cases by physical separation of intake and discharge structures and by taking advantage of natural topography and prevailing current directions. The problem is more difficult in tidal waterways, due to the reversal of the current direction in the receiving water. If the temperature distribution in the vicinity of the intake has a fairly well-developed vertical stratification, recirculation can be prevented by the design of a skimmer wall intake which will prevent the upper layer from being drawn into the intake channel.

A schematic diagram of a skimmer wall intake is shown in Figure 15. The important quantity in the skimmer wall design is the vertical distance, h_s, from the bottom of the intake channel to the elevation of the assumed interface between the upper heated layer and the ambient receiving water. The vertical opening of the skimmer wall should be $h_s/2$, and the width b_i of the opening is given by

$$b_i = \frac{Q_c}{\sqrt{g \frac{\Delta\rho}{\rho}\left(\frac{2}{3} h_s\right)}},$$

where Q_c is the condenser water flow rate and $\Delta\rho$ is the density difference due to the temperature difference in the upper and lower layers. Detailed development of design procedures are given by Harleman (1969).

The basic concepts of surface heat dissipation which are of importance in the far field analysis are given in the following section.

Heat Transfer at a Water Surface

A body of water gains heat by absorption of short-wave solar radiation and long-wave atmospheric radiation. Heat is lost by evaporation and by long-wave back radiation, and heat may be either gained or lost by conduction, depending on the difference between water surface and air temperatures. A thorough discussion of the energy balance for a body of water is given by Edinger and Geyer (1965) and is illustrated in Figure 16.

Equilibrium Temperature

If a water body at a given initial temperature is exposed to a given set of constant meteorological conditions, it may either warm by gaining heat or cool by losing heat. The approach to a final steady-state temperature occurs exponentially, and that temperature is called the equilibrium temperature T_E. At equilibrium, the heat gained by absorbing radiation exactly balances the heat losses by back radiation, evaporation, and conduction.

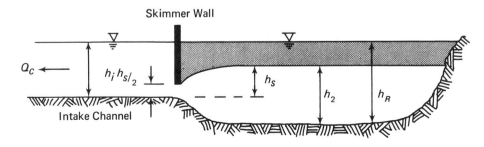

FIGURE 15 Skimmer wall at entrance to intake channel.

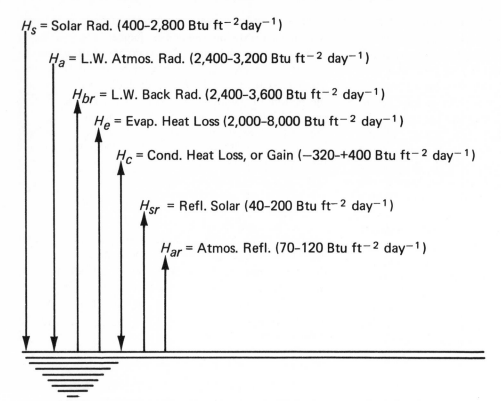

H_s = Solar Rad. (400–2,800 Btu ft^{-2}day^{-1})

H_a = L.W. Atmos. Rad. (2,400–3,200 Btu ft^{-2} day^{-1})

H_{br} = L.W. Back Rad. (2,400–3,600 Btu ft^{-2} day^{-1})

H_e = Evap. Heat Loss (2,000–8,000 Btu ft^{-2} day^{-1})

H_c = Cond. Heat Loss, or Gain (−320–+400 Btu ft^{-2} day^{-1})

H_{sr} = Refl. Solar (40–200 Btu ft^{-2} day^{-1})

H_{ar} = Atmos. Refl. (70–120 Btu ft^{-2} day^{-1})

NET RATE AT WHICH HEAT CROSSES WATER SURFACE

$$\Delta H = (H_s + H_a - H_{sr} - H_{ar}) - (H_{br} \pm H_c + H_e) \text{ Btu ft}^{-2} \text{ day}^{-1}$$

H_R Temp. Dependent Terms

Absorbed Radiation $H_{br} \sim (T_s + 460)^4$
Independent of Temp. $H_c \sim (T_s - T_a)$
                               $H_e \sim W(e_s - e_a)$

FIGURE 16   Mechanisms of heat transfer across a water surface.

Observed water surface temperatures in lakes, rivers, or the ocean are rarely equal to the equilibrium temperature because the equilibrium temperature is continually changing in response to local meteorological conditions. In practice, equilibrium temperatures are usually computed on the basis of daily or monthly averages of the meteorological parameters. Actual water surface temperatures are usually less than equilibrium in the spring when temperatures are rising and are greater than equilibrium in the fall when they are falling.

Surface Heat Exchange Coefficient   The net rate of heat transfer at an air–water interface is proportional to the product of the water surface area and the difference between the temperature of the water surface and the equilibrium temperature. In equation form,

rate of heat dissipation [Btu/day] $= -\rho c_p k_h A (T_s - T_E)$,

where

ρ = density of water, 1.94 slugs/ft^3
c_p = specific heat of water, 32.2 Btu/slug − °F
k_h = kinematic surface heat exchange coefficient, ft/day
A = water surface area, ft^2
T_s = water surface temperature, °F
T_E = equilibrium temperature, °F

The parameters $\rho c_p k_h$ are usually lumped together into a surface heat exchange coefficient, $K = \rho c_p k_h$. Thus,

rate of heat dissipation [Btu/day] $= -KA(T_s - T_E)$,

where the units of K are Btu/ft^2-day-°F.

Typical values of K range from 100 to 200 depending upon wind speed and other meteorological variables.

Evaluation of K and T_E   Brady *et al.* (1969) prepared the chart shown in Figure 17 on the basis of the analysis

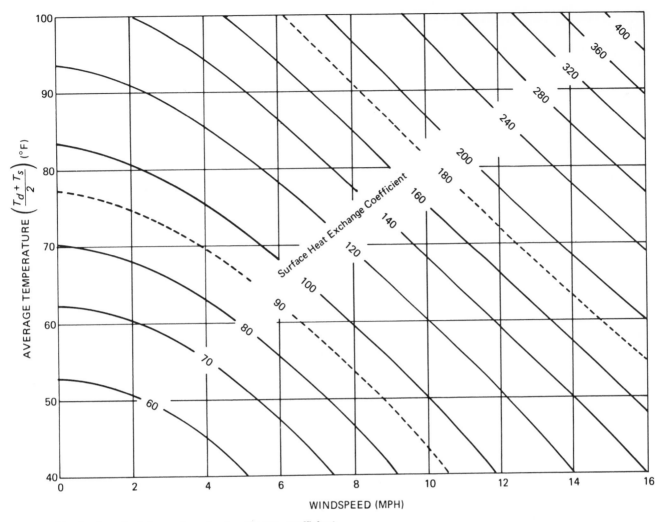

FIGURE 17 Design chart for surface heat exchange coefficient.

of extensive field data. Values of the surface heat exchange coefficient K may be determined for various combinations of wind speed, dewpoint temperature T_d and water surface temperature. The dewpoint temperature T_d is the temperature at which air would become saturated with water vapor when cooled at constant pressure. It can be measured directly by means of an electronic instrument or calculated from wet and dry bulb air temperatures. The mean dewpoint temperature during the month of July in the northeastern part of the United States is about 60°F.

In many applications the water surface temperature is the quantity to be determined. In this case an iterative procedure should be used in which an initial value of K is based on an assumed value of T_s. This process usually converges very rapidly.

Brady *et al.* (1969) have developed an approximate method for determining the equilibrium temperature. The recommended equation is given in terms of the dewpoint temperature, the mean daily solar radiation, and the surface heat exchange coefficient:

$$T_E = T_d + \frac{H_s}{K}.$$

The mean daily solar radiation H_s in July in the northeast is approximately 2,000 Btu/ft^2-day. Thus, if $K = 150$ Btu/ft^2-day-°F and $T_d = 60$°F, $T_E = 73$°F.

Heat Dissipation in a Uniform River Consider a river of constant cross-sectional area in which the discharge is constant both spatially and temporally. The one-dimensional heat conservation equation for a well-mixed river may be written in terms of the heat content per unit volume of water, $\rho c_p T$, as

$$\frac{\partial(\rho c_p T)}{\partial t} + U\frac{\partial(\rho c_p T)}{\partial x} = -\frac{\rho c_p k_h b \Delta x(T - T_E)}{bh\Delta x}$$

where U is the section average velocity, b is the width, and h is the depth. Longitudinal dispersion is neglected, and heat transfer is considered only across the water surface. Since ρc_p are generally assumed to have a constant value corresponding to the mean temperature of the system, the heat conservation equation becomes

$$\frac{\partial T}{\partial t} + U\frac{\partial T}{\partial x} = -\frac{k_h}{h}(T-T_E) = -\frac{K}{\rho c_p h}(T-T_E).$$

In the above equation, T is the river temperature due to a combination of natural and artificial heating, such as from a power plant. Solutions, giving the temperature distribution in the downstream direction, may be obtained for specified rates of heat addition for a power plant discharging heated water at $x=0$. Frequently, it is desirable to be able to predict the temperature increases above the existing ambient river temperatures before such a power plant is constructed. The conservation of heat equation can be written for ambient river conditions as

$$\frac{\partial T_a}{\partial t} + U\frac{\partial T_a}{\partial x} = -\frac{K}{\rho c_p h}(T_a - T_E),$$

where T_a is the natural or ambient river temperature, which is not necessarily equal to the equilibrium temperature. If the latter equation is subtracted from the former, the heat transport equation can be written in terms of the temperature excess ΔT, where $\Delta T = T - T_a$.

$$\frac{\partial(\Delta T)}{\partial t} + U\frac{\partial(\Delta T)}{\partial x} = -\frac{K}{\rho c_p h}(\Delta T)$$

This equation is similar to the one-dimensional conservation-of-mass equation for a substance undergoing a first-order decay,

$$\frac{\partial c}{\partial t} + U\frac{\partial c}{\partial x} = -K_d c.$$

The heat transport equation, written in terms of the temperature excess ΔT, is more amenable to physically meaningful steady-state solutions than the preceding equations. Under conditions of constant plant heat input, the term $\partial(\Delta T)/\partial t \to 0$.

The steady-state solution of

$$U\frac{\partial(\Delta T)}{\partial x} = -\frac{K}{\rho c_p h}(\Delta T)$$

under conditions of constant values for U, K, h and a uniform rate of heat addition of HR [Btu/sec] at $x=0$ is

$$\Delta T = \frac{[HR]}{\gamma Q}\left[\exp -\frac{xK}{U\rho c_p h}\right] = \frac{[HR]}{\gamma Q}\left[\exp -\frac{xKb}{\rho c_p Q}\right],$$

where $\gamma Q = $ lb/sec of water flowing in river and $Q = $ ft^3/sec river discharge. Thus the excess temperature ΔT approaches the ambient temperature exponentially in the downstream direction.

Jaske (1968a, 1968b) has developed computer programs for heat dissipation in rivers in which both river flow and heat additions are functions of distance and time. Numerical methods for heat dissipation in estuaries and coastal embayments are discussed by Pritchard and Carter (1965) and by Storer et al. (1971).

Supplemental Cooling

Supplemental cooling systems are once-through systems that include cooling towers or cooling ponds tied in with the discharge system. They are designed both to discharge condenser cooling water directly to the receiving body and to allow partial heat rejection to the atmosphere through the supplemental cooling facilities, when required. In the former case, the temperature at the point of discharge is essentially the same as at the condenser outlet; in the latter case the temperature is reduced by an amount dependent on the capacity of the supplemental cooling facility and on atmospheric conditions.

In general, atmospheric wet bulb temperatures have wide fluctuations in comparison to natural water temperatures and at times may even exceed them. Since a supplemental or "helper" wet cooling tower can reduce the temperature of the warm water discharge only to within a certain approach to the wet bulb temperature, there are inherent limitations in its cooling performance and in its effectiveness in controlling thermal effects. Similarly, the performance of a supplemental cooling pond has inherent limitations, since it is controlled by the approach to the equilibrium temperature of the water which may exceed, at times, the natural temperature of the receiving water body.

Closed-Loop Cooling

Closed-loop systems require no warm water discharge to the receiving water body and are designed to dissipate into the atmosphere the entire amount of waste heat carried by condenser cooling water. The heat transfer to atmosphere occurs in cooling towers, cooling ponds, or spray ponds.

Cooling towers can be of the wet, evaporative type or the dry, nonevaporative type. Depending on the means used for causing air circulation, cooling towers can be of the forced or induced mechanical draft type or the natural draft type.

Presently there are no technological limitations in the use of a wet cooling tower for any kind of condenser cooling water, including seawater. However, the use of seawater is impractical due to the effects of salt-laden drift on the land environment. For example, a wet cooling tower for a 1,000-MW(e) nuclear power plant using seawater for make-up would result in the deposition of approximately 100 tons of salt per day even if the concentration factor did not exceed 2.0 and the tower were equipped with the most efficient drift eliminators presently available.

Under certain circumstances, other limiting factors in the use of wet cooling towers may be (a) consumptive use of water; (b) restrictions on blowdown discharges; (c) fogging and icing hazards.

The total make-up water requirement for wet cooling towers includes evaporative losses, blowdown, and drift losses. The evaporative losses are approximately 1 percent of the cooling water flow for each 10°F of cooling effected in the tower. Blowdown is a nonconsumptive use and depends on the allowable concentration of water constituents to meet either water quality restrictions in the receiving water body or nonscaling requirements in the circulating water system. Drift, or windage losses, are currently estimated at 0.1 to 0.2 percent of the cooling water flow, but it is anticipated that they may be reduced in the near term to 0.05 percent by means of improved drift eliminators. For a 1,000-MW(e) nuclear unit, the consumptive loss of water at full load is approximately 32 cfs, and the total make-up water requirement is on the average approximately 40 cfs.

Whenever necessary, blowdown can be treated prior to discharging into a receiving water body. In certain cases, the cooling tower blowdown can be discharged into a closed reservoir or lagoon where solids are precipitated and then removed.

The use of dry cooling towers is presently limited to relatively small generating units, but technological progress in this field is anticipated. Presently they impose a penalty on thermal efficiency, and their cost is estimated to be three to five times larger than for wet cooling towers.

Cooling ponds are extensively used in geographical areas where the availability of large land areas and low humidity provide effective surface cooling. Heat transfer to the atmosphere occurs by evaporation, radiation, and conduction, and their magnitude depends on the temperature of the water surface. If there are no regulatory restrictions on temperatures in the pond, approximately 1,500 acres of water surface area are required for a 1,000-MW(e) nuclear power plant. Where temperature restrictions apply to ponds, much larger surface areas are necessary. Consumptive evaporative losses depend on the selected approach to natural water temperature. For example, for a 1,000-MW(e) nuclear power plant, using a cooling pond designed for a 15°F approach, average losses due to forced evaporation in summertime are approximately 18 cfs. In

addition, there are losses due to natural evaporation from the pond surface; these vary considerably with geographical region and season of the year. In evaluating total consumptive losses for a cooling pond, consideration must be given to a possible reduction of evapotranspiration losses due to creation of the pond. In addition to evaporative losses, make-up water requirements for a cooling pond include blowdown, which is necessary to control the increase in the concentration of chemical substances due to water evaporation.

Spray systems can be added to cooling ponds to increase the heat dissipation per unit surface area and reduce the pond size. Such systems have been used to date only for relatively small generating units.

In assessing the effects of cooling ponds, consideration must also be given to recreational and other possible uses.

Cooling Towers

Major Sources of Information

Though a wide variety and large number of articles have been published on devices for cooling condenser water, the two best and most detailed technical references in the English language are *The Industrial Cooling Tower,* by K. K. McKelvey and Maxey Brooke (Elsevier, 1959), and *Evaporative Cooling of Circulating Water,* by L. D. Berman, which is a 1961 translation by Raymond Hardbottle for Pergamon Press of the Russian book published in 1956. Though quite different in its outlook, one should also note the excellent study *Water Demand for Steam Electric Generation,* (1965, Resources for the Future, by Paul H. Cootner and George O. G. Lof). For a basic understanding of the integrated concept of electric power generation at an individual station comprising boilers, generators, condensers, and cooling tower, and their effects on efficiency and costs, it is unsurpassed. What the book does not do is to compare alternate systems at alternate sites. While run of the river cooling water may, in general, be cheaper than cooling towers, if the run of the river plant is not at the load center or at the fuel source, the integrated cost for delivered power may be less for a plant with a cooling tower at the load center or mine.

For a basic understanding of the workings of a central electric power generating station, *Power Station Engineering and Economy,* by Bernhardt G. A. Skrotski and William A. Vopat (1960), is recommended; and for a more detailed examination of the generation and production of steam from fossil fuels, the in-house publication by the Babcock and Wilcox Company *Steam —Its Generation and Use* (1963) is recommended.

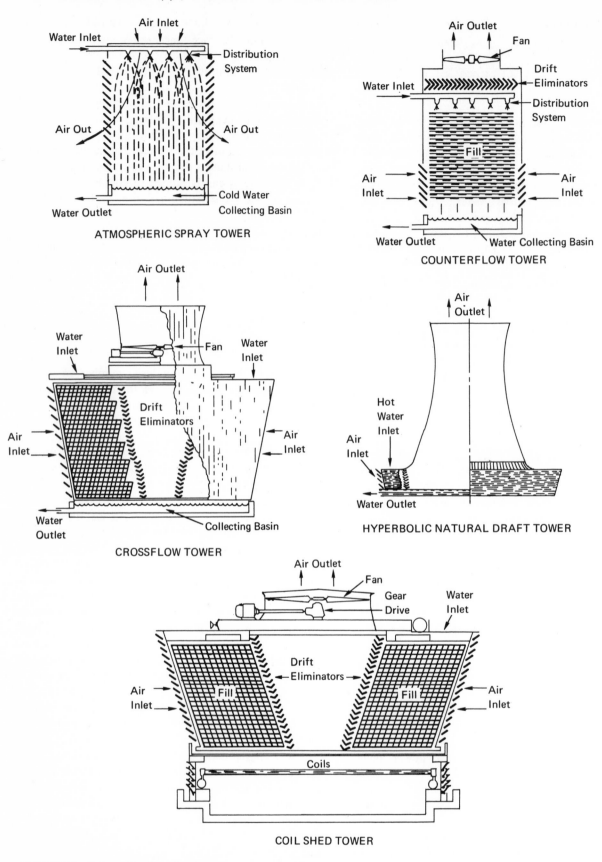

FIGURE 18 Schematic of cooling towers.

Also useful as reference works are the house publications of the major manufacturers of cooling equipment and the Cooling Tower Institute. Representative publications include: *Cooling Tower Fundamentals and Application Principles,* Marley Company (1967), and *Cooling Tower Performance Curves,* Cooling Tower Institute (1967). Up-to-date articles on individual topics in the cooling water field are ubiquitous, and only the articles deemed to be most important are included herein.

A good pictorial introduction to the field, outlining basic components and concepts, is the special report "Cooling Towers" by Steve Elonka in the March 1963 issue of *Power Magazine.* See also Figure 18.

Definitions

Certain specialized definitions are required to read the literature in this field:

Cooling range or range: The number of degrees water is cooled from inlet to outlet of device.

Approach: The difference in temperature of the cold water leaving and the wet bulb temperature of the air entering. Without special treatment, water cannot be cooled in a water-cooled device to a temperature lower than the wet bulb ambient temperature. It has been suggested by Agnon and Young (1952) that it is possible to reduce the wet bulb temperature of the incoming air below ambient by passing the incoming air through an air cooler or heat exchanger before it is admitted to the tower proper. The coolant used was part of the chilled water from the tower.

Heat load: Amount of heat dissipated per unit time in the cooling device. It is equal to the weight of water circulated per unit time times the approach. The heat load can be increased by

1. Increasing the velocity of the air in contact with the water surfaces.
2. Increasing the water surface exposed to the air.
3. Lowering the atmospheric pressure.
4. Raising the entering water temperature.
5. Reducing the vapor content of the inlet air.

Performance: The measure of the device's ability to cool water. Usually expressed in terms of cooling a quantity of water from a specified hot water temperature to a specified cold water temperature at a specific wet bulb temperature.

It should be noted and stressed that circulating cooling water device selection is intimately associated with condenser and turbine design and operation.

Though one can make approximate statements on the effect of decreased or increased pressure, initial temperature, increased reheat temperature, and increased feed water temperature, the efficiency of the turbine and the resultant heat rate (efficiency—Btu's required to produce 1 kilowatt-hour of electricity), for any specific installation for the most economic solution it is necessary to take into account the interrelationships of the following factors (Devereaux, 1966):

1. Average annual duration of temperature and humidity at the site
2. Condenser heat loads and vacuum correction curves for the turbines desired
3. Thermal performance of a number of cooling towers
4. Rough calculation of the pump and pipe network to determine costs and hydraulic characteristics
5. Performance of a number of condensers designed for the heat loads and cooling ranges under study
6. Costs of each component and thereby the amortization charges, fuel costs, plant lifetime load factor, and incremental capability credit

As LeBailly (1951) noted, "There is no practical and simple formula to compare directly the combined performance of different combinations of condensors and cooling towers operating under the same heat load and wet bulb temperature."

Atmospheric Towers

This discussion will be limited to the types of cooling towers that are used with large modern central electric generating stations. It should be noted, however, that in 1967, electric utilities cooling tower requirements were only $20 million out of a total $70 million annual sales of cooling towers (*Fluor-o-scope,* Fall 1967). It should also be noted that in the time period 1962–1967, over 85% of the cooling towers installed in United States power generating stations were of the induced cross flow design (Hansen and Parker, 1967). Atmospheric towers are no longer used because they cannot provide the large cooling capacity required.

Mechanical Draft Wet Cooling Towers

Theory All wet cooling towers are essentially described by the Merkel formulation (total heat method) (Merkel, 1925):

$$\frac{K_G dA}{L c_s} = \frac{dti}{i_i - i_G},$$

where

K_G = mass coefficient of heat transfer
A = surface area

C_s = specific heat of the air/water vapor mixture
L = weight of water circulating per unit time
t_i = interface temperature between water and air
i_i = enthalpy at interface temperature (t_i)
i_G = enthalpy of moist air at temperature t_G
t_G = air temperature

The total heat transfer per unit area per unit time is a function of the difference between the air/water interface and the air temperature and the difference in concentration of water vapor in the saturated air at air/water interface temperature and the humidity of the main air stream. The first difference takes into account the sensible heat transfer and the second, the evaporative heat transfer. The two heat transfer forces can be summed into the single equation shown where the difference in entering and leaving interface water temperature is a function of the differences in enthalpy between the interface air and the moist air. To maximize heat transfer, the difference in enthalpy must be maximized.

In addition, the heat dissipated can be increased by

1. Increasing the velocity of the air past the water surface, as that increases both the sensible heat transfer and the evaporative cooling
2. Increasing the surface area of water exposed to the flowing air again, which increases sensible heat transfer and evaporative heat transfer
3. Lowering the atmospheric pressure, thus increasing the evaporation rate
4. Raising the entering water temperature, which will increase the sensible heat loss and evaporation rate
5. Reducing the vapor content (humidity) of the inlet air to increase the evaporation rate

Advantages and Disadvantages of Mechanical Draft Wet Cooling Towers Mechanical draft towers are divided into two categories: forced air flow or induced air flow. The advantages and disadvantages of both are listed below (McKelvey and Brooke, 1959):

a. *Forced Draft Towers*
   Advantages

1. Absolute control over air supply—correct quantities of air and suitable air velocities may be selected
2. Close control of cold water temperature
3. Small ground area required
4. Generally low pumping head
5. Location of the tower not restricted
6. More packing per unit volume of tower
7. A closer approach and longer cooling range possible
8. Capital cost less than for a natural draft tower

Disadvantages

1. Horsepower required to operate the fans
2. Subject to mechanical failure
3. Subject to recirculation of the hot humid exhaust air vapors into the air intakes
4. Maintenance costs high
5. Operating costs high
6. Performance varies with wind intensity
7. Not suitable for all climatic conditions

b. *Induced Draft Towers* Induced draft towers are subdivided into crossflow and counterflow towers.

i. *Counterflow Induced Draft Tower*
   Advantages Coldest water contacts the driest air and the warmest water contacts the most humid air. Maximum performance is thus obtained.

   Disadvantages

1. Greater resistance to inlet flow of air increases the fan horsepower
2. Greater resistance to upward flow of air through falling water requires a greater fan horsepower than in cross draft towers
3. Uneven distribution of air velocities through the filling
4. Moisture eliminator area restricts air flow
5. High pumping head necessary
6. Water load capacity is limited
7. Hot water distribution system is inaccessible for ready maintenance
8. High inlet velocities are liable to suck airborne trash and dirt into the plant

ii. *Crossflow Draft Tower*
    Advantages

1. Low pumping head
2. Low static pressure drop on the air side
3. Convenient arrangement of the distribution system
4. Possible to clean the distribution system while tower in operation
5. The fill height is approximately equal to the tower height
6. More air per fan horsepower
7. More wood fill cooling surface per cubic foot of tower volume
8. Higher water loadings are possible for a given height
9. The water temperatures in the basin vary from the center to the edges according to a definite pattern, so that water may be drawn off at a selected temperature
10. Larger diameter fans can be used so that fewer cells are required for a given capacity

    Disadvantages

1. An insufficient pressure head on the dis-

tribution pans to keep the orifices from becoming clogged

2. Entire water feed exposed to the air which favors growth of algae

3. A substantial crossflow correction factor needs to be applied to the driving force, particularly where long range and close approach performances are required. In such cases for some pumping heads, a crossflow tower may need more ground area, and more fan horsepower than a counterflow cooling tower.

The major characteristics of induced draft towers—crossflow and counterflow—have been compared (Marley, 1967):

a. Plan area—counterflow units use less plan area than crossflow. Closer approaches can be obtained by simply increasing the height of the tower if space is a problem.

b. Low draft loss—for a specified applied horsepower, a greater air flow is possible in crossflow towers because of lower static losses, and therefore the crossflow tower is more efficient.

c. Thermal performance—crossflow towers can usually attain better thermal performance (i.e., heavier water loadings, longer ranges, and closer approaches) because the air flow distance is independent of fill height and static losses are not increased as length of air travel increases.

d. Low pumping head—static head is constant for crossflow towers but increases as depth of fill increases for counterflow towers. In addition, counterflow units which use spray nozzles require sufficient pressure to break up and distribute the water properly across the tower.

e. Recirculation—recirculation is less likely in crossflow towers than in counterflow towers. When it does occur in crossflow towers, only the upper section of the tower is usually affected. In counterflow towers, the recirculation is likely to be more uniformly distributed, hence decreasing efficiency.

f. Multiple cold water temperature withdrawal—in crossflow towers, the temperature of the water in the towers decreases from the edge to the center of the tower. By dividing the tower, it is possible to extract cold water at various temperatures.

g. Winter operation—wide louvers inclined to the vertical enable warm water to be diverted to the sides and to keep the air flow spaces open.

For the above reasons, crossflow towers are most commonly used.

Though forced air flow towers would seem to have overwhelming advantages over induced flow towers such as fewer vibrational problems (fans located at intakes at lower levels), less corrosion since the fans contact the ambient (relatively dry) air, lower noise levels, and slightly more efficiency (since some of the velocity pressure is extracted in useful work), their drawbacks are even greater. Recirculation of the moist, warmed air to the air intakes of the tower is greater than in induced air flow; uniform air distribution through the forced air draft tower is also more difficult to obtain in the large sizes, and fan size is limited in comparison to the size of fan utilized in induced air towers. Consequently, for the uses discussed here, induced draft towers are much favored over forced draft towers, and crossflow induced towers over counterflow induced draft towers.

Because of the competition from hyperbolic towers and the ever-increasing demand for higher efficiencies, many recent improvements have been made in mechanical draft cooling towers (Dickey, 1964). The use of single flow towers where space is a problem or where the wind flows predominantly in one direction has increased efficiencies more than 10%. The single flow arrangement can also be useful in preventing freezing by having the northern face solid. Fan and fan cylinder design has been advanced in recent years. Fan blade shape has come closer to the theoretical, and cylinders have incorporated energy recovery shapes. Various towers have been built to maintain their initial efficiency and to handle water temperatures up to 200°F (Anonymous, 1969).

Natural Draft Wet Cooling Towers

Natural draft towers have been used in Europe as far back as 1907. The first concrete hyperboloidal shell for a natural draft tower was installed in 1916 at the Emma Pit, Heerlen Coal Mine, in the Netherlands (McKelvey and Brooke, 1959). Though these towers have been popular in Europe for many years, only recently have any been built in the United States.

Over one half of the generating capacity in the construction program for Great Britain following World War II utilized natural draft cooling towers (Chilton, 1952). The first natural draft tower was built in the United States in late 1962 at the Big Sandy Plant of the Kentucky Power Company. Possibly the reluctance to build natural draft towers in the United States was due to the fact that they appear, in general, to be best suited economically for areas with high humidities. Such areas in this country, eastern and northwestern U.S., also have had abundant water and loose temperature rise restrictions and consequently have used once-through river cooling. Now with greater demands on the available water and tighter water temperature rise restrictions, new attention has been given to the natural draft towers. Among the advantages of natural draft towers are long-term maintenance-free operation (no moving parts and a concrete shell), smaller amount of ground space for multiple towers (no circulation of warm air from one tower to another), reduced piping costs (towers can be located adjacent to plant), no electricity required for operating fans, fewer electrical controls, and less mechanical equipment. Disadvantages include a decreased ability to design as precisely as for mechanical draft towers and inability to control outlet temperatures as closely as in mechanical draft towers.

McKelvey and Brooke (1959) indicate the following advantages and disadvantages for the hyperbolic natural draft towers.

1. Advantages

   a. They produce cooling effects similar to those provided by mechanical draft towers without the mechanical parts and the power required to run them

   b. Maintenance costs are negligible

   c. They can practically never break down

   d. Compared with atmospheric towers, they are independent of wind velocity

   e. They can cope with tremendous water loads

   f. They use comparatively small ground areas

   g. The stream of air is in the opposite direction to that of the falling water with the coldest air meeting the colder water first, which insures no loss in efficiency

2. Disadvantages

   a. Resistance to air flow must be kept at a minimum, hence grid sections must be shallow unless some form of film flow is used

   b. The great height necessary to produce the draft

   c. Inlet hot water temperature must be kept hotter than the air dry bulb temperature

   d. Exact control of outlet water temperature is difficult to achieve

Natural Draft Wet Cooling Tower Theory

The basic theory of natural draft cooling towers was restated by members of England's Central Electricity Generating Board (Rish and Steel, 1959). They point out that the reason for the hyperbolic shape is aerodynamic and structural rather than thermodynamic; also substantially less material is required for the hyperboloid shape than for an equivalent straight cylindrical tower (Jones, 1968). The shell is designed by membrane theory and results in extremely thin sections, as little as 5 inches of reinforced concrete, for towers over 300 feet in height. Details of the analysis, including wind measurements, movement of the structure as it

becomes saturated, icing, makeup, and integration with the rest of the equipment in the power station, are given.

Chilton (1952) had pointed out that the Merkel total heat equation was applicable to counterflow cooling towers in which the air flow is known. When the air flow is more variable and a function of climate and loading, such a direct solution is not possible. Wood and Betts (1950) have provided a trial and error solution to the problem, but the method requires knowledge of constants which are difficult to evaluate. Chilton has derived an approximate solution which is valid if the air becomes saturated during flow through the tower packing. The duty coefficient of the natural draft tower (D), defines the overall capabilities of a tower under all operating conditions. It is essentially a constant over the normal range of operation and is a function of total heat of the air passing through the tower (Δh), the change of temperature of the water passing through the tower (ΔT), the water load in lb per hour (W_L), and the temperature difference between the dry bulb and wet bulb air temperature (Δt):

$$D = \frac{W_L}{90.59 \frac{\Delta h}{\Delta T} \sqrt{\Delta t + 0.3124 \, \Delta h}}.$$

The draft, then, is due to the difference between the density of the air leaving the tower and the density of the air entering the tower and to the aerodynamic "lift" of the wind passing over the top of the tower.

Cranshaw (1963–1964), in a comprehensive series of tests, has more recently studied the variation of performance of natural draft towers and found

1. That the air temperatures should be measured at both top and bottom of the fill

2. Frictional resistance of the fill and the water to air loading are strongly affected by their change, especially at half loadings where the difference can be as great as 4 to 5°F

3. Wind has a lesser effect at full loading than at half loading on tower performance and high winds cause a 2°F drop in the cooling range at half load

4. Performance is negligibly affected by barometric changes

5. Air stability conditions do reduce the cooling range—an inversion by 1.2°F and a neutral lapse by 0.7°F

6. Increase in loading, cooling range, and humidity all tend to improve cooling tower performance

It should be noted that all of these results were obtained under English climatic conditions.

Stern (1967) outlines some important items for consideration in choosing natural draft towers:

1. Temperate climate with above average humidity
2. Lesser ground area for hyperbolic towers
3. Height of effluent exhaust will mean less air pollution problems

Technical factors to be considered are

1. Tower spacing—at least one and one half times base diameter
2. In crossflow towers, utilizing splash cooling, the fill is outside the shell and increases the diameter of the cold water basin. In the counterflow towers the fill is inside the tower and utilizes film cooling
3. Packless, or spray-filled, towers seem to hold some promise

Jones (1968) has noted that the market for natural draft cooling towers is almost equally divided between crossflow and counterflow towers. For a specific cooled water temperature, higher ambient wet bulb conditions can be tolerated at higher humidities. For a given tower diameter, as the approach narrows, the water flow must decrease. The height of towers increased dramatically in the last 10 years, from approximately 300 feet to close to 450 feet. Evaporation losses average about 1.5% of the circulating water, and drift will not exceed 0.2% of the circulating water.

Crossflow towers utilizing wood fill are slightly cheaper to build than counterflow towers. However, the maintenance costs are higher, since the wood will eventually have to be replaced, and the operating costs will also be higher, since 10 to 15% higher heads are required than in an equivalent counterflow design.

New Designs for Cooling Towers

Experimental designs for more efficient cooling towers are under development. Space requirements for both hyperbolic and induced draft towers are so great that combining the best features of both would reduce the space required, improve the efficiency, and reduce the variability of performance. Therefore, mechanical draft has been added to hyperbolic towers to achieve this higher efficiency. The Central Electric Research Laboratories, England, has completed such a design for an assisted-draft tower raising the capacity of a natural draft tower from 250 to 660 megawatts (Stubbs, 1968). In addition, model designs of ellipsoidal form have been tested to separate the airflow and to help create a high draft velocity. Such a slope will also help prevent downdrafts into the tower. Such towers would reduce aerodynamic interaction between closely spaced shells (which was partially responsible for the Ferrybridge

collapse), would be less conspicuous, and would occupy less land (Anonymous, 1967).

Dry Cooling Towers

The discussion above pertains primarily to wet cooling towers (i.e., those towers where the major mode of heat dissipation is evaporation). For certain cooling conditions such as very high water temperatures, insufficient water, and problems of blowdown disposal, systems that depend primarily upon convection and use air as the transport medium may be preferable. (Dry cooling towers may be either of the natural draft or mechanical draft types.) McKelvey and Brooke (1959) list the following advantages and disadvantages of a dry cooling tower:

1. Advantages
   a. It can be used where fluids to be cooled are at a high temperature
   b. It eliminates water problems such as availability, chemical treatment, corrosion, spray nuisance, freezing hazard, and fouling
   c. There is no upper limit to which air can be heated
2. Disadvantages
   a. Normally a dry cooler is less economical than a cooling tower of the ordinary evaporative type
   b. The specific heat of air is only one-fourth that of water
   c. Maintenance costs (e.g., the prevention of corrosion) are high

Dry Cooling Tower Theory

Though air cooling has been used for many years for industrial cooling in the petroleum industry and in automobiles, it was only with the recent work of Heller and Fargo (1956, 1962) that the theory advanced sufficiently to permit larger scale application to central electricity generating stations. As early as 1939, Gesellschaft für Luftkondensation (GEA) had installed an air-cooled condenser on a high-vacuum stationary steam turbine. The design was the familiar finned tubes with air cooling (GEA, 1969).

In the unique design of Heller, the exhaust steam from the turbine is sent to a direct contact spray-type jet condenser using cooling water of feed water quality. As shown in Figure 19, the condensate is then circulated to the air-cooled tower. The entire system is kept under pressure to avoid air leakage.

In addition to a 16-MW plant in Hungary, two 6-MW plants in the Soviet Union, two 7.5-MW plants in Pakistan, and a 120-MW plant in Rugeley, England,

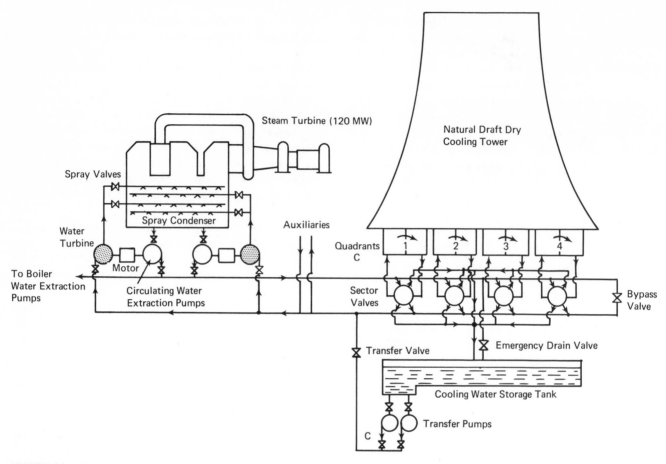

FIGURE 19 Heller-type dry cooling tower system.

have been constructed. The Rugeley plant has been in successful operation for over 7 years (Fiehn, 1969). The turbine operates at a steam pressure of 1,500 psig and steam temperature of 1,000°F and reheat to 1,000°F, and the tower dissipates 587×10^6 Btu/hr with a water savings of 1.8×10^6 gallons per day (Christopher, 1965). The application of the spray condenser to steam turbine was the idea of Heller, but the success of the system depended upon having a highly efficient and cheap heat exchange. This was designed by Fargo. To prevent excessive corrosion, 99.5% pure aluminum was used and the joints were mechanical rather than welded or brazed. The tower was the largest in the world at that time, reaching 350 feet and having a base diameter of 325 feet. The tower and condenser went into full service in July 1962. Makeup water in the first three years (for boiler flow-down and other losses) averaged about 2 percent. The system has been judged to be a complete success. Designs have been proposed for stations up to 4,000-MW capacity, and the English Electric Company expects the dry towers to become a common feature of power stations in the future (Anonymous, 1965).

Recently a 150-MW unit at Ibbenburen, West Germany, has been placed in operation, and a 200-MW turbine and two 100-MW turbines at Gyongyos, Hungary, are being installed (Jaszay and Tomcsanyi, 1968).

Research Required

1. Though cooling tower design has evolved over a long period of time and has reached a high state of development, it would still be useful to study from a basic principle point of view the heat transfer mechanism in a cooling tower. It may be that the utilization of tortuous surfaces would result in longer flow-through times and consequently increase evaporation for a given air flow.

2. The interference of multiple towers on each other versus the costs of longer towers should also be studied.

3. The physical and economic limit of towers should be studied. It may be that the efficiencies of scale of prime movers are completely negated by increased tower and pollution costs.

4. Dry cooling towers have only been built in small

sizes. The physical and economic problems of extrapolation to larger sizes should be investigated.

5. The shape of towers is not aerodynamically efficient with respect to air flow across the towers. Changes in shape to reduce the recirculation and improve air flow out of the towers should be studied.

6. The design and efficiency of fan-assisted natural draft towers should be investigated to determine the possible savings in tower size and ground space. The increase in the rise of the tower plume should also be studied.

Cooling Tower Problems

Though in many respects cooling towers seem to offer overwhelming advantages and to be an absolute necessity in some instances, they are not without their own problems. The wet towers, both mechanical and natural draft, have perhaps more problems than the dry towers, but, in general, they are also less costly. Some of the problems are described in the following section.

Climatic Effects of Cooling Towers

The use of the atmosphere as a sewer to carry off our wastes is well established and is favorably looked upon even by professional meteorologists, provided that it is done under the proper conditions and with no ill effects (Smith, 1963). For dry heat injection this feeling is even more pronounced, and professional meteorologists indicate that this discharge into the environment is acceptable and preferred to discharges to the hydrosphere or lithosphere (personal communication, Thom, July 1969).

Among the first problems to be answered is the question of major climatic changes due to waste heat rejection to the atmosphere. It has been shown that the amount of heat rejected from our major cities is a not negligible fraction (3%) (Brooks, 1967) of the solar heat reaching the earth's surface for that region. At present over the entire earth's surface the yearly production of man-made energy is about 1/2,500 of the radiation balance of the earth's surface. It could equal the surface radiation balance if compounded annually at 10% for 100 years (Fletcher, 1969). Its present growth rate is 4% per year. Therefore, even if we alleviate local thermal pollution problems now, the global problem will be upon us in a few decades. In the meantime the following general conclusions have been reached (Fletcher, 1969):

1. It does appear to be within man's engineering capacity to influence the loss and gain of heat in the atmosphere on a scale that can influence patterns of thermal forcing of atmospheric circulation.

2. Purposeful use of this capability is not yet feasible because present understanding of atmospheric and oceanic dynamics and heat exchange is far too imperfect to predict the outcome of such efforts.

3. Although it would be theoretically more efficient to act directly on the moving atmosphere, engineering techniques for doing so are not presently available.

4. The inadvertent influences of man's activity may eventually lead to catastrophic influences on global climate unless ways can be developed to compensate for undesired effects. Whether the time remaining for bringing this problem under control is a few decades or a century is still an open question.

5. The diversity of thermal processes that can be influenced in the atmosphere, and between the atmosphere and ocean, offers promise that, if global climate is adequately understood, it can be influenced for the purpose of either maximizing climatic resources or avoiding unwanted changes.

The second problem is the effect of cooling towers on regional climatology. The professional opinion is that cooling towers will not have a measurable effect on a region's climatology. This opinion is based upon the effect of Lake Meade and similar large scale man-made systems upon the local climatology (Personal communication, Holzman, July 1969).

This was recognized as long ago as 1937, when Holzman stated "that the moisture for precipitation in the United States was derived mainly from the oceans and transported by maritime air masses" (Holzman, 1941).

This is not to imply that cities do not have a measurable effect on their climate. Landsberg (1961) showed that cities did have a profound influence, as illustrated in Table 7. Landsberg (1961) notes "that the influences of the city on precipitation are not easily unraveled. We can say with reasonable confidence, however, that most of them tend to increase precipitation."

The main causes for increased precipitation are additional water vapor from combustion processes and factories, added nuclei of condensation and freezing, thermal updrafts from local heating, and updrafts from increased friction turbulence. Chagnon (1961) notes that the urban effect on precipitation is most pronounced in the colder half year of humid continental climates. Though Chagnon (1961) notes increases in precipitation over the Champaign–Urbana region, he concludes "whether a given city produces, or can produce, a 5, 10, or 15 percent increase in rainfall and number of rainy days in portions of its urban area is a question that neither meteorology nor climatology can at present answer accurately."

TABLE 7 Climatic Changes Produced by Cities

Element	Comparison with Rural Environs
Contaminants	
Dust particles	10 times more
Sulfur dioxide	5 times more
Carbon dioxide	10 times more
Carbon monoxide	25 times more
Radiation	
Total on horizontal surface	15 to 20% less
Ultraviolet, winter	30% less
Ultraviolet, summer	5% less
Cloudiness	
Clouds	5 to 10% more
Fog, winter	100% more
Fog, summer	30% more
Precipitation	
Amounts	5 to 10% more
Days with 0.2 in.	10% more
Temperature	
Annual mean	1 to 1.5°F more
Winter minima	2 to 3° F more
Relative humidity	
Annual mean	6% less
Winter	2% less
Summer	8% less
Wind speed	
Annual mean	20 to 30% less
Extreme gusts	10 to 20% less
Calms	5 to 20% less

More recently, Czapski (1968) has suggested that the climatic effects "observed" near cities may be due more to the effect of the added heat than the condensation and freezing nuclei to which they are usually attributed. Czapski foresees severe consequences of large latent heat and thermal emissions:

1. Rainfall will be increased downwind for a considerable distance
2. Cumulus clouds will prevail most of the time downwind
3. Severe thunderstorms and even tornadoes can be caused in very unstable weather by dry and clear heat

Thom (personal communication, September 1969) has pointed out that most of the estimates about climatic effects upon cities are speculative, and the interpretation of many of the observations about climate are subject to question.

Hall, Russell, and Hamilton (1958) have noted that even for major ecological changes such as forests, the most noticeable effect on the microclimate occurs within the forest itself. For shelter belts, the microclimate of the adjacent open areas is most affected.

There is no question that man can modify his local

rainfall temporarily as seen in the rain after the fire bombings in Tokyo and Hamburg. It does not seem that the energy and moisture released from cooling towers has yet reached that level, but it is a fruitful area for research.

Fog and Plumes from Wet Cooling Towers

Localized ground fog does occur around many cooling towers when the cooling air water vapor saturation exceeds the capacity of the ambient air for water vapor (Buss, 1968). It is suggested that the physical location of the cooling tower be adjusted so that the fog does not obscure roadways or other industrial buildings. One method used to reduce ground fog was to increase the height of the discharge stack to enable the fog to diffuse before reaching the ground. In addition, individual discharge stacks were used for each fan cell to reduce the obstruction to air flow and, consequently, to reduce downdrafts (Baker, 1960). Another ingenious method was to heat the cooling tower exhaust air so as to increase its velocity and heat content and to promote greater turbulence so that the mixtures were below saturation (Hall, 1962).

The prediction of the size and path of the fog from cooling towers is possible by analogy with smoke plumes. The angle of the spread of the plume usually varies from 18 to 24 degrees, and the length of the visible plume is

$$Xp = 5.7 \left(\frac{Vg}{102.\pi.Vw} \right)^{\frac{1}{2}} \cdot \left(\frac{tge - tgi}{tp - tgi} - 1 \right)^{\frac{1}{2}},$$

where

  tg = air or plume temperature (°C)
  tp = temperature at end of visible plume (°C)
  Vw = wind speed (ft/sec)
  Vg = total rate from tower (m³/hr) (N.B.—All gas volumes refer to 20°C and 1 atm pressure)
  Xp = visible plume length (ft)
  i = inlet
  e = exit

Longer and lower plumes occur when air temperature is low, humidity high, and wind speed moderate to high (Baker, 1967).

Noise

Large high-speed rotating machinery and enormous quantities of air moving through restricted spaces cause noise. Consequently, around mechanically induced draft cooling towers, there are going to be high sound levels. A three-step procedure is followed to evaluate noise problems (Miller and Long, 1962). First, using established values, determine the sound levels that will be acceptable to the neighboring installations. (In most instances, the building served will be insulated from the outside noise.) Background noises may mask the tower noise. Second, estimate the noise level produced by the cooling tower installation and received at the neighboring site. Make sure that all electrical and water connections are isolated from the tower and the only noise is one from the tower operation itself. Third, if the sound level at the neighboring site is excessive, reduce it by any or a combination of the following methods:

1. Increase the wall density or thickness, close windows, close air vents, seal cracks, and barrier walls at the site.

2. Change orientation of the tower so that sound is projected in another direction; reflecting walls can be used.

3. Move the tower further away from the critical site.

4. Run fan motors at lower speed at night when load is lowest, but when ambient noise levels may also be lowest.

5. A more expensive but more effective method is to use a lined plenum chamber. Miller and Long (1962) present a very complete set of curves and tables on acceptable sound levels, noise reduction provided by various walls, background noise, and distance correction values.

In a similar paper (Seelback and Oran, 1963), values of noise levels for fans of different horsepower are given. It is also pointed out that fan noise carries further and is more noticeable than water noise. Outlet noise is also greater than inlet noise, and satisfactory noise level reduction can sometimes be obtained by baffling the outlet only.

For more complete details on noise and its control, reference should be made to *Noise Reduction,* by L. L. Beranek (1960), and to *Handbook of Noise Control,* edited by C. M. Harris (1957).

Chemical Wastes from Cooling Towers

To promote long serviceable life for cooling towers, algicides and fungicides must be used to prevent damage to the wooden components of cooling towers, as must corrosion inhibitors to prevent oxidation of the metal parts. These chemicals are a potential pollution hazard if discharged directly to the stream.*

---

* The major references to these problems are (1) National Association of Corrosion Engineers *Primer on Cooling Water Treatment,* Houston, Texas, 1966; (2) papers presented at the Cooling Tower Materials and Water Treatment Symposium at the 1965 meeting of the American Chemical Society, Illinois State Water Survey Circular 91, 1966; and (3) Berg *et al.,* 1964, *Water Use and Related Costs with Cooling Towers.* This paper was adapted from the Illinois State Water Survey Circular No. 86.

Scale and foulants adhere to heat transfer surfaces or settle out in low flow areas and reduce the efficiency of the cooling towers. Common types of scales are calcium carbonate (most common), calcium sulfate, silica, magnesium compounds, iron and manganese compounds, and phosphate compounds. Scale prevention in enhanced by (1) softening the water (but this also increases its corrosiveness); (2) pH adjustment, which is not too satisfactory since conditions vary widely throughout the cooling tower; (3) chemical treatment of the water by the addition of polyphosphates which slow the rate of precipitation of the scale forming materials; polyphosphates with organics which primarily make the scale crystals nonadherent and organics alone (occasionally for water extremely high in scaling tendencies).

The rate of corrosion can be reduced by the use of anodic and cathodic inhibitors. Typical inorganic anodic inhibitors are sodium and potassium chromates, phosphates, silicates, nitrites, ferrocyanides, and molybdates. The chromates are most frequently used at high concentrations more than 200 ppm Na_2CrO_4 due to their relatively low cost and superior corrosion inhibition frequency.

Cathodic inhibitors may be salts of zinc, nickel, manganese, and trivalent chromium. They do not appear to be as successful as anodic inhibitors.

Organic inhibitors are sometimes used, but their chief use appears to be in conjunction with polyphosphate and chromates to improve their corrosion inhibition.

There are also a wide variety of proprietary compounds which are used to inhibit corrosion. They are mostly based on chromates and polyphosphates with some additives.

Biological fouling not only plugs screens and restricts flow in lines; through the photosynthetic process it can release oxygen to accelerate the corrosion process. The most commonly used agent to control biological growth is chlorine in shock loadings. However, the free chlorine residual in the hot return line to the tower should be less than 1 ppm to reduce chlorine attacks on the tower lumber. For cooling tower sumps where chlorine is not available, a wide variety of toxicants is available.

Cooling tower lumber, to retain its strength, should not be exposed to any strong oxidizing agent and can be protected from biological attack by using wood pressure-treated with chromates and zinc. High temperatures across the wooden surfaces should be avoided to reduce changes in structure and loss of wood material (National Association of Corrosion Engineers, *Primer on Cooling Water Treatment,* 1966).

A vast literature details treatment of these problems by proper material control, operation, and chemical treatment. The following are among the more useful works: Dalton (1965), giving a table of chemicals used for cooling water treatment, including purpose, usual dosage, relative effectiveness, waste problems, and relative cost. Dinkel *et al.* (1967) details the advantages of antifoulants in extending operating periods. High molecular weight polymers prevent deposition by flocculation, and organics such as lignins and tannins disperse the potential fouling materials. Berg *et al.* (1964) showed that chemical costs could be reduced if the mineral concentration in the cycled water could be increased 5 to 10 times rather than the common practice of 1 to 2 times. Recommendations for increasing the mineral concentration had previously been made by Denman (1961). Berg *et al.* (1964) gives estimates of all the various costs associated with cooling towers, and for comparison purposes with once-through cooling. It is these data which are used in Cootner's and Lof's *Water Demand for Steam Electric Generation* (1965). Details of the proper maintenance of cooling towers are given by Willa (1966). The right materials for use in cooling towers, with information on their service under various exposure conditions, are presented by Nelson (1966) for normal conditions, cast iron, modular iron, and galvanized steel for major mechanical and support parts; cast aluminum for blades; wrought aluminum for splash bars; copper alloys for nails; redwood or pressure-treated douglas fir for fill splash bars and framing; douglas fir plywood for fan decks and hot water basins; glass fiber reinforced polyester for fan blades and cylinders; and polypropylene and high-density polyethylene for orifices and spray nozzles.

Water effects of cooling towers are due primarily to the chemicals used for corrosion and fouling control. If the blowdown material cannot be rejected directly to the stream, ion exchange resins may be used to remove the chromates on streams and thereby may even eliminate blowdown altogether (Anonymous, 1964). A report by Hoppe (1966) indicates that it may be possible to use a zinc and biodegradable organic without phosphate and chromate and that the blowdown can be discharged to the stream.

Water Quality Effects of Cooling Towers

In passing through wet cooling towers, the temperature is reduced, the water is oxygenated, the BOD reduced, and the chlorine demand lowered (Dalton, 1965). Perhaps the most comprehensive paper on water quality changes in cooling towers (Davies, 1966) has been based upon studies at English power stations ranging from 120 MW to 600 MW. It points out that for a large, modern power station, blowdown can be as much as one million gallons per hour and this can be a sizable

fraction of a stream. Since mostly pure water is evaporated, it would be expected that the concentrations of inorganics such as chloride, sodium, sulfate, and totally dissolved solids increase directly as a function of a number of recirculations, as does conductivity. This was found to be true in most cases for chlorides, in half the cases for sodium, and in only one third of the cases for sulfate and conductivity. The hydronium ions (pH) at 3 of the 5 stations tested were concentrated slightly more than the inorganic constituents, apparently because of the loss of carbon dioxide to the air while passing through the tower. Other possible reactions which would lower the concentration of the hydronium ion below the recirculation concentration are oxidation of ammonia to nitric acid, hydrolysis of molecular chlorine to hydrochloric acid, and oxidation of sulfur dioxide. At three of the stations, hardness follows chloride concentrations, but at one station there appears to be precipitation of calcium carbonate.

The main chemical changes in water passing through cooling towers, then, are the loss of carbon dioxide, consumption of some of the oxidizable matter, and nitrification of ammonia. These processes are accelerated by the heated waters over their natural rate of occurrence in streams.

Discharge of Condenser Cooling Waters to the Ground

Though few large central station power plants reject their condenser cooling waters to the ground, because of the magnitude of the flows, smaller units do. The Dutch have taken a preliminary look (Dam, 1969) at the problem and concluded that fresh water is too valuable for their country and would not be used. Saline and brackish water can be used but will eventually raise groundwater to temperatures at which they cannot be used for cooling purposes. The Dutch also conclude, for their country, that the costs of recharging, maintenance, etc., are so great that, if possible, other methods of cooling should be used (Bardet, 1968).

The practice of using groundwater for cooling is quite old in the United States. By 1937, over 22 million gallons per day were being recharged to the ground formations in Kings and Queens counties in New York (Leggette and Brashears, 1938) at an average temperature rise of 10°F to 15°F. In 1937, a rise of 9°F in the groundwater supply due to recharge operations was noted (Leggette and Brashears, 1938). By 1940, recharge had reached 30 million gallons per day and groundwater temperatures had risen as much as 20°F at some of the pumping wells (Brashears, 1941). By 1941, 60,000,000 gallons per day were being recharged and were causing gradual temperature rises in the groundwater. These rises, however, were confined to areas near the recharge wells. In Nassau County, Long Island, New York, though high water temperatures have been found near some recharge wells, no widespread thermal pollution has been detected (DeLuca *et al.*, 1965).

It is interesting to note that a recent European paper on the influence of central electric generating stations on the groundwater regime near rivers does not even mention thermal pollution as a problem (Stundl, 1967).

A beneficial use of the condenser cooling water discharged to the ground has been noted by Granby (1966), who suggests a modification of production periods by creating a microclimate of several thousand hectares.

Tower Failures

The most spectacular problem with cooling towers was the failure of three natural draft cooling towers at the Ferrybridge C Power Station in England. The prime mode of failure was a vertical tensile failure within the lower part of the structure (Hannah, 1967). The design of the structures was deficient in three areas:

1. One-minute maximum mean wind speeds were used, whereas the structure was vulnerable to winds of shorter duration.
2. No safety margin was left for uncertainties in steady or dynamic wind loadings.
3. It was not appreciated how much the design could be changed by minor changes in the pressure coefficient distribution. The safety factors were applied to the strength of materials rather than to the load.

No defect or deficiency in the material or workmanship on towers was found which could be considered to have contributed significantly to the collapse. It was also found that insufficient attention had been paid to the effect of adjacent towers or the building itself in changing the wind loads (Anonymous, 1966).

Research Required

1. Practically no information is available on the climatic effect of the discharge of large quantities of heat and water vapor into the atmosphere. The effects of dry cooling towers may be even more adverse than those of wet cooling towers. These changes should be compared to the changes produced by the heat, water vapor, and particulate matter emitted by an urban complex.
2. The size and path of fog from cooling towers has not been sufficiently well established. When the size and location are determined as a function of wind speed, air

temperature and humidity, and approach, then a non-interfering location can be specified.

3. Means of preventing fog from becoming a nuisance need to be more thoroughly investigated. The degree of heating, induced turbulence, etc., that might be required needs to be detailed.

4. More satisfactory means of handling the blowdown from cooling towers need to be found. The maximum number of recirculations compatible with the heat transfer properties, life of components, and treatment need to be determined. If the blowdown contains many impurities, it may be denser than the fluid into which it is discharged and would then become a negatively buoyant flow.

5. Little information is available concerning negatively buoyant flows in flowing streams and under stratified flow conditions. The effect of the angle of discharge, the initial momentum, the flow of the ambient fluid, etc., need to be investigated to determine what conditions will induce complete mixing.

6. There appear to be a number of improvements in the water quality of the cooling water passing through cooling towers. Some information on changes occurring in towers in England is available, but a more detailed study of conditions in the United States needs to be made. It may be possible, by suitable alterations of operating techniques, to substantially improve the quality of the discharged water.

7. The effects of multiple towers upon the wind stresses induced need to be elucidated. The Ferrybridge inquiry needs to be broadened to take into account the change in wind loadings due to their towers and the power station itself.

8. The possible mixture of water vapor from cooling towers and sulfur dioxide from stacks needs to be considered. If the plumes mix, then an acidic condition can result, with deleterious effects. The effects of tower and stack spacing, height of plume rise, structure of the atmosphere, wind speed, etc., all play some role in this phenomenon.

Comparison of Cooling Methods

Climatic Requirements

Though one frequently sees statements limiting the use of mechanical draft and natural draft towers to certain climatic conditions, these represent, in part, an economic judgment on the part of the writer. For example, McKelvey and Brooke (1959) state, "On the other hand, with a mean wet bulb temperature of 40°F there is not *economic* application at all for mechanically ventilated towers, however low the exhaust heat loading. Canada or countries in North Europe are examples of places where natural draft towers will consequently tend to predominate. With mean wet bulb temperatures of 70°F and over, natural draft towers are not suitable, countries where such temperatures are habitual may therefore require mechanical equipment."

Fluor Corporation's house organ (*Fluor-o-scope,* Fall 1961) states, "Because of this (air moves in a natural draft tower as a result of a chimney effect created by the difference in density between the moist air inside the shell and the denser air outside), natural draft towers are economically best suited for areas with high humidities, such as Western Europe and part of the eastern and northwestern United States."

It is obvious that more than climatic conditions are involved when these statements are made and assumptions of the cost of labor, materials, power, land, etc., are also factored into the economic decision. It should also be noted that the temperature of the returned condensate has ramifications in the choice of the condenser and even the turbine itself. We restrict ourselves here to the climatic influence on cooling tower choice.

It should be noted that the method of dissipating heat in both natural draft and mechanical draft wet towers is the same—primarily evaporation. Therefore, for wet towers the minimum temperature to which the condensate can be cooled is the wet bulb temperature. For dry towers, the dry bulb temperature is the minimum temperature to which the condensing water can be cooled.

Cooling Tower Operation

Possibly the best review of current operating experience with mechanical draft cooling towers in the United States is that offered by Waselkow (1969), which details 60 years of experience of the Public Service Company of Colorado. Seventy-six percent of Public Service steam generating capability is on cooling towers. Waselkow also presents data on the experience of other utility companies in the mountain and southwestern states with cooling towers. He cautions about the difficulties in specifying tower performance since it is so intimately related to weather conditions.

The following figures reflect their experience:

1. One half to three quarters of a gallon of water consumptively used per kilowatt-hour. (For nuclear plant use, 0.9 gallons per kilowatt-hour.)

2. Drift loss is less than 0.1 percent of the circulating water.

3. Lake cooling is most trouble-free operation, followed by tower operation, and most troublesome is once-through stream flow.

4. Deterioration of redwood fill does not follow any consistent pattern.

Natural draft tower experience must be gleaned from European data since only limited American experience is available. Jones (1968a) estimates evaporative losses at 1½% of the circulating flow and drift at less than 0.2 percent. Blowdown will vary depending upon water quality but should be less than two or three percent of the flow even under extreme conditions.

Jones (1968b) also notes that natural draft towers, because of their high initial cost, can only be justified where the cooling demand is extremely great. He judges that the smallest individual generating unit which will use natural draft towers will be on the order of 500 megawatts.

Dry towers have much smaller losses. At Rugeley, total makeup has been about 1.5 percent of the boiler feed makeup because of faulty pump glands and cooler leakage (Anonymous, 1965). Tower makeup of 1.8×10^6 gallons per day is, of course, not required (Christopher, 1965).

It should be noted that if the existing tower is poorly designed, real estate is at a premium, and repairs are needed on the existing tower, it may be possible to increase its capacity by as much as 60% (McCann, 1966).

Up to 20% increase in capacity can be gained by newer types of fill material which offer greater surface area. The use of positive pressure sprays can increase capacity by 15%. New designs for drift eliminators can give up to 5% better performance. Up to 10% more capacity can be gained by varying air movement through changes in fan blade angle and in speed and size of motor. A properly designed velocity recovery stack can increase capacity as much as 70%. Partitioning of individual fan cells will help to reduce recirculation.

Economic Comparison of Cooling Towers

Can we make an economic comparison of these various alternatives? It is extremely difficult to do so on any objective basis. Jones (1968a) has pointed out that the physical size and cost of an optimum tower for a generating unit of a given size can differ in cost by a factor of two because of location, fuel costs, load factors, capability penalties, and capitalization rates. Because of these considerations, there are wide discrepancies though there have been a fairly large number of comparisons made. They are valid only for the conditions assumed and are not generally applicable.

Possibly the most recent and one of the best compari-

TABLE 8   Economic Comparison of Cooling Methods

Heat Rejection Method	Temperature Rise (°F)	Inlet Temperature (°F)	Condenser Back Pressure	Capacity ($/kW)
1. Run of river cooling	17	55	1.2	5
2. Bay/lake cooling	—	—	1.2	6
3. Natural draft cooling towers—run of river makeup	28	70	1.5, 2.16	7.5
4. Natural draft cooling towers—makeup reservoir	—	—	1.5, 2.16	11
5. Cooling pond	—	65	—	10
6. Dry cooling towers	—	—	—	22

sons was that of Shade and Smith (1969), where they examine six different methods of heat rejection for a modern power plant consisting of two 900-MW units with fossil-fuel-fired boilers using present-day costs. They emphasize that this is only for general comparison, since the costs for any project are greatly influenced by local conditions. The results are shown in Table 8.

F. A. Ritchings and A. W. Lotz (1963) studied the economics of closed vs. open cooling water cycles and likened a 200-MW(e) open cycle and three closed cycle systems for natural and induced draft dry cooling towers to a mechanical draft fin tube heat exchanger of American design. Results indicate 10% greater investment for closed-type systems in comparison to an open cycle wet cooling tower. Annual costs, including fixed charges on investment, are 10% more. Where water costs are high or where conservation must be practiced, dry towers may be useful.

These systems were compared against a standard American-made redwood design induced-draft tower. The economic optimum was adopted for each type. These are listed in Table 9. The water requirements, land areas, operating costs and annual charges are also listed.

Other broad-gauge studies of this type by various consulting firms include that of Steur (1962), who for a

150-MW 1,800 psig, 1000/1000°F unit of 70% capacity factor gives comparative capital costs of the various items needed for different types of cooling systems (Table 10).

Weir and Brittain (1962), of still another consulting firm, compare the alternatives of various cooling systems. Though they go through many calculations, they basically make a direct comparison only for induced draft towers and run-of-river cooling. Surprisingly, for the Mississippi River water used as cooling, where the bed was constantly changing, the cost of the structure was so great as to overshadow the normal savings in run-of-river water cooling. In that case, for combined total costs, an inland tower to serve two 160-MW units was $707,000 cheaper than a run-of-river plant, and $1,529,000 cheaper than a tower at the river site. Rysselberge (1959) compared natural draft with mechanical draft towers for a 450-MW plant in the Illinois–Indiana–Ohio area and reached the conclusion that though the first costs of the natural draft tower were 1.25 times the cost of the mechanical draft tower, the former was more economical when operating costs were taken into account. Rysselberge maintains that on a cost accounting basis only, capital costs of natural draft towers can exceed mechanically induced draft towers by a factor of 1.6 to 1.7 and still be competitive overall.

TABLE 9   Economic Comparison of Cooling Towers

	Induced Draft Wet Tower	Natural Draft Wet Tower	Natural Draft Dry Tower	Induced Draft Dry Tower	Mechanical Draft Fin Tube Exchanger
Circulating water flow GPM	91,000	95,000	115,000	160,000	95,000
Cooling range, °F	21	21	17	12	20
Design approach, °F	20	20	38	33	25
Wet bath temperature, °F	65	62	—	—	—
Dry bath temperature, °F	—	—	95	95	95
Land area, square feet	18,600	32,000	69,000	22,500	60,000
Total investment cost—millions of dollars	28,715	29,580	31,905	32,305	30,980
Backpressure, in. Hg (44% station load)	1.3	1.4	1.5	1.9	1.5
Capability loss, $1,000	105	69	448	430	292
Annual costs plus capability loss, $1,000	6,453	6,476	7,217	7,283	6,979

TABLE 10 Equivalent Fuel Costs

Costs	River	Lake	Mechanical Draft Cooling Tower	Natural Draft Cooling Tower
Capital costs, $/kW	1.0	3.5	3.2	7.2
Typical fuel costs to give equivalent annual costs, ¢/million Btu	25.0	24.4	24.2	23.5

Smith and Bovier (1963) showed that for Keystone Station in western Pennsylvania, though the cooling tower had an economic penalty of $0.62/kW for capital and operating costs, the transmission and fuel cost credits for this nine-month plant were so overwhelming, $18.54/kW, that it not only was the choice, but there was no alternative at that location since sufficient cooling water would not otherwise be available.

Ravet (1963) compared costs for a 400-MW unit for natural draft and induced draft towers for investor-owned and publicly owned utilities. The interest charges for the higher initial cost natural draft tower can make an appreciable difference in the yearly costs of a project. For optimum towers with an 18.8°F approach for a natural draft tower and 15°F approach for an induced draft tower, the natural draft tower is $140,000 lower in cost. For public power utility with an approach of 16.5° natural draft and 13.5° induced draft, natural draft has an advantage of $440,000.

It should be noted that the last two economic analyses have been made by Hamon engineers, who have a patented process for constructing hyperbolic natural draft cooling towers.

Probably the most objective figures are the average prices actually paid for the various types of cooling towers over the past five years. Marley Company, which manufactures both mechanical draft and natural towers, states (Furnish, 1969):

For mechanical induced draft cooling towers, the average cost per kilowatt of the twelve largest towers purchased in the last five years was $1.67/kilowatt. This cost is a delivered and erected cost but does not include the concrete basin, the electrical controls and wiring for same or the piping external to the tower. This cost figure is a weighted average and the spread can be over twice this figure for smaller stations to somewhat less than this figure for the very large stations. Plant design conditions and design ambient temperatures will also affect the cost.

For eight large natural draft installations over the last five years, we have an average delivered and erected price of $3.75/kilowatt. This price includes the concrete basin and the cooling tower but does not include piping to the tower. Again, this cost is a weighted average; however, the swing is much narrower.

These average prices do not indicate the costs at a particular site. More important, however, they do not reflect the cost of delivered power since distance to load center and to fuel supply are not represented in the cost figures.

Estimates have been made of the cost of the various types of cooling towers (Kolflat, 1968) and the costs to the year 2000 (Bergstrom, 1968) using conventional wet induced and natural draft technology and extrapolations for dry cooling towers. The results are shown in Table 11.

Conclusions

As in many environmental problems, the choice of the proper cooling device is not a simple process, nor is there a unique result. Because of the relationship between condenser, turbine, and cooling device, a system analysis study must be done to explore a variety of choices. In addition, the other components of the production of power, transmission, fuel supply, water supply, etc., all play major roles in determining the best choice. One must agree with Jones (1968b), manager of cooling towers for Hamon-Cottrell, who stated, "I am very much against attempting to make statements which indicate the relative costs of various forms of condenser cooling. I have seen studies resulting in numbers which

TABLE 11 Unit and Total Costs of Cooling Devices

Cooling Device	Area Required (Normalized)	Water Required (gal/kW/yr)	Additional Unit Costs ($/kW)	Cost to Year 2000 ($ billions)
Open cycle	—	500,000	—	—
Pond	500–1,000	—	—	—
Mechanical draft				
1. Wet	1–2	4,000–8,000	7	11
2. Dry	—	70	27	—
Natural draft				
1. Wet	1	4,000–8,000	11	16
2. Dry	—	70	25	60

could be used to indicate any conclusion desired." The figures quoted in this chapter, including the comparison of river cooling, show the validity of the above statement. For planning purposes, the average costs of delivered towers as indicated by Furnish, might best be used.

Research Required

The interrelationship of the turbine, condenser, and cooling device and their relationship to the geographical location and meteorological conditions make a generalized choice selection methodology impossible. However, it would be useful for a number of typical electric loads, environments, distances to load center, etc., to set up a systems analysis methodology which would detail approximate tradeoffs. For example, how many miles removed from a load center can one be as a function of transmission costs and tower costs in comparison to run-of-river cooling at a nearby site and its attendant thermal pollution costs?

These comparisons are highly dependent upon site conditions and must in essence be made for each site. It might be appropriate to fashion a standardized list of all considerations to be taken into account in comparison of this sort in order to make such judgments on a more objective basis. Possibly most useful would be an analysis after the fact of how close to the design criteria the actual facility operates.

COMPARISON OF HEAT DISSIPATION ALTERNATIVES

From an environmental standpoint, the following factors must be included in a comparison of once-through and closed-loop cooling water systems: (1) effects on water and air quality; (2) consumptive use of water.

Effects on Water and Air Quality

The most obvious effects of once-through cooling systems on receiving water bodies and existing ecosystems are of a thermal nature, and they have been discussed at length in the preceding sections. When these effects are detrimental, closed-loop cooling systems afford the only alternative which eliminates completely the problem of thermal discharges. However, in comparing once-through and closed-loop systems in general, the overall effects on water and air quality must be evaluated.

In coastal areas where fresh water is not available for makeup, the application of wet cooling towers should be restricted at this time, since salt-laden drift can have very harmful effects on vegetation. However,

except in shallow embayments, tidal and wind-driven currents are generally available in coastal waters for rapid dispersion and dissipation of the added heat.

In certain bodies of water, the addition of heat, by judiciously designed cooling water discharges, can create new circulation patterns which may enhance water quality. For example, in dead-end channels or embayments with little natural flushing, promoting a stratified flow condition by means of a thermal discharge may substantially increase flushing effects and reduce the overall pollution level. Similarly, in naturally stratified large reservoirs, including some originally constructed for hydroelectric development, the addition of waste heat from steam electric plants may, in certain cases, be beneficial for water quality by promoting reaeration of hypolimnetic waters through new circulation patterns.

The effects of cooling tower blowdown on natural water bodies must be evaluated and, whenever required, they must be controlled by adequate treatment. Fogging effects due to cooling towers and cooling ponds must be considered in any environmental evaluation of closed-loop systems. The extent and configuration of vapor plumes should be investigated for the entire range of meteorological conditions at the site. In evaluating the environmental effects of cooling ponds, due consideration should be given to multipurpose utilization of water resources, including recreational value.

Consumptive Use of Water

Evaporative water losses may vary widely, depending on whether once-through or closed-loop cooling systems are used and on the type of cooling facilities used in a closed-loop system. For example, evaporative losses for an 850-MW nuclear power plant located on a river were found to be 4 to 5 times less in a once-through system controlled by a 5°F permissible temperature rise in the river than in a closed-loop system on cooling towers.

CONSTRAINTS DUE TO REGULATORY REQUIREMENTS

In certain cases, across-the-board application of temperature criteria may discourage thorough investigation of site conditions and may not be conducive to the best engineering concept from a total environmental standpoint. To better define criteria applicable to a given site and ensure that all necessary site studies are conducted, the following procedures for site selection and investigation are recommended.

Before a new power plant site is definitively selected, it is necessary to demonstrate the suitability of the site, based on preliminary field investigations. The scope and extent of such investigations will depend on the characteristics of the aquatic environment at the site, but in principle they should determine the physical, chemical, and biological parameters of the area potentially affected by plant operation, and ascertain the interactions of these parameters.

It is desirable that the scope and duration of the preliminary field investigation be determined jointly by the owner of the proposed power plant and competent regulatory agencies. In general, the objectives of these studies should be as follows:

1. To demonstrate that construction and operation of the proposed power plant will not have unacceptable adverse effects on the aquatic environment and ecosystems at the site.

2. To enable regulatory agencies to define applicable criteria for the site that can be used in the design of plant facilities

3. To better define the scope of preoperational and postoperational investigations aimed at determining the actual effects of power plant operation on the environment, if any

The adoption of the above procedure for demonstrating site suitability and setting criteria for the selected site would ensure a more thorough knowledge of site characteristics and nature of design requirements at the earliest stage of a project. It would largely avoid delays due to unexpected regulatory requirements once the site has been finally adopted and the project started. It would allow greater flexibility in determining design criteria of all the characteristics for a given site under consideration.

ENGINEERING DESIGN CONSIDERATIONS

Condensers

The condensers used in steam-electric plants are essential components designed to remove the latent heat of vaporization from the turbine exhaust steam. In stations with once-through circulating water systems, condensers are surface-type shell and tube heat exchangers, with the heat exchange between steam and circulating water occurring through the thin walls of numerous small-diameter tubes. There are also other types of condensers, such as jet, spray or barometric type condensers, but they are used in closed loop nonevaporative systems.

Surface condensers are built with one or two passes. In general, a two-pass condenser requires less circulating water flow and has a higher temperature rise than a single-pass condenser. In many cases single-pass condensers are used for once-through systems and two-pass condensers for closed-loop systems with cooling towers or cooling ponds.

The amount of waste heat transferred from steam to the circulating water in a surface-type condenser depends on the size of the generating unit and fuel used. Typically, the temperature rise in the condenser ranges between 10°F and 30°F. Upper limits of temperature rise are determined by limitations in turbine backpressure. Temperature rises lower than the normal range are possible in principle but are practically restricted by the increase in cooling water flow which affects the entire circulating water system and by physical space limitations between turbine supports. In certain cases, a more practical means for reducing the temperature rise at the point of discharge into a receiving water body is dilution pumping after the condenser outlet.

The outside diameter of condenser tubes normally ranges between 5/8 inch and 1 inch. Water velocities in condenser tubes depend on tube material and water conditions. They typically range between 7 and 10 ft/sec.

The metallurgical makeup of condenser tubes and water boxes depends on whether fresh water or seawater is used, on the degree of industrial pollution of the water, and on local ecological factors. For seawater applications, materials generally used include stainless steel, aluminum brass and 90/10 copper–nickel alloy.

Algae and bacteria present in water used for cooling form slime on condenser tubes, which can severely reduce thermal efficiency. To control these effects, biocides, such as chlorine or, in some cases, sodium hypochlorite, are added periodically to the cooling water. The residual chlorine at the point of discharge into the receiving water body is usually kept at 0.1 to 0.5 ppm, depending on the characteristics of the water and the time of the year.

In certain cases, the condenser can be designed to allow periodic backwashing with flow reversal to remove, by means of thermal shock, slime or crustaceans building up on condenser tubes or circulating water pipes.

Alternative methods for cleaning condenser tubes consist of using sponge rubber balls or captive brushes. The sponge rubber balls continuously circulate through the condenser tubes, being retrieved from the condenser outlet box and released again into the inlet water box. The captive brushes are attached to the tubes and effect cleaning by flow reversal in the condenser. These alternate cleaning methods can be used only for condenser tubes and may have to be supplemented by other antifouling procedures for the rest of a circulating water system.

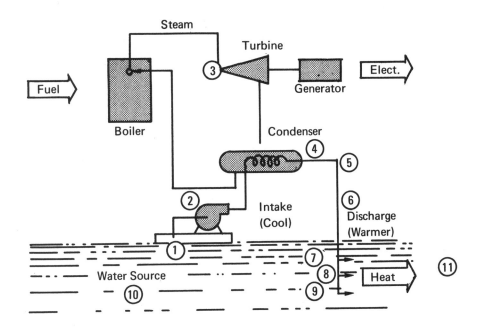

DESIGN PARAMETER	GENERAL PREFERENCE	ECOLOGICAL BASIS
1. Intake design	Behaviorally avoidable or provide safe return to environment	Poorly designed intakes trap fish, crabs, etc.
2. Volume of water pumped	Low (but site dependent)	Numbers of organisms affected
3. Turbine backpressure	Lowest feasible heat rates	Lowest backpressure permit low temperature discharges to environment (7) highest feasible efficiencies
*4. Temperature rise	Site and season dependent	Temperature–time relationships of effects
5. Length of cooling water piping in plant	Short (minimum transit time)	Temperature–time relationships of effects on entrained organisms
6. Length of transit to receiving waterway (canal or pipeline)	Short (minimum transit time)	Temperature–time relationships of effects, fish entrapment
7. Discharge location	Beyond littoral contact	Shoreline abundance of organisms (may be seasonal)
8. Discharge depth	Semistratified plume	Keep highest temperature water away from resident bottom organisms
9. Turbulence (exit velocity, port size or number)	High	Temperature–time relationships and areal extent of effects
10. Dilution (near field)	High	Plume entrainment, temperature–time relationship
11. Circulation (far field)	High	Temperature buildup for recirculation may change overall species composition

*Subject to mutual trade-offs at specific sites.

FIGURE 20   Summary of cooling system design needs.

Organisms entrained through condensers are also subjected to pressure changes. In the circulating water lines ahead of the condenser, the pressure may be 25 to 30 psia. In the condenser outlet water box, the pressure may be as low as 2 to 3 psia for extreme operating conditions. In some cases the drop from maximum to minimum pressure may be experienced in only 5 to 10 seconds.

Pumps

Pumps used in condenser cooling water systems of steam-electric plants typically range in capacity from about 20,000 gpm to 250,000 gpm, and there are usually two to four pumps for each generating unit. Circulating water pumps normally have axial or mixed flow impellers and are of either the wet pit or the dry pit type. Smaller pumps used in steam-electric plants may be of the centrifugal type.

Rotating speeds may normally range from 150 rpm for the large, low head pumps to 900 rpm for the lower capacity range. In once-through systems, total dynamic head may range between 20 and 50 feet. In closed-loop systems with cooling towers, higher pumping heads are required. The pump setting and design must be such as to avoid cavitation for all operating conditions. Water velocities at the pump discharge may range between 8 and 12 ft/sec.

Due to the existence of traveling screens ahead of circulating water pumps, organisms that are allowed to pass through the pumps are smaller than the clearances within the pumps and therefore should be subjected to less damage than would have been the case with larger organisms.

Intake Structures

Intakes are major components of once-through condenser cooling water systems. Their design can have a marked effect on the entrainment of aquatic organisms into the circulating water system. Design features which are the most significant in this respect include location of the intake; overall intake arrangement; velocities; curtain walls and bottom barriers; fish protection facilities.

Provisions to minimize impact on the aquatic environment as related to these features include

1. Locating the inlet as flush as possible with the bank to avoid fish traps. The location of the intake depth should be selected, whenever possible, considering the vertical distribution of aquatic species. Seasonal considerations are important also.

2. Control of approach flow patterns by means of overall intake arrangement to discourage fish from entering the intake.

3. Controlling approach velocities to the inlet racks and, within the intake, to the traveling screens. In view of the impact these velocities can have on organism entrainment on the one hand, and on the engineering and total cost of intake structures on the other hand, permissible approach velocities should be set carefully and be based on the characteristics of each type of aquatic environment, *viz.*, cruising speed capabilities of local finfish species.

4. Controlling warm water recirculation by means of curtain walls in conjunction with control of inlet velocities, when feasible. Curtain walls can also be used to prevent access into the intake by fishes residing in upper water layers, when required.

5. Preventing access of bottom fishes into the intake by means of bottom barriers, when necessary.

6. Fish protection facilities designed to minimize mechanical damage on the screens and, when necessary, providing a means for fish escape or recovery from the intake.

Impingement of entrained organisms against traveling screens in intakes depends on approach velocity. Normally, the velocity based on gross areas is approximately 1 ft/sec for mean water level and may reach 1.2 to 1.5 ft/sec for extreme low water levels. At these velocities, impingement is small. More recent designs attempt to maintain 0.5-ft/sec velocities at the intake structure.

Engineering design considerations for cooling systems are summarized in Figure 20.

4   METHODOLOGY AND DATA ACQUISITION

BACKGROUND ECOLOGICAL INFORMATION

Adequate regional environmental surveys, providing proper information for power generation site selection, are generally not available. Lack of such information is due (1) partly to the relatively small service area of most utilities, (2) to past lack of interest in developing such information, and (3) to the fact that other regional environmental surveys had different objectives. To assist proper siting, therefore, it is crucial that an approach be developed between resource agencies and the utilities to determine as quantitatively as possible the species composition, distribution, and ecology of the biota and its commercial and recreational usage over wide areas that are likely to be considered for siting of electric generation facilities. Such background data should be accrued for at least one annual cycle, with two to five years of data considered more desirable [see Coyote and North (1971) and Mihursky (1963) as two examples of attempts at preoperational surveys].

This information should be collated and evaluated in a manner that will assist in determining environmental vulnerabilities and flexibilities to generating facilities. For example, where are the migration pathways, spawning grounds, concentrations of important resident species such as oysters? Where is heavy fishing pressure, both sport and commercial? What are the seasonal dynamics of abundance and location of important species?

One possible avenue for accomplishing the above objectives is to establish an administrative structure and a research fund for this purpose to be obtained from a surcharge tax on kilowatts of electricity produced in the given region. Such a system will go into effect in the State of Maryland beginning in 1972. A tax of 0.1 mill/kW will be levied with the stipulation that the program may continue to 1985, with 0.3 mill/kW established as the maximum taxable rate (Goodman, 1971). This program also provides funding for engineering research and development as well as for site acquisition (Tawes, 1971).

RECONNAISSANCE SURVEYS

The objectives of the reconnaissance studies in the aquatic environment are

1. To make a hydrographic survey of temperature, water chemistry, and other factors that may influence the dilution and dispersion of aquatic discharges from the plant, with due regard to seasonal, hydrographic, and temporal changes.

2. To develop qualitative descriptions of the biotic community near the plant site, with emphasis on the seasonal species composition and distribution of the principal macrofloral and faunal communities.

3. To relate, insofar as practicable, the above biotic and hydrographic surveys to similar surveys made by other agencies in the same general area.

4. To make preliminary predictions on the extent of the thermal discharge from the power plant under different hydrographic conditions, using several different intake and discharge configurations.

5. To predict, insofar as possible, the probable effect

of the thermal discharge on the principal ecological communities, using field data, laboratory data, thermal biotic predictive models, available literature, etc. (see Coyote and North, 1971; Cronin and Mihursky, 1970; Mihursky *et al.,* 1971).

ANALYSIS OF CRITICAL BIOTIC COMPONENTS

Background ecological data and reconnaissance surveys should eventually provide the following types of detailed biological information needed to determine biological utility of an area, to assist in plant design considerations, and to properly evaluate preoperational and postoperational data.

1. Food chain components
   a. Important for commercial and/or recreational value
   b. Important due to large biomass component
   c. Important due to major role in energy flow in system
   d. Important due to uniqueness
2. Major life history patterns of above species
   a. Seasonal aspects
   b. Highly localized
   c. Extended geographically, i.e., migratory
   d. Apparent sensitive or vulnerable life history stages
3. Attempt to design operations to preserve the most sensitive species and will then automatically preserve the remainder

Species-diversity indices (McErlean and Mihursky, 1968) may also be employed as a guide to whether a stressed environmental condition exists as shown by a loss in species. Ultimately, however, it must be demonstrated whether a shift in species composition is or is not impeding energy flow to useful biological components.

STANDARDIZATION OF APPROACH AND TECHNIQUES

Review of present practices in site surveys has revealed a striking dissimilarity of approach and techniques that makes meaningful comparisons of the resulting data virtually impossible. Sites differ, and therefore some latitude must be allowed in order to pattern the studies to the specific site-related questions. This necessary freedom has apparently been taken as a general license for poor definition of study goals, inadequate sampling (either type or number), incomplete data analysis and presentation, or undue emphasis on the technical specialty of the investigator. Principal investigators should be conversant with the potential problems from power

plants in order to set study priorities, yet a survey of current projects and the technical backgrounds of principal investigators (Ulrikson and Stockdale, 1971) revealed shocking inadequacies. Technical problems remain unresolved, such as biased sampling gear and progressive increases in proficiency of workers during the course of a study. Many of these needs could be met through periodic workshops (held under multiple agency sponsorship—see *Ches. Sci.* 10[3–4]) designed specifically to iron out technical differences and through publication of standard guides to survey practice which indicate techniques, minimum length on surveys, measurements to be made including physical, chemical, and biological, and methods of evaluation resulting from workshop sessions. These tasks are of very high priority.

There is a distinct need for high-level, *cooperative supervision* of surveys and predictions by agencies with diverse interests in electric generation station effects. Agencies with both "resource protecting" and "resource utilizing" mandates should serve on joint supervisory committees or review boards. This is of unusual importance to assure competence and fairness in the development of all major predictions and detailed research projects. It will also serve as a route for rapid dissemination of information to all interested parties. For instance, the State of California has a Resources Agency Task Force on Thermal Power Plant Siting which coordinates the planning of site investigations with the utilities. An Ocean Environment Task Force of the Siting Committee is responsible for aquatic environmental studies. Representatives of federal agencies such as the Federal Water Quality Administration, the National Marine Fisheries Service, and the Bureau of Sport Fisheries and Wildlife also participate in the task force.

Maryland has also established a Committee on Power Plants and the Environment and a Chesapeake Bay Cooling Water Studies Group, both with objectives similar to the California task forces. Federal, state, industry, and university representatives serve on the Maryland committees.

Utilities and other agencies not having broad expertise in ecological aspects of thermal effects need a source of qualified investigators or technical advisors to direct their site selection and monitoring programs. Federal and state agencies, under joint supervision, should provide guides to investigators with pertinent professional experience.

ECOLOGICAL EFFECTS LITERATURE

Available Literature—Thermal

The biological and ecological effects of temperature change are many and diverse, but the literature is readily located. The titles in the comprehensive bibliographies and literature reviews indicate a thorough coverage of the pertinent literature through February 1970.

Kennedy and Mihursky (1967, 1969) listed approximately 1,500 references published through 1966. Few English-language papers of biological importance were missed, and many papers concerned principally with physical effects are included. Their bibliography covered only published material, and many relevant "in-house" staff reports or agency documents were not included. Raney and Menzel (1969) published a bibliography containing 1,870 references on heated effluents and their effects on aquatic life, particularly fishes. All titles were permuted and indexed alphabetically on each significant or key word—a feature that makes the compilation particularly useful. Many references are duplicated from the reviews by Kennedy and Mihursky and from earlier versions of their own list. Literature relevant to assessing the effects of thermal discharges into aquatic environments has been reviewed annually for the years 1967, 1968, and 1969 by Coutant (1968, 1969, 1970). While the review for 1967 was not as complete (37 references), the reviews of 1968 and 1969 literature (111 and 200 references, respectively) are quite comprehensive. Each paper was abstracted to indicate the scope of the findings.

Foreign-langauge literature available in English-language abstracting services and staff and agency reports on progress of research are included in this continuing series of reviews published by the Water Pollution Control Federation.

In addition to bibliographic and annual summaries, several recent "state-of-the-art" symposia and documents relate directly to the ecological effects of power-plant siting. Battelle-Northwest (1967) prepared an extensively researched document for the U.S. Bonneville Power Authority which analyzed the many interdisciplinary problems associated with nuclear-power-plant siting in the Pacific Northwest. Appendixes and the text of the 547-page report summarize ecological, regulatory, geological, meteorological, heat-dissipation, power-transmission, public-understanding, and economic considerations. Representative power-reactor sites in the Northwest are analyzed, but the principles applied are relevant elsewhere in the country.

The U.S. Federal Water Pollution Control Administration sponsored preparation of a status report (Parker and Krenkel, 1969) on thermal pollution that includes considerations of water quality, aquatic organisms, waste assimilation, beneficial uses, heat dissipation, mechanisms of heat discharge, modeling of thermal discharges, cooling ponds, cooling towers, and a literature review specifically on Pacific salmon. Consideration of all biological effects was not extensive, however.

The proceedings of a National Symposium on Thermal Pollution were published in two volumes, one principally on biological aspects (Krenkel and Parker, 1969) and the other principally on physical and engineering aspects (Parker and Krenkel, 1969). Each volume included several reports by people with administrative or research responsibilities bordering on their respective topics. The biological volume cannot be considered comprehensive, for several schools of research on thermal effects and several important field studies of power-plant sites were not included.

The Proceedings of the 2nd Thermal Workshop of the U.S. International Biological Program were published as a special issue of *Chesapeake Science* [10 (3/4), 1969]. Many areas of research were reported that were not included in the above volumes. In addition, there are summaries of workshop discussion sessions on fisheries and management and a series of papers on standardized techniques for the assessment of thermal effects on primary productivity. The English Central Electricity Research Laboratories (1971) have recently released the results of a symposium on "Freshwater Biological Research and Electrical Power Generation." Topics included current problems of water use heat dispersal, assessment of thermal effects and fishery considerations; ecological studies dealing with bacteria and plankton and a summary of relevant findings on Lake Trawsfynydd; laboratory studies on effects of temperature and other factors on reproduction, physiology, and development.

The U.S. Senate Committee on Public Works (1968) has published transcripts of hearings held to determine the extent to which environmental factors are considered in selecting power-plant sites, with particular emphasis on the ecological effects of the discharge of waste heat into rivers, lakes, estuaries, and coastal waters. This testimony not only reveals the scientific basis for concern over "thermal pollution" but affords an understanding of public attitudes as well.

The phase of activity involving bibliographic searches other than annual reviews of current literature would appear over, and additional effort in this direction (other than updating) is superfluous. On the other hand, there are few in-depth, critical reviews of even our present ability to control the ecological effects of power-plant waste heat (although see Coutant, 1970; Levin *et al.,* 1970). Pursuant to such an overall critique is the necessity for detailed reviews in specialized problem areas. Such reviews are generally not available despite the urgent demand. A number of popular articles on the topic of thermal effects have served admirably to arouse public concern. Unfortunately, many have been overly speculative, some have been ill-formed, and few have emphasized evidence at operating stations.

Other Available Literature

Although bibliographies have been completed covering thermal considerations applicable to electric generation, there remain subject areas in which detailed information is not available or is scattered. Examples include the ecological aspects of pressure changes, turbulence, chlorination, blowdown wastes, mechanical devices, and heavy metals. Properly documented data on fish behavior in response to seasonal thermal release patterns, turbulence, and velocity and in response to various intake–discharge structures are generally not available in the published literature.

Computerized Information Center

There is a need for a major *clearinghouse or information center* for material relevant to the electric generation effects question. *Computer simulations* and up-to-date data should be readily available to persons in utilities or regulatory agencies that must efficiently evaluate power plant sites.

Current methods of collection and dissemination of information are generally archaic, slow, and often unresponsive to real needs. The poor suitability of traditional published literature for resource management decisions has been recognized by most fields for years. From frustration with this system has come an even more haphazard collection of agency reports, progress summaries, newsletters, prepublication drafts, symposium proceedings, and, more recently, computerized information centers. There is simply no orderly flow of information from investigator to user and no orderly return of research (information) needs. This situation is critical in areas of fast-moving public opinion and management decisions in electric generation effects. Partly because of this information bottleneck, decisions are being made on the basis of reaction, speculation, poor information, and in some cases, false information.

Existing information centers may provide important solutions to short-range needs. The existing literature system is worth retaining for historical retrospect. The principal drawback to present information centers is their apparent inability to keep abreast of current information and to have this information available for "instant playback." In many cases entire reports (or copies thereof) must be retrieved for critical appraisal, and information centers often cannot provide this service. Return of a three-sentence abstract and citation may facilitate literature searches, but it hardly affects the arduous task of obtaining reports and locating pertinent data.

It is clear that ecologists and managers requiring up-to-date information must be involved in the establishment and maintenance of advanced, computerized information centers if they are to obtain useful information from them. The topical nature of the center must be apparent to all potential users (not hidden as part of another topic) and must be readily accessible to them. The services should include operation of existing simulations on a cost basis for customers not able to establish their own operating capability. Periodic work-

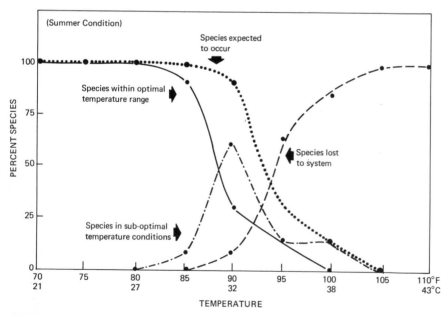

FIGURE 21 Thermal–biotic predictive model for an estuarine system.

shops and seminars should be held to acquaint planning, design and monitoring personnel with the ecological questions pertinent to electric generation effects.

BIOLOGICAL PREDICTIVE MODELS

Predictive modeling of biological responses to proposed environmental modifications is a relatively new endeavor in the field of interaction of electric generation and the environment. Mihursky and his colleagues (Mihursky, 1969; Mihursky, McErlean *et al.*, 1969; Mihursky, McErlean, and Kennedy, 1970; Mihursky *et al.*, 1971) have developed thermal biotic predictive models for an estuarine system, the Chesapeake Bay. Using laboratory data on temperature effects on mortality, respiration, and activity, they have produced two graphical models. Figure 21 gives a line plot of the percent of total Bay macrofauna assumed to be living within optimal temperature conditions, with all life processes functioning at an adequate rate of efficiency. These temperature levels, however, could be near upper levels that may reduce efficiencies. A second line plot gives percent of total Bay macrofauna still expected to be found under various possible summer temperature conditions but which may be living under suboptimal thermal conditions.

Surface water temperatures may naturally reach 30° to 32°C in the Bay system during extremely hot and humid periods, generally in July or August. At such times the surface water temperatures will cool slightly during evening hours. Under such high temperature conditions, temperature-sensitive mobile species must seek out deeper and cooler water of the Bay system. Immobile, benthic species living in shallow shelf zones must therefore be able to tolerate high temperature conditions either by high temperature tolerance or by ability to tolerate suboptimal conditions for a short period of time. If the temperature elevation persists as a chronic condition for immobile, temperature-sensitive species, then mortalities can be expected. This apparently occurred with the temperature-sensitive soft-shell clam during 1965 and 1966 in the Bay system. This is a northern genus *Mya,* living in the Chesapeake Bay at the southern range of its distribution on the east coast.

It appears from Figure 21 that chronic 30°C temperature are not likely to endanger native Chesapeake Bay fauna. Chronic 32°C temperature will cause a shift in population structure; uncertainties exist concerning food chain efficiencies and biomass production at these altered temperatures, since a large percentage of the species will be living under suboptimal conditions. Persistent 35°C temperature would definitely reduce

species. Under both the 30° and 32°C conditions, the tolerant species present might increase in biomass due to lack of competition from other species. The net effects cannot be fully assessed, since it is not known whether production at these elevated temperatures will yield biologically useful foodstuffs. At 38°C, a few species would remain under suboptimal conditions, while at 41°C all usual macrofauna would apparently disappear.

Figure 22 is an attempt to produce a predictive model applicable to all seasons. It gives minimum and maximum LD_{50} mortality slopes obtained from an array of Chesapeake Bay species and plotted according to the method of McErlean, Mihursky, and Brinkly (1969). The line marked M.A.T.E. (maximum allowable temperature elevation) is an extrapolated contour that would protect 100% of the species tested, based on laboratory data.

In effect, it states that if the allowable temperature increase is to $2+°C$ (4°F) below the LD_{50} points for the most temperature-sensitive species studied, the temperature change will then presumably have no adverse effects on this most temperature-sensitive species. Also, it assumes that if the most temperature-sensitive species is so protected, more temperature-tolerant species will also be protected. Thus the area below the M.A.T.E. line identifies the maximum increase in temperature that will still permit optimum biotic functioning and production of the estuarine system throughout the year.

The thermal biotic predictive models above are among the first models developed and must be refined and checked against new data; however, they do provide us with a useful start on making predictions and identifying the role of temperature from an ecological viewpoint. By indicating the optimal seasonal tempera-

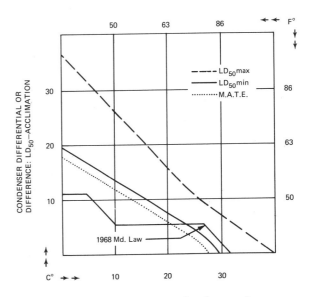

FIGURE 22 Ambient or acclimation temperatures.

ture levels it is possible now to focus on and recognize the suboptimal temperature conditions that may not kill directly but in essence are ecological death zones. This latter category recognizes that temperature levels can prevail that would reduce energy conversion efficiencies, result in reduced physiological condition, and alter predator–prey, host–parasite, and host–disease relationships (Mihursky, McErlean, and Kennedy, 1970).

Bloom, Levin, and Raines (1969) released a report entitled "Mathematical Simulation of Ecosystems—A Preliminary Model Applied to a Lotic Freshwater Environment." The paper attempts to apply a mathematical approach to determining biological effects of thermal effluents from power plants. They concluded the following:

1. The results of this study indicated that the response of aquatic biota to waste-heat additions can be simulated mathematically and that mathematical models can be developed to predict changes in an ecosystem that would result from the thermal discharges of a power plant.

2. In the test cases studied, it was shown that the extent of the effects on aquatic biota depended on the trophic level under consideration and on the geographical location of the biota. The importance of considering the interrelations among the various trophic levels in the food web rather than limiting the examination to the particular species of interest, e.g., the top carnivore such as a sport fish, was also evidenced.

3. The study highlighted the requirements for appropriate field data-collection programs to provide the needed input to the models to allow confident predictions for all cases of interest.

Several steps were identified which are required in the development of a comprehensive model for predicting the effects of effluents from industrial or power-plant operation on environments, particularly the effects on the biota within a given geographical area. Following a review of existing information on the environment of interest, a preliminary mathematical model can be formulated from the literature data and estimates of unknown parameters. This preliminary model can then be used to guide the acquisition of field and laboratory data. It is important that the data-collection programs be coordinated with the mathematical developments so that optimum use of the field efforts can be made and so that the required predictive capabilities can be obtained. Complementary laboratory data are needed on biomass transfer rates as a function of heat or material additions. With sufficient data, a comprehensive mathematical model can be developed, and predictions of the response of biota which can be subjected to tests by field experiments can be made. The model, thus developed and validated by field experiments, can then be used to predict effects on the environment for the applications of interest.

Jaske, Templeton, and Coutant (1969) published a paper entitled "Thermal Death Models," which predicts the mortality and equilibrium loss of young salmon under conditions of fluctuating but lethal temperatures, such as those occurring in many power discharges. They state: "These models should prove useful in the initial planning stages of industrial and municipal release structures (e.g., ability to predict the loss of fish as a result of a given discharge design). It will also be possible to examine other aspects of ecological loss besides that of fish. Ultimately such models will permit the delineation of boundary conditions for plant design, not only as regards thermal releases but also chemical toxicants, as similar resistance patterns have been shown by fish exposed to various toxic substances."

Thus, attempts are being made at producing biological predictive models applicable to electric generation facility design and site location. These are our first models and need confirmation and field verification. Presently, a number of assumptions are being made that need clarification and additional data. The approaches are valid, however, and considerably more work of this type should be undertaken.

RESEARCH NEEDS IN ECOLOGICAL AND ENGINEERING DISCIPLINES

To obtain reliable information necessary for continuous review and updating of water quality criteria affecting power plant siting and engineering, a nationally coordinated long-term program is necessary. Its objective should be to evaluate in a consistent manner the impact of selected electric generation facilities on the aquatic environment. More specifically, the program would assess the total near-field and far-field effects of all aspects of power plant operation on ecosystems and, in particular, the effects of thermal discharges.

The proposed program should include and expand some of the investigations currently carried out by electric utilities, universities, and state and federal agencies. It would extend over a period of several years, with progress reports being released annually and discussed in broad public and professional forums.

The program would include the following steps:

1. Selection of a number of representative existing power plant sites, so that it is possible to define a wide range of discharge conditions and a diversity of aquatic environments.

2. Setting up an investigation program for each of the selected sites to include both long-term measurements and short-term surveys. These would be aimed at defining effects on physical, chemical, and biological characteristics of water bodies by comparing conditions in selected control areas and in areas directly affected by power plant operation.

3. Whenever possible, studies already under way at certain existing power plant sites would be incorporated in the proposed long-term investigation program, provided any required modifications can be made in the scope of such current studies to ensure a consistent approach in the over-all program.

4. The various regional programs must employ a combination of field and laboratory studies. Field studies deal with the real work; however, the variations of natural systems make it extremely difficult to sort out what factors are actually limiting or stimulating growth. Laboratory studies, properly designed with the aid of available field data, permit focusing on one variable at a time. Agreement of field and laboratory data gives greater confidence in evaluations and gives a firmer basis for knowledgeable recommendations for alternative designs.

5. The selection of representative power plant sites and aquatic environments, as well as the scope and extent of the proposed long-term investigation program, should be thoroughly discussed and reviewed to ensure wide acceptance of methods and objectives.

6. Data obtained from the investigation program, together with pertinent power plant operation parameters related to the effects under investigation, would be analyzed and evaluated in a consistent manner for all selected sites. The results would serve as a basis for a rational and objective review of water quality criteria and their applicability to various types of aquatic environments.

7. The research programs must permit the development of predictive models, physical, chemical, and biological.

Understanding of the interaction between electric generation and the environment results in the realization that there are a number of specific areas where improvement in engineering designs and environmental understanding is needed. The following list includes some of the more important areas of research needs.

1. Devise methods other than the use of biocides for the control of fouling organisms.

2. Devise better screens and methods for the safe removal of organisms from intake waters.

3. Develop water intake structures, condensers, and discharge structures which reduce the stress and mechanical injury to entrained organisms.

4. Devise better methods for the cooling and recirculation of cooling waters.

5. Devise effective and efficient dry cooling facilities and methods.

6. Devise, develop, and test methods for the utilization of waste heat.

7. Develop nontoxic materials for the production of condenser tubes.

8. Improve pump and turbine design so as to reduce damage to organisms passing through them.

9. Determine more effectively and accurately the effects of thermal electric plants on entrained organisms.

10. Develop facilities, equipment and methods on a national and regional basis to determine the effects of thermal electric plants on the production of aquatic life resources in receiving waters.

One of the basic and prime needs arising from the production of thermal electric power is the determination of safe and favorable water temperatures for aquatic life in the areas of discharge. Special problems exist because of the great diversity in water temperatures and conditions from north to south and east to west. Problems and requirements differ in the various biotic zones, and studies should be made in each zone. The following is an example, from the coastal zone, of the kind of national plan necessary to really answer existing needs.

Proposed Coastal Biotic Zones

1. Canadian border to Cape Cod
2. Cape Cod to Cape Hatteras
3. Cape Hatteras to Ft. Lauderdale, Florida
4. Fort Myers to the Mexican border
5. Mexican border to Point Conception
6. Point Conception to Canadian border
7. Coast of Alaska—probably should be 2 or 3 zones
8. Tropical islands and tip of Florida south of a line from Fort Lauderdale to Fort Myers

Proposed Research Plan

1. For each biotic zone determine or estimate the organisms which comprise about 90% of the biomass in each of the following groups: fishes, arthropods, molluscs, worms, zooplankton, and phytoplankton. List first those species unique to each area and then divide among the areas species common to two or more areas so each area does not have an inordinate number of species.

2. In each zone select from the species listed with that area those which are of economic and recreational importance, those which are important as food for

recreational and economic species, and those species that are important in energy flow in the system.

3. Devise facilities, equipment, and methods for holding and rearing these organisms through their entire life history.

4. Devise equipment and methods for rearing these organisms so that any desired life history stage can be secured at any time of the year in numbers needed for bioassay studies.

5. Devise, develop, and test short-term bioassay procedures for determining the relative sensitivity of the life stages of these organisms to specific toxicants or environmental factors.

6. Carry out short-term bioassays with the selected important species to determine the species in the various groups and regions which are most sensitive to temperature and temperature changes.

7. Carry on long-term studies for two generations with the selected sensitive species to determine the range of favorable temperature and the maximum temperature which is not harmful in the normal diurnal and seasonal cycle under conditions of continuous exposure.

8. On the basis of all data secured, determine the temperature requirements of the most sensitive species in each zone. This is done on the basis that protection of the most sensitive, important species will protect the entire biota with a certain degree of safety.

9. Develop thermal–biotic predictive models for each zone.

10. Set water temperature standards based on environmental water temperature requirements for each zone or for smaller coastal areas.

11. Monitor to determine if the standards are being met.

5 CONCLUSIONS AND RECOMMENDATIONS

1. Wherever possible, environmental enhancement on a broadly based scale should be a principal consideration of all future utility planning, and interim measures which, for economic reasons or inelasticity of public policy, are adopted as expedients without progress towards a long range, constructive goal, should be rejected as soon as such measures are revealed to be incorrect or misleading.

In the pursuit of a common purpose, the leadership of the various contributing techniques and disciplines should find a common ground for creating a basis for maintaining an open, credible, and competitive dialogue without resort to measures which reduce the credibility of technology in the public eye.

2. Water management should be combined with innovative ventures in the utility field which permit decentralization of metropolitan concentration and improve decision making for the overall public good in proper development of empty lands for new cities, recreational lakes, commercial fisheries and aquiculture, and of new environmentally compatible technologies.

3. Under the combined auspices of concerned public agencies and the investor-owned utility industry, regional planning centers should be established which permit the accumulation, study, and distribution of conceptual information vital to the orderly use of resources in a coordinated and informed way. On a regional basis, sufficient model technology presently exists to develop the framework for a system of such centers which would provide a social basis for the use of energy technology as an instrument in socioeconomic affairs.

4. The systematic organization of constraints in rela-

tion to the array of available alternatives of both equipment type and site location is very complex and requires extensive refinement in terms of graphics, words, and mathematical modeling in order to display a reasonable decision system which will yield a socially optimum result. Such trade-off options as standards, design alternatives, ecological values, and energy policy need improved quantification and technical justification. *It is essential* that any adversary proceedings involved in determining public courses of action enhance the credibility of the technological process needed to resolve the courses of action.

5. The systematic use of systems technology and modeling for derivation of a predictive basis for siting of electric generation with integrated and overlapping effects is preferable to the use of arbitrary criteria in accomplishing the effective use of water as an adjunct of siting technology.

6. The use of power and energy by society is projected to become so large that associated climatological modifications may affect water resources. Climatology may have to be included in the systems evolution of national siting policy as related to energy policy.

7. Additional private and public resources should be arranged according to a priority listing compatible with the principal conclusions of this study for the purpose of improving the rate of development of new ideas and reducing the difficulties now faced by utilities in the adoption of innovative technology on a broad scale. This program of increased emphasis on research and equipment development would put primary emphasis on the determination of a credible basis for the use of

predictive technology with full response to the political process.

8. The use of water in large volumes as a medium of dissipation of thermal energy has a definite place in the engineering of electric generation and should not be arbitrarily eliminated as a usable mechanism for thermal dissipation. Rather, decisions should be made on the basis of environmental compatibility.

9. A basis exists for improved coordination of water resources and electrical energy technology in order to provide more systematic use of water for low flow augmentation, recreational development, commercial fisheries enhancement, irrigation, and new cities development, including interbasin diversion.

10. A review of published information indicates that in water-short areas, unless supplemented by water sources outside of a given basin, an optimum number of reservoirs can exist beyond which the impounding of streams for single or multipurpose duty is undesirable. Stated alternatively, the incremental resources losses from evaporation or by disturbance of the existing fishery are minimized by extensive development of existing reservoirs rather than the creation of new ones. Resolution of this trade-off is an example of a distinctly local or regional issue and needs to be considered at all times in this context.

11. The evaporation of extensive quantities of water [22,000 acre-ft/year/1,000 MW(e)-nuclear] from wet cooling towers represents a serious diversion of water from water-short areas or areas where waters are subject to withdrawal rates regulated by international treaties. The planning of a regional power system must of necessity take into account these modifications of the normal hydrology and the relative merits of other cooling systems which evaporate materially less water.

12. Waters of low biological productivity or poor quality should receive primary consideration for SES use. Examples are treated sewage, acid mine water, other already polluted water that will not be improved in the immediate future, and the deep, low-oxygen water of the central Chesapeake during the summer months.

13. The use of coastal and offshore areas for the siting of very large plants should be encouraged as a means of minimizing the thermal loading of areas of the country with limited freshwater resources.

14. Options available to the power industry in terms of prime mover flexibility, efficiency improvement incentives, and general inelasticity in the power industry equipment field are too limited by convention and lack of fundamental research. Equipment lines appear to be based primarily on quick response, near-term considerations, and regulatory requirements rather than the long-term objectives. Apparently for reasons of overhead control, the cost of special features or design options,

including the cross compound machine which was optimized on an operational efficiency basis, have become overpriced so as to standardize production runs of relatively identical single shaft units. As a result, options of importance in minimizing environmental effects (such as low heat rate, flexibility in condenser arrangement for transit time minimization, innovative concepts such as "helper" or postcondenser cooling towers or ponds, smaller units, and other concepts for making significant efficiency improvements) have become impracticable. This can effectively inhibit individual leadership by utilities.

15. The state-of-the-art of discharge system design and supporting technology seems to be presently capable of satisfying many criteria for the use of water as a heat receiving medium. These criteria can be specified by the present state of knowledge of the ecology of the affected areas. Modeling research, however, remains to be done to better define the thermal residual of highly diluted effluents with peripheral isotherms less than 1°F, *viz.,* in large lakes with low natural circulation.

16. Insufficient attention is being given to the development of environmentally inert processes and materials such as nonfouling surfaces.

17. As a general principle, the use of designs which emphasize short travel and exposure times of biota passing through cooling circuits is a desirable alternative to other systems approaches.

18. As a special case because of importance to designs currently under consideration, examination of applicable tradeoffs reveals that high condenser rises (ΔT) are not intrinsically bad. If exposure times are short and rapid enough, and effluent dispersion is carried out, high ΔT may be within limits acceptable to the key species involved. A suggested conceptual guide is to keep the product of travel time in seconds times delta T in degrees C less than 2000. That is: $(t)(\Delta T) < 2000$, where $t =$ exposure time in seconds and $\Delta T =$ temperature rise across the condenser. In the future, criteria of the type $< (t)^a (\Delta T)^b < C,$ where a, b, and C are experimentally determined constants for each species of interest, and t and Δ are as above, are a preferred approach to the quantification of exposure.

Volume of water pumped at a site should be as low as possible to minimize the number of organisms passing through the cooling system, but sufficient to eliminate a high temperature exposure that would have detrimental effect on organisms passing through the system. Volumes are season and site dependent.

19. Siting of thermal electric plants on aquatic systems must avoid overlapping their zones of environmental impact.

Siting designs must not create thermal barriers to migratory organisms, i.e., adjacent plants on a river system should locate on the same side of the river; on lakes, plants should avoid locations at the mouths of tributary streams used for spawning purposes by migratory species. In general, siting should avoid those bottle-neck areas through which migratory stages of important species must pass.

20. Siting should avoid areas of restricted water supply or circulation.

21. Aquatic plants and animals have specific and seasonal temperature requirements that must not be exceeded if the system is to remain biologically productive.

22. High temperatures, toxic substances, and various other aspects of present electric generating station designs have sometimes exceeded survival limits for local flora and fauna, resulting in kills of crabs, fish, and food organisms, including microscopic plant and animal life.

23. Regional environmental differences appear to exist with regard to ability of the resident organisms to tolerate increased temperature levels. For instance, organisms in northern latitudes seem to have more flexibility with regard to slight temperature increases than those in southern latitudes, especially during the summer period. Careful SES site selection is critical in order to minimize harmful environmental effects.

24. As a part of the confirmation of site desirability, and to serve as baseline data for determining and comparing ecological effects, the performance of preoperational surveys of careful design are an absolute necessity. While regional characteristics are significant in such experimental design, wherever possible, standardization of techniques and compilation are desirable.

25. In order to verify the models used for both physical and biological (ecological) prediction, postoperational sampling and correlation at every new site are an intrinsic part of the full, systematic evaluation and development of siting and regulatory technology.

26. Siting should avoid small or moderate-size streams and small lakes or reservoirs unless offstream cooling is provided and return water represents less than 10% of the low water flow.

27. Plants sited on lakes other than the Great Lakes which have populations of salmonid fishes in the hypolimnion should not take cooling water from below the thermocline. There should be no discharge of heated water into that area.

28. Plants sited on large streams and using once-through cooling should not use more than 20% of the low water flow of the stream.

29. Deep water intake below the thermocline in oxygen-poor water or from water of low biological utility is desirable in the warmer seasons.

30. Shallow water intake in the coldest zone is preferable in the colder seasons when deep water quality conditions change and organisms tend to congregate in deep waters for overwintering.

31. Intake designs should enable mobile species to avoid being trapped or pulled into the intake system. Improved revolving screens with hydraulic flushing and sluiceways for return of biota to the receiving water are desirable. Screening flush to the shoreline is preferable to recessed-intake screening.

32. Chlorine and other biocides should not be used for the control of fouling organisms.

33. A great deal of additional research, design development, and prototypical testing of intake devices and systems needs to be done to minimize the attraction and involuntary circulation of organisms in the circulating and makeup water systems of all types of power plants.

34. The extent of knowledge of the physiology of species involuntarily transported and recycled is insufficient for sophisticated predictive modeling with respect to time–temperature and other time exposure relationships involved within and outside cooling system components.

35. There is insufficient knowledge of, or attention being placed on, the active behavioral response of mobile species in order to model the interaction of the niche of these species with modifications of the ecology, *viz.*, intake structures, discharge velocities, thermal patterns, etc.

36. The blowdown from wet cooling tower for heat dissipation systems of large size has received insufficient attention with respect to chemical content, thermal content, biocide content, and general water resources impact. A related problem of even greater scope is the disposal of condensate and other wastes from geothermal systems which do not recycle all of the water used in the plant steam cycle.

37. A review of testing methods, environmental survey procedures, and regulatory standards reveals a lack of standardization of sampling and measurement technology. This makes meaningful interpretation of the extent of compliance with standards or the design of experiments to model and verify compliance difficult, if not impossible.

38. Constructive use of waste heat has sound theoretical foundations. At present, however, industrial production of waste heat presents a number of limiting factors that must be eliminated in order to meet aquatic biological and commercial requirements. Although some work is being done in this field with encouraging results, no economically self-supporting program is yet in operation.

REFERENCES

Abraham, G., July 1965, "Horizontal Jets in Stagnant Fluid of Other Density," Proc. ASCE, *J. Hydraulics Division, 91, HY4,* pp. 138–54.

Agnon, S., and Chia-Yung Young, 1952, "Cooling of Water Below Wet-Bulb Temperature of Ambient Air by Cooling Tower." *Heating, Piping, and Air Conditioning, 24,* 139.

Anonymous, 1964, "Systems Recover Chromate from Cooling Water Discharge." *Chemical and Engineering Use, 43.*

Anonymous, 1965, "Experience at Rugeley Establishes Design and Use of Dry Cooling Towers." *Power Engineering, 69,* 62.

Anonymous, 1966, "What Lessons from Ferrybridge." *Engineering, 201,* 401.

Anonymous, 1967, "Ellipsoidal Cooling Tower Promises High Efficiency." *Power, 95.*

Anonymous, 1969, "Ceramic Cooling Towers," Ceramic Cooling Tower Company, Fort Worth, Texas.

Babcock and Wilcox, 1963, "Steam—Its Generation and Use." Babcock and Wilcox, New York, N.Y.

Baker, D. R., 1960, "Unique Cooling Tower Overcomes Ground Fog." *Oil and Gas Journal, 58,* 78.

Baker, K. G., 1967, "Water Cooling Tower Plumes." *Chemical and Process Engineering, 48,* 56–59.

Bardet, J., 1968, "Technical Aspects of Groundwater Recharging." *De Ingenieur,* A758.

Beranek, L. L., Ed., 1960, "Noise Reduction." McGraw-Hill, New York, N.Y.

Berg, B., Lane, R. W., and Larson, T. E., 1964, "Water Use and Related Costs with Cooling Towers." *Journal of American Water Works Association, 56,* 311.

Bergstrom, R. N., 1968, "Hydrothermal Effects of Power Stations." Paper presented at American Society of Civil Engineers, Chattanooga, Tennessee.

Berman, L. D., 1961, "Evaporative Cooling of Circulating Water." Pergamon, New York, N.Y.

Brady, D. K., Graves, W. L., and Geyer, J. C., "Surface Heat Exchange at Power Plant Cooling Lakes," Edison Electric Institute, Publication No. 69-901, New York, 1969.

Brashears, M. L., Jr., 1941, "Groundwater Temperature on Long Island, New York, as Effected by Recharge of Warm Water." *Economic Geology,* 811.

Brooks, N. H., 1967, "Man, Water, and Waste," 110. California Institute of Technology, Pasadena, California.

Buss, J. R., 1968, "How to Control Fog from Cooling Towers." *Power, 112,* 72.

Camp, T. R., and S. D. Graber, February 1968, "Dispersion Conduits," Proc. ASCE, *J. Sanitary Engineering Division, 94,* SA1, pp. 31–39.

Cederwall, K., February 1967, "Jet Diffusion: Review of Model Testing and Comparison with Theory," Chalmers Institute of Technology, Goteborg, Sweden, Division of Hydraulics.

Chagnon, S. A., Jr., 1961, "A Climatological Evaluation of Precipitation Patterns over an Urban Area," 45. Public Health Service, Cincinnati, Ohio.

Chilton, H., 1952, "Performance of Natural-Draught Water Cooling Towers." *Proc. Institution Electrical Engineers, 99.*

Christopher, P. J., 1965, "The Dry Cooling Tower System at the Rugeley Power Station of the Central Electricity Generating Board." *English Electric Journal, 20,* 22–33.

Cooling Tower Institute, 1967, "Cooling Tower Performance Curves." Cooling Tower Institute, Houston, Texas.

Cootner, P. H., and Lôf, G. O. G., 1965, "Water Demand for Steam Electric Generation." Johns Hopkins Press, Baltimore, Maryland.

Cranshaw, C. J., 1963–1964, "Investigation into Variations of Performance of a Natural-Draught Cooling Tower." *Proc. Inst. Mechanical Engineers.*

Czapski, U. H., 1968, "Possible Effects of Thermal Discharges to the Atmosphere," 19. Proceedings Fifth Annual Environ-

EDITOR'S NOTE: Many of the citations that appear throughout the report of the Working Group on Water have been left incomplete by the authors and thus are not included in this list of references.

mental Health Research Symposium—Thermal Discharges, New York State Department of Health.

Dalton, T. F., 1965, "Cooling Towers: Problems and Some Treatment Practices." *Water Works and Wastes Engineering, 2*, 53.

Davies, I., 1966, "Chemical Changes in Cooling Water Towers." *Air and Water Pollution, 10*, 853.

Deluca, F. A., Hoffman, J. F., and Lubke, E. R., 1965, "Chloride Concentration and Temperature of the Waters of Nassau County, Long Island, New York, U.S. Geological Survey, 5.

Denman, W. L., 1961, "Maximum Reuse of Cooling Water." *Industrial and Engineering Chemistry, 53*, 817.

Devereaux, M. B., 1966, "Selecting Integrated Cooling Tower-Condenser-Turbine Combination." *Am. Power Conference Proc., 28*, 457–68.

Dickey, Joe Ben, Jr., 1964, "Large Mechanical Draft Towers Show Many Recent Improvements." *Power Engineering*, 30.

Dinkel, C. C., Beecher, J. S., and Corwin, S. H., 1967, "Anti-Foulants in Cooling Water Treatments." *Industrial Water Engineering*, 17.

Elonka, S., 1963, "Cooling Towers." *Power, 107*, S1–S16.

Edinger, J. E., and J. C. Geyer, "Heat Exchange in the Environment," Edison Electric Institute, Publication No. 65-902, New York, 1965.

Fan, L.-N., "Turbulent Buoyant Jets into Stratified or Flowing Ambient Fluids," California Institute of Technology, W. M. Keck Laboratory of Hydraulics and Water Resources, Report No. KH-R-15, 1967.

Fiehn, A. J., 1969, "Cooling Tower Applications in Power Cycle Design." *Journal Power Division American Society of Civil Engineers, 95*, 55.

Fletcher, J. O., 1969, "Managing Climatic Resources." Rand Corporation, Santa Monica, California, 3.

Fluor-o-scope, 1961, Fall, Fluor Corporation, Los Angeles, California.

Fluor-o-scope, 1967, Fluor Corporation, Los Angeles, Calif.

Furnish, A. G., 1969, February 26, Personal communication.

GEA, 1969, "Air Cooled Steam Condenser." GEA, Bochum, Germany.

Granby, A., 1966, "The Heating of Soils Using the Cooling Cycle of a Power Reactor." *Energie Nucleaire, 8*, 193–4.

Hall, H., Russell, V. S., and Hamilton, C. D., 1958, "Trees and Microclimate." Climatology and Microclimatology, UNESCO, Paris, 259.

Hall, W. A., 1962, "Elimination of Cooling Tower Fog from a Highway." *Journal Air Pollution Control Association, 12*, 379.

Hannah, I. W., 1967, "I—Resume of the Report on the Failures." *Natural Draught Cooling Towers—Ferrybridge and After*, Proceedings of the Conference held at the Institution of Civil Engineers, 5.

Hansen, E. P., and Parker, J. J., 1967, "Status of Big Cooling Towers." *Power Engineering*, 38.

Harleman, D. R. F., 1969, "Mechanics of Condenser Water Discharge from Thermal Power Plants," Chapter 5 in *Engineering Aspects of Thermal Pollution*, F. L. Parker and P. A. Krenkel, Editors, Vanderbilt University Press, Nashville, Tennessee.

Harleman, D. R. F., Stolzenbach, K. D., and G. Jirka, June 1971, "A Study of Submerged Multi-Port Diffusers with Application to the Shoreham Nuclear Power Station,"

M.I.T., Department of Civil Engineering, Ralph M. Parsons Laboratory for Water Resources and Hydrodynamics, Technical Report.

Harris, C. M., Ed., 1957, "Handbook of Noise Control." McGraw-Hill, New York, N.Y.

Heller, L. I., and Fargo, L., 1956, "Operational Experiences with the Condensation System of a Power Plant With Air-Cooled Water Circuit and Potential Improvements." *Fifth World Power Conference*.

Heller, L. I., and Fargo, L., 1962, "Recent Operational Experiences Concerning the 'Heller' System of Air Condensation for Power Plants. Latest Results of Developments." *Sixth World Power Conference*.

Holzman, B., 1941, "The Hydrologic Cycle in Climate and Man," U.S. Government Printing Office, Washington, D.C.

Holzman, B., 1969, Personal communication.

Hoppe, T. C., 1966, "Coping with Cooling Tower Blowdown." *Industrial Water Engineering*, 27.

Illinois State Water Survey Circular 91, 1966, "Selected Papers on Cooling Water Treatment," Urbana, Illinois.

Jaske, R. T., April 1968, "A Test Simulation of the Temperature of the Illinois River and a Prediction of the Effects of Dresden II and Dresden III Reactors," AEC Research and Development, Pacific Northwest Laboratory.

Jaske, R. T., and J. L. Spurgeon, February 1968, "A Special Case, Thermal Digital Simulation of Waste Heat Discharges," Presented at 63rd National Meeting of the American Institute of Chemical Engineers, St. Louis, Missouri.

Jaszay, T., and Tomcsanyi, G., 1968, "A Survey of Air-Cooled Condensing Plants 'System Heller.'" *Transelektro News, 9*, 32–33.

Jones, W. J., 1968a, "Natural Draft Cooling Towers." Paper presented at Cooling Tower Institute Meeting.

Jones, W. J., 1968b, July 8, Personal communication.

Kolflat, R., 1968, statement at hearings before the Senate Subcommittee on Air and Water Pollution of the Committee of Public Works, U.S. Senate, 90th Congress, 2nd Session, 22.

Landsberg, H. E., 1961, "City Air—Better or Worse." Public Health Service, Cincinnati, Ohio, 3.

LeBailly, A. R., 1951, "Some Economic Factors in the Selection of Cooling Towers." *Transactions American Society Mechanical Engineers, 73*, 1021.

Leggette, R. M., and Brashears, M. L., Jr., 1938, "Groundwater for Air Conditioning on Long Island, New York." U.S. Geological Survey, 413.

Liseth, P., November 1970, "Mixing of Merging Buoyant Jets from a Manifold in Stagnant Receiving Water of Uniform Density," University of California, Hydraulic Engineering Laboratory Report 23-1, Berkeley, California.

Marley Company, 1967, "Cooling Tower Fundamentals and Application Principles." Marley Company, Kansas City, Mo.

McCann, J., 1966, "Here Are Six Ways You Can Increase Cooling Water Capacity." *Power, 110*, 67–70.

McKelvey, K. K., and Brooke, E. M., 1959, "The Industrial Cooling Tower." Elsevier, Amsterdam, Netherlands.

Merkel, F., 1926, *Zeitschrift des Vereines Deutscher Ingenieure, 70*, 123.

Miller, L., and Long, F. M., 1962, "A Practical Approach to Cooling Tower Noise." *Heating, Piping, and Air Conditioning, 34*, 141–142.

National Association of Corrosion Engineers, 1966, "Cooling Water Treating." NACE, Houston, Texas.

Nelson, J., 1966, "Selecting Materials, Coatings for Water Cooling Towers." *Heating, Piping, and Air Conditioning, 38,* 109.

Parker, F. L., and P. A. Krenkel, December 1969, "Thermal Pollution: Status of the Art," Vanderbilt University, Department of Environmental and Water Resources Engineering, Report No. 3, Nashville, Tennessee.

Pritchard, D. W., and H. H. Carter, February 1965, "On the Prediction of the Distribution of Excess Temperature from a Heated Discharge in an Estuary," Johns Hopkins University, Chesapeake Bay Institute, Report No. 33.

Ravet, L. C., 1963, "Natural Draft Cooling Towers—Shape of Things to Come." *Power Engineering, 50,* 67.

Rawn, A. M., Bowerman, F. R., and N. H. Brooks, 1961, "Diffusers for Disposal of Sewage in Sea Water," *Trans. ASCE, 126,* Pt. III, pp. 344–388.

Rish, R. F., and Steel, T. F., 1959, "Design and Selection of Hyperbolic Cooling Towers," *ASCE—Proc. (J. Power Div.), 85,* 89–117.

Ritchings, F. A., and Lotz, A. W., 1963, "Economics of Closed Versus Open Cooling Water Cycles." *Proceedings of the American Power Conference,* 416–431.

Rysselberge, J., 1959, "Hyperbolic Towers Deserve Study." *Power Engineering, 96,* 63.

Seelbach, H., Jr., and Oran, F. M., 1963, "What To Do about Cooling Tower Noise." *Sound, 2,* 38.

Shade, W. R., and Smith, A., 1969, "Economics of Cooling Water Use." *Engineering Aspects of Thermal Pollution,* ed., Parker, F. L., and Krenkel, P. A., Vanderbilt University Press.

Skrotzki, Bernhardt, G. A., and Vopat, William A., 1960, *Power Station Engineering and Economy,* McGraw Hill, New York, N.Y.

Smith, Alexander F., III, and Bovier, Ralph E., 1963, "Hyperbolic Cooling Towers with Reservoir Storage of Makeup to Serve the Proposed Keystone Generating Station." *Proceedings of the American Power Conference, XXV,* 406–415.

Smith, Maynard E., 1963, "The Use and Misuse of the Atmosphere." Brookhaven Lecture Series, 24, Upton, N.Y.

Stern, W. M., 1967, "Hyperbolic Cooling Towers," *Power,* 59–62.

Steur, W. R., 1962, "Artificial Cooling Can Broaden Choice of Sites for Today's New Steam Stations." *Electrical World, 157,* 42.

Stolzenbach, K. D., and D. R. F. Harleman, February 1971, "An Analytical and Experimental Investigation of Surface Discharges of Heated Water," M.I.T., Department of Civil Engineering, Ralph M. Parsons Laboratory for Water Resources and Hydrodynamics, Technical Report No. 135.

Stover, J. E., Huston, R. J., and W. D. Bergman, January 1971, "Mathematical Modelling of Thermal Discharges into Shallow Estuaries," TRACOR Report No. T70-AV-7425-U, Austin, Texas.

Stubbs, C. T., 1968, "Cooling Tower Designs Improved." *Electrical World, 108.*

Stundl, Karl, 1967, "L'Influence des Centrales Electriques sur le Régime des Eaux Souterraines, à Proximité des Rives des Fleuves." *La Tribune du Cebedeau, 20,* 74–81.

Thom, C. S., 1969, June 23, Personal communication.

Thom, C. S., 1969, September 18, Personal communication.

Van Dam, J. C., 1969, "The Availability of Groundwater for Cooling Purposes." *De Ingenieur,* A743.

Viganer, S., Elder, R. A., and N. H. Brooks, February 1970, "Internal Hydraulics of Thermal Discharge Diffusers," *Proc. ASCE, J. Hydraulics Division, 96,* HY2, pp. 509–527.

Waselkow, C., 1969, "Design and Operation of Cooling Towers." *Engineering Aspects of Thermal Pollution,* ed., Parker, F. L., and Krenkel, P. A., Vanderbilt University Press, Nashville, Tennessee.

Weir, George E., and Brittain, James F., 1962, "Economic Features in the Selection of Circulating Water Supplies for Electric Generating Stations." *American Power Conference,* Chicago, Illinois.

Willa, J., 1966, "Cooling Towers: Step-Children of Industry." *Heating, Piping, and Air Conditioning, 58,* 153.

Wood, B. R., and Betts, P., 1950, "A Temperature–Total Heat Diagram for Cooling Tower Calculations." *The Engineer, 189,* 455–457.

Working Group I(c)

Environmental Protection: Radiological Engineering Aspects of Power Plants and Their Fuel Cycles

152

INTRODUCTION

In the past few years the public has expressed increasing concern about the impact of electric-generating facilities on the environment. This concern has been focused primarily on three environmental effects areas: radiological, air pollution, and thermal effects. The latter two areas are being reviewed by other Working Groups and are treated in their reports. The objective of this Working Group's report is to define those radiological aspects which have become issues in power-plant siting and to evaluate the adequacy of present technology to deal with these questions, both in terms of the engineering solutions and in terms of the standards which these solutions must meet. The Working Group limited its study to nuclear power plants, since radioactivity considerations from other types of power plants are negligible.

In the course of preparing this document, the Working Group benefited from discussions with members of the nuclear and public health community other than those represented on this group. It was concluded that the total environmental impact of nuclear power generation cannot be measured by considering the power plants alone, but that the radiological aspects of the remainder of the fuel cycle, which is necessarily a part of nuclear power generation, must also be taken into account. In the same fashion, it was proper for an evaluation of the impact of fossil fuel generation to include the health, safety, environmental, and aesthetic effects of coal mining and transport in addition to the normal environmental considerations of generating stations themselves.

The document that follows treats several basic areas. First, the questions involving current radiation protection standards and criteria, their application to nuclear facility siting, and control of discharges to the environment are discussed. In this regard, it is noted that the National Academy of Sciences is making a comprehensive review of the biological-effects literature, in an attempt to evaluate the risks associated with exposure to radiation at the low doses characteristic of environmental exposures. Basic radiation protection criteria have been reaffirmed by the National Council on Radiation Protection and Measurements as recently as January 1971. Therefore, although the development of nuclear energy application in this country may have reached a stage where it becomes wise to consider an allotment or apportionment of these basic protection criteria to specific uses of radiation, there is no basis for anticipating substantial changes in the criteria currently in effect.

1 RADIATION PROTECTION STANDARDS AND CRITERIA

The nuclear power industry is expanding rapidly. Construction and startup of nuclear power reactors and associated fuel reprocessing plants are proceeding at a fast pace, with the number of operating power reactors in the United States increasing from 21 as of March 1971 to about 90 by the end of 1975, and fuel reprocessing plants from 1 to 4 in this same period. Since these facilities cannot operate without discharging small quantities of processed radioactive wastes into the atmosphere and hydrosphere, and since no absolute assurance can be given that significant inadvertent releases will never occur, segments of the public have become increasingly concerned about the radiological aspects of nuclear facilities. Although initial concern for preserving the environment has sometimes been expressed about nuclear plant transmission rights-of-way and thermal discharges (which are common to all steam electric generating stations, and considered by other Working Groups), the radiological features of these stations have usually been the principal focus of public interest.

Nuclear facility operators are required by the Atomic Energy Commission, as a condition of their license, to monitor effluents from their plants for radioactivity and to control such releases well within standards set in the interests of public health and safety. In some instances, state health agencies separately survey individual facilities for environmental radioactivity levels. The Atomic Energy Commission also obtains independent information on radioactivity discharges and environmental levels through its own offsite surveillance programs. In no case has human exposure in the environment of these facilities exceeded, or even approached, current radiological protection criteria. In the vast majority of instances, changes in the human radiological environment have not been observable in the "noise level" of natural background radiation.

The increase in the number of nuclear power facilities, however, presents two situations which require additional consideration in applying protection criteria. The first, which already exists, involves the location of several units on the same site, under the same management, resulting in a potentially larger local discharge of wastes to the atmosphere and water than for a single unit. The second situation involves the location of nuclear facilities, under different management, sufficiently close to each other so that the same population group is potentially exposed to radioactive discharges from both sites into the same air mass or watershed. Such considerations are included in the operating restraints set as conditions for licensing of reactors. For example, the releases from all facilities at a single site are added after the different limits based on effective plume heights, etc., have been taken into consideration, and together must be less than the limits which would apply if only one plant were on the site. Although the fraction of the U.S. population theoretically exposed to radioactivity will increase as the number of nuclear plants increases, this fraction will remain extremely small, although not negligible, for the foreseeable future (the next generation).

A small minority of the radiological protection community has, within the past two years, asserted the need for a substantial and immediate reduction in current radiation protection criteria and standards, claiming that their application will lead to a substantial numerical incidence of malignancies and genetic disorders. This interpretation of both the radiological data and the application of present standards has been subjected to considerable criticism by the majority of official and unofficial standards-setting (or recommending) bodies and by most knowledgeable individuals. However, since an understanding of the radiation protection criteria and their application to nuclear facilities is vital to both the broader professional community involved in power plant siting and plant design and to the public at large, this section of the report is devoted to a discussion of the implications and application of these criteria and standards. The subject of emergency situations—which are in fact more crucial to nuclear facility site acceptability than the normal operational factors—is reviewed in the following section.

RADIATION EFFECTS AND RISKS

Although man has developed in a radiation environment, only within the last 75 years has he become aware of its presence and of the biological effects of exposure to levels of radiation higher than those present in the natural environment (as, for example, from x-ray and radium applications in medical and industrial practice and the recent industrial, research, and military applications of nuclear energy and its by-products). These effects can be categorized as acute responses to high short-term exposures, or as delayed reactions to chronic low-level exposures.

Several groups have played dominant roles in the establishment of current radiation protection standards. On an international scale, the International Committee on Radiological Protection (ICRP) has been recommending protection standards since the early 1930's. The U.S. counterpart of ICRP is the National Council on Radiation Protection and Measurements (NCRP), which has also provided continuing scientific guidance in the radiation protection area for almost forty years. On several occasions, the National Academy of Sciences has provided intensive reviews of the state of knowledge in radiation protection and biological effects [National Academy of Sciences–National Research Council (NAS–NRC), 1960]. A similar function has been performed on a continuing basis by the United Nations Scientific Committee on the Effects of Atomic Radiation.

On a statutory basis, the Federal Radiation Council (FRC, the functions of which are now being continued by the Division of Criteria and Standards, Office of Radiation Programs, Environmental Protection Agency) has, since 1960, provided radiation protection guides to be used by all federal agencies. As related to power plant siting, these have been expressed as regulations governing licensees of the Atomic Energy Commission [Code of Federal Regulations, Title 10, Part 20 (10 CFR 20), 1968], and in more limited form, as part of the Public Health Service Drinking Water Standards (1963).

The radiation effects which must be considered in connection with the normal operation of nuclear facilities are the delayed responses to very low-level long-term radiation exposure. It is customary to categorize these effects either as *somatic* effects (damage occurring in the exposed individual) or *genetic* effects (damage passed on to future generations). These biological responses have been reviewed and described in detail in many documents, and the interested reader is referred to these for supplementary information. Among the most comprehensive of these reviews are the series of reports by the United Nations Scientific Committee on the Effects of Atomic Radiation (1958, 1966, 1969). A somewhat abbreviated discussion is contained in the recent NCPR Report No. 39, *Basic Radiation Protection Criteria* (1971). The following material is necessarily only an outline of the vast literature in this area.

Somatic Effects

Somatic effects are known to result from high doses of radiation. High doses may cause the potential induction of various types of neoplastic diseases (cancer), including leukemia and thyroid and bone cancers; cataracts; fetal (prenatal) abnormalities; and nonspecific life-span shortening. Of the neoplasms that may result from near whole-body irradiation, the risk of leukemia is greatest. The combined incidence of all other neoplasms is cited (International Commission on Radiological Protection, 1969) as from two to ten times higher than the leukemia incidence, with a factor of five now generally accepted. It should be noted in this context, however, that the data relating dose to this general response in adults are all derived from exposures to high doses (on the order of 100 rads or greater) delivered at the high dose rates characteristic of the experience of the Japanese bomb survivors and of medical x-ray therapy. At very much lower exposures and exposure rates, such as might be experienced in the vicinity of nuclear power facilities, the validity of extrapolation of incidence rates at the high dose and dose rates for these neoplasms has raised much speculation as to the dose-effect response; that is, whether there is a threshold below which radiation has no influence on incidence, or whether the neoplastic response increases as a linear function of dose, or as a more complex function of dose. The present working assumption is that no threshold exists and that the response increases linearly with dose. To quote the NCRP (1971):

In the particular case of leukemia, where the consequences of induction are severe and the uncertainty of the data high, the *assumption* of a conservative dose-response relationship seems the only proper course. It seems possible, however, that leukemia induction at low dose and low dose rate will eventually be proved to be lower than estimates derived on the basis of the present linear, no-threshold assumption. In numerical terms, this implies that the leukemia incidence at the levels of our dose-limiting recommendations cannot be more closely characterized than to say that it ranges from zero to a possible yearly upper limit of about one to two cases per million persons per rad.

Irradiation of the embryo during diagnostic procedures has been implicated in the increased incidence of childhood leukemia and other neoplasms. These result from relatively low doses (on the order of a few tenths to a few rads) delivered at high rates (via x-ray exposures) during a limited period of pregnancy. The number of excess cases appears to be linearly related to exposure.

Abnormalities have been observed in children following irradiation *in utero* at relatively large doses. Mental retardation and reduced head size have also been observed in Japanese children who were exposed during the first prenatal trimester to doses above 50–100 rads.

Partial body or specific organ irradiation can result in a variety of tumors, depending on the site exposed. Tumors have been observed in thyroids inadvertently exposed to radiation during the treatment of infants' thymuses to x-radiation doses on the order of 20 rads and higher; such tumors have also occurred in Marshall Islanders exposed to high fallout levels.

Radium deposition in the skeleton has been related to bone cancers, and occupational exposures of the extremities have resulted in skin lesions and malignancies. Dr. Robley D. Evans in his testimony at the Rulison trial* stated that, on the basis of his experience, a cumulative exposure greater than 1,000 rads to the skeleton could be interpreted as one form of threshold dose. Below this dose, the skeleton shows no effect that could be discerned.

Cataract, or vision-impairing opacity of the lens of the eye, has been observed in man to result only after exposure to x rays of the order of 600 R. Much lower doses on the order of 1–5 rads of high-LET radiations (e.g., neutrons) can produce cataracts. For those radiations to which the public may be exposed, a "practical threshold" of a few hundred roentgens is widely accepted as a value which might seriously affect the eye.

The final somatic effect considered to be of potential significance is nonspecific life-span shortening, possibly attributable to the residual, or nonrecoverable, injury done by radiation. Although not producing a specific disease, this may reduce the vitality or reserve capacity of the cells or organs affected. *There is no basis at present for quantitatively relating life shortening and dose in man.* Again, according to the NCRP (1971):

In a general way, existing knowledge from past and present experience with radiologists and with those employed in atomic energy programs indicates that life shortening of persons whose exposure is maintained within presently recommended *occupational* limits* would be too small to detect in the presence of so many other variables. In fact, were it not so, some *measurable* experience would be shown in current epidemiological studies.

In summary, a variety of somatic effects associated with radiation exposure have been observed in humans and other mammals. However, with the exceptions of childhood leukemia and other neoplasms resulting from prenatal exposure to x rays, these effects have not been observed to result from very low radiation doses permitted by current protection criteria. In the case of the exceptions noted, the doses were delivered at high rates during a restricted period of the mother's pregnancy. Despite the lack of observations at dose levels of interest to population protection, the working assumption has generally been adopted that incidence of neoplasms at low levels is linearly related to their incidence at dose values (and dose rates) several orders of magnitude greater. The only somatic risks which the ICRP felt were quantifiable on this basis were those related to neoplasms (ICRP, 1969). Other scientific groups such as the UNSCEAR have felt that assigning risks to exposure at these very low doses is a scientifically meaningless exercise, not supported by the data at hand. The practical outcome of the working assumption of linearity is a radiation protection practice that stresses that any exposure to man-made radiation should be kept as low as is practicable.

The difficulty of quantitatively determining the somatic effects of radiation exposure at very low dose levels should be obvious from the foregoing. First, the radiation dose levels of interest in this discussion would be calculated to produce effects which, in the absence of any other cause, could be detected, if at all, only by statistical methods applied to extremely large population groups. Second, since the same types of effects occur in any population much more frequently from other unidentified causes, the presence of this "background"

---

* The Rulison litigation (*Crowther et al.* v. *Dr. Glenn T. Seaborg et al.,* Civil No. 1702), U.S. District Court, Denver, Colo., January 12–18, 1970, p. 888.

* These limits are 10 times the maximum limit set by regulations to be permitted any member of the public and roughly 100–1,000 times what living next to a nuclear plant might involve.

makes the determination of quantitative relationships at very low doses essentially impossible. This is not to imply that specific cellular changes cannot be observed at very low doses. Such changes have been seen, both *in vitro* and *in vivo*. It is intended to indicate that the production by these dose levels of sufficient irreparable cellular damage to be expressed as an observable somatic effect in the exposed individual is extremely rare, if it occurs at all.

Genetic Considerations

The mutagenic nature of radiation has been known since the mid-1920's. These effects of irradiation may result from gene mutations or chromosomal aberrations, both of which contribute to hereditary damage. Quantitative expression of this damage as a function of dose is more difficult than in the case of somatic effects, because of the variable nature and significance of the genetic effects. For example, the mutation may be dominant, in which case its expression would be in the first-generation offspring. Or it may be recessive, in which case its expression may take many generations to occur. Further, the significance of a mutation may vary widely, from a predisposition to a fatal ailment to a relatively minor personality characteristic. More specifically, the significance can range as follows:

1. Mutations of severe effect and high social cost, e.g., severe diseases with onset in childhood or early maturity
2. Mutations of severe effect and low social cost, e.g., lethals that might occur during embryogenesis or in early fetal stages
3. Mutations whose effects may be trivial, e.g., electrophoretic variants of normal proteins, where function may, perhaps, be only slightly altered

As related to radiation protection, it is generally accepted that some genetic damage might be produced by any exposure, no matter how low, although the linearity of overall response is still open to question. It has been demonstrated that genetic effects are sensitive to dose rate, although no "credit" is claimed for this effect in evaluating protection criteria. Finally, unlike somatic injury, genetic injury is a function of population dose, rather than individual dose.

The most comprehensive method commonly used to date for estimating the total genetic detriment to man is the concept of "genetic death." It is important that the term "genetic death" be understood, and not confused with the actual death of individuals due to radiation.

A genetic death is defined as the extinction of a gene lineage through the premature death or reduced fertility of some individual carrying the gene. By way of explanation, consider a large population group starting out with some given number of harmful mutant genes. In such a group, the number of mutant genes will decrease with each generation, since the carriers of the mutants have a smaller chance of survival and reproduction. This process of loss will continue until all the mutant genes are eliminated from the population. Thus, each harmful mutation leads on the average to one extinction or "genetic death."

The ICRP has estimated (ICRP, 1965) that the exposure of a parental generation of one million persons to one unit of radiation (one rad) may cause 1.9×10^3 genetic deaths over the first ten generations and 8.5×10^3 if projected to infinity.* In relation to genetic risk, the ICRP states in summary:

In the present state of knowledge any evaluation of genetic risks from radiation is beset with uncertainty, largely because it is necessary to rely on much indirect evidence obtained from experiments with animals. In view of this, and because of the importance of avoiding the inadvertent underestimation of risks, many assumptions which may in due course prove to be unduly pessimistic, have been made.

While this "genetic death" concept of estimating the total genetic detriment to man has been used extensively in the past, it is not favored at the present, largely because the experiments with laboratory mammals given large doses in each generation fail to show the accumulated damage predicted by a model equating the extinction of an individual mutant to a single "genetic death." To some, the approach advanced by geneticists such as Dr. J. Lederberg appears more reasonable. In this, a fraction of the present rate of morbidity is assigned to the effects of recurrent mutation. This is then used as the basis for estimating the effects of some fractional increase in mutation rates as a result of increase in exposure to radiation.

CURRENT NORMAL DOSE CRITERIA

With this brief background on radiation effects and risks, we can now examine the current criteria for radiation exposure and their application to nuclear facilities. The current effective recommendations and standards by recommending organizations and agencies, ICRP, AEC, NCRP, etc., are largely, but not entirely, in agreement. The basic AEC radiation protection standards

---

* Projections of power growth to the year 2000 include estimates that the average exposure of people in the United States from nuclear power plant operations might be on the order of 1 mr or 1/1,000 of a rad.

TABLE 1 Dose-Limiting Recommendations

	NCRP Report #39	FRC Report #1	ICRP Publ. #9 AEC Part 20
Maximum permissible dose equivalent for occupational exposure:			
Combined whole-body occupational exposure			
Prospective annual limit	5 rems in any one year	—	5 [a]
Retrospective annual limit	10–15 rems in any one year	—	12
Long-term accumulation to age N years	$(N-18) \times 5$ rems	$(N-18) \times 5$	$(N-18) \times 5$
Bone		0.1 μg Ra^{226} (or equiv.)	30 rems
Skin	15 rems in any one year	30	30
Hands	75 rems in any one year (25/qtr)	75 (25)	75
Forearms	30 rems in any one year (10/qtr)	75	75
Other organs, tissues and organ systems	15 rems in any one year (5/qtr)	15 (5)	15
Thyroid		30 (10)	30
Fertile women (with respect to fetus)	0.5 rem in gestation period	—	1.3 rems per quarter
Dose limits for the public, or occasionally exposed individuals:			
Individual—whole-body	0.5 rem in any one year	0.5	0.5 [a]
Skin, bone, thyroid		—	3 rems in any year [b]
Hands, forearms, feet, ankles		—	7.5 rems in any year
Other single organs		—	1.5 rems in any year
Students	0.1 rem in any one year	—	—
Population dose limits:			
Genetic	0.17 rem average per year	0.17 average of suitable sample	0.17
Somatic	0.17 rem average per year		—

[a] Gonads and red blood marrow.
[b] 1.5 rems to thyroid of children up to 16 years of age.

agree substantially with the recommendations of the ICRP and are entirely consistent with the FRC protection guidance, but differ somewhat in a number of specific areas with the most recent recommendations of the NCRP, published in January 1971.

Most differences lie in the protection criteria applicable to radiation workers. The recent NCRP report also recommended the application of a single dose value for individual members of the general public of 0.5 rem per year, eliminating those differences in specific organ doses which still remain in ICRP, FRC, and AEC standards. Recommendations for population group dose limits, however, remain unchanged by NCRP at 170 mrem per year. Current recommendations and standards are summarized in Table 1.*

Specifically with regard to nuclear power plant siting, an understanding of the application by the AEC regulatory staff of current dose criteria to review and control of normal operation at nuclear facilities is very impor-

tant in assessing the margin of safety which exists in practice. AEC regulatory standards are defined in Parts 20 and 50 of Title 10, Code of Federal Regulations, and, in common with many other similar standards, represent a compromise between strict scientific validity and administrative practicality. For example, the so-called "MPC's," or maximum permissible concentrations, as promulgated by the AEC, are numerically almost identical to those recommended by ICRP and NCRP. These MPC's are not applied as the intake-value limits intended by these organizations. They are used instead as discharge-limit values, particularly for liquids, prior to dilution beyond the plant boundary. This is an administrative simplification which applies a factor of conservatism and thus avoids the necessity of evaluating environmental dilution between the point of release and the nearest point of use.

The AEC proposed, on June 4, 1971, an amendment to 10CFR50 "Licensing of Production and Utilization Facilities" which adds an Appendix I to provide numerical guidance on design objectives for light water-

* The natural background radiation in the United States averages between 100 and 125 mrem/year.

cooled power reactors to meet the requirements that radioactive material in effluents released to unrestricted areas be kept "as low as practicable."

The proposed guides are based upon the fact that existing technology makes it possible to design and operate light water-cooled nuclear power plants which meet the requirements concerning radioactive effluents. Conformance with the design objective guides is expected to achieve the following results:

1. Provide reasonable assurance that annual exposures to individuals living near the boundary of nuclear power plant sites from the radioactivity released in liquid and gaseous effluents from all reactors at the site (5 mrem/yr) will generally be less than about five percent of average exposures from natural background radiation. This level is about one percent of federal radiation protection guides for individual members of the public.

2. Provide reasonable assurance that annual exposures to sizeable population groups from radioactivity released from all power reactors on all sites in the United States, for the foreseeable future, will generally be less than about one percent of exposures from natural background radiation. This level of exposure is also less than one percent of federal radiation protection guides for large groups of people. These levels of exposure would be indistinguishable from exposures due to variation in natural background radiation. They are so low that they would not be measurable with existing techniques, and would be estimated using effluent data from nuclear power plants by calculational techniques.

Under the President's Reorganization Plan No. 3 of 1970, the Environmental Protection Agency (EPA) is responsible for establishing generally applicable environmental radiation standards to protect the general environment from radioactive materials. The AEC is responsible for the implementation and enforcement of EPA's generally applicable environmental standards.

EPA has under consideration generally applicable environmental standards for these types of power reactors. AEC has consulted EPA in the development of the guides on design objectives and limiting conditions for operation to control radioactivity in effluent releases. If the design objectives and operating limits established should prove to be incompatible with any generally applicable environmental standard subsequently established by EPA, the AEC will modify these objectives and limits as necessary.

The design objectives and limitations set forth in the proposed amendments to AEC regulations are specifically applicable to light water-cooled nuclear power reactors. These are the only type of power reactors being installed in relatively large numbers and on which there is sub-stantial operating experience in this country. Numerical guides for other types of nuclear power reactors, such as gas-cooled and fast breeder reactors, will be developed as more design and operating experience is acquired.

The guides are not appropriate for regulating levels of radioactivity in effluents from other types of nuclear facilities such as fuel reprocessing plants, fuel fabrication plants, or radioisotope processing plants where the design characteristics of the plant and the nature of the operations involve different considerations. The AEC will give further consideration to amendments to its regulations to specify design objectives and operating limitations to minimize levels of radioactivity released from these other types of plants.

This does not mean, however, that environmental factors are not considered. License applications must describe environmental features, including existing dilution and reconcentration mechanisms, and their significance in terms of potential restrictions on plant discharges. Additional discharge restrictions may be imposed if these factors appear to be governing in terms of a "critical path" or "critical population segment," or discharges from multiple sources.

Utilities are required to demonstrate that even under unfavorable fuel and plant performance, and conservative meteorological conditions, doses to hypothetical individuals at the most exposed locations on or beyond the site periphery do not exceed the established limits. This is done by providing detailed information on system designs for control of effluents and estimates of expected quantities and concentrations in these effluents to show that their rate will be as low as practicable. Under normal operating conditions, these discharges would be expected (and required) to be substantially below the limits of Part 20.

However, substandard fuel performance and outages of radioactivity control systems or other plant systems have in the past increased (and can be expected to in the future) these discharge values for short periods of time. Under these abnormal (in terms of discharges extended over a period of years) conditions, exposures are still confined by the limits of Part 20. Under these limits a hypothetical individual who is assumed to reside outdoors on the downwind fencepost 24 hours a day for 365 days per year, or who obtains all his drinking water and fish from the plant discharge before dilution in the receiving body of water, must not receive more than 500 mrem per year from plant discharges. Since, in fact, no real individual can fulfill these assumed conditions of occupancy, the maximum individual dose would be substantially lower, even assuming the worst fuel or other plant system performance. Further, since dilution occurs in both the water and atmospheric environments, exposures of large numbers of people are usually several orders of magnitude less than the maximum individual dose calculated. In fact, it has been shown for a number of power plant facilities that the ratio of maximum hypothetical individual exposure to average population exposure lies in the range of about 10^2 to 10^3, depending upon site conditions (Goldman, Calley, et al., 1969; Parsont and Goldman, 1970; Rogers and Gamertsfelder, 1970).

Thus, the restriction imposed by the maximum hypothetical individual exposure provides an automatic assurance that population exposures cannot exceed, at maximum, a few percent of the population dose criterion. As a corollary, the exposure of a substantial population group to a significant fraction of the population dose criterion would imply a gross violation of individual dose limits applied at the plant perimeter. Further, these limits are site values, rather than unit limits; thus the addition of units at an existing site requires that waste management capability be adequate to insure that, for the total station, existing limits are complied with.

ADEQUACY OF CURRENT RADIATION STANDARDS

As mentioned previously, in the past few years a small segment of the scientific community has questioned the adequacy of current radiation protection standards. A few have attacked the entire foundation of the radiation protection standards recommended by the Federal Radiation Council on the basis that these permit unduly hazardous exposure. These concerned scientists have concluded that if the population of the United States were exposed at the average level permitted by the FRC's Radiation Protection Guides, more than 30,000 additional cancer deaths would occur per year.

Statements to this effect have been presented by individuals before the Congressional Joint Committee on Atomic Energy (1969) and the Subcommittee on Air and Water Pollution of the Senate Committee on Public Works (Senate, Public Works, 1969). These claims have been examined and largely refuted (Gamertsfelder, 1971; Bond, 1970; Stannard, 1970) as technically indefensible, both in terms of the "risk per rad" estimates used, and the requirements for physically impossible conditions, in delivering an average population dose of 170 mrem per year without exceeding the maximum individual limits of 500 mrem per year, as discussed earlier. Further, charges have been made of reactor operations having already caused increased infant and fetal mortality. These charges have invariably been disproved by epidemiologists of the Environmental Protection Agency and state health departments.

Nevertheless, this controversy emphasizes the obligations of responsible regulatory agencies: first, to continually review radiation protection standards, as has been done by responsible scientific organizations such as UNSCEAR (1969), ICRP (1969), and NCRP (1971); and second, to report regularly and clearly to the public the results of these reviews. Largely because of what has become an emotional public concern, the Secretary of Health, Education, and Welfare, as Chairman of the Federal Radiation Council, asked the Federal Radiation Council in January 1970 to institute a careful independent review and evaluation of the relevant information that has become available in the past decade. The review committees are considering pertinent scientific information and professional judgments, including the reports prepared by such organizations as the National Council on Radiation Protection and Measurements, the International Commission on Radiological Protection, and the United Nations Scientific Committee on the Effects of Atomic Radiation. The biological effects review is being carried out by the National Research Council–National Academy of Sciences Advisory Committee to the Federal Radiation Council. The NRC review will attempt to quantify the risk of death and of nonfatal injury to the U.S. population from radiation exposure. This committee (and its four subcommittees) and the NCRP (which is providing dosimetry models for population and occupational exposure) will address themselves specifically to the "risk" side of the equation.

Anticipating the results of this review—at least the biological effects aspect of standards as they may apply to nuclear facilities—some insight may be provided by statements of the Chairman of the NAS-NRC Advisory Committee to the FRC (Senate, Public Works, 1969):

Recently the adequacy of radiation protection standards has been questioned. Allegations have been made that insufficient attention has been paid to human data that have become available in the past few years, and that as a result risks to the public are being grossly underestimated, and that maximum permissible levels should therefore be reduced immediately.

Radiation Protection Standards are formulated by several independent national and international bodies, namely, the NCRP, ICRP, and FRC. In addition, periodic scholarly reviews of pertinent data are provided by UNSCEAR. Recent reviews by these groups (ICRP, 1966, 1969; UNSCEAR, 1964, 1966, 1969) have considered in depth essentially all the available data relevant to the setting of standards. These bodies have found no evidence that warrants a downward revision of the basic radiation standard of 5 rems per 30 years or 170 mrem per year to the general population.

Pertinent data have been under continuous review by the NAS–NRC Advisory Committee to the FRC. This Committee has specifically reviewed the statements presented before Congressional Committees and elsewhere to support the allegations referred to above and concludes that these statements contained no data that would significantly alter the base upon which current standards were established. There is no evidence available to the Committee that exposure to the public will increase at a rate that would in any way justify an emergency revision of the existing standards.

Because of the allegations and widespread public concern the Committee feels it must plan further consideration of the interpretation of data relative to estimating risks associated with low levels of radiation exposure and the utilization of such interpretations for establishment of radiation standards.

And by the NCRP (1971):

The NCRP last reduced the basic MPD in 1957, at which time it undertook a continuing review of the whole state of our knowledge of the effects of ionizing radiation on man, especially with respect to its influence on radiation exposure for the population and standards for occupational conditions. During this same time period intensive reviews of the progress of research on the biomedical effects of radiation have been carried out by the International Commission on Radiological Protection, the United Nations Scientific Committee on the Effects of Atomic Radiation, and the National Council on Radiation Protection and Measurements. Continuous review of all of these efforts has led the NCRP to the conclusion that in spite of the enormous amount of such research, very few new facts had been discovered that would strongly influence the concepts of MPD as developed in the 1949 and 1957 reports. Instead, the facts and concepts used in earlier reports were essentially confirmed, although not in every detail. The descriptions of biological factors and specific radiation effects given in this report, not all of which are interpreted alike in the scientific fraternity, represent a balanced fusion of the reviews of Committee members and of many knowledgeable reviewers of early drafts of this report. For this reason, they are not presented as a scientific treatise with full references to the original scientific data.

On the other hand, it became increasingly evident that, in the absence of positive knowledge, it was desirable to continue to make certain conservative assumptions in such areas as dose-effect relationships at low doses, and the existence or not of thresholds and genetic effects in the general population. Positions on these had been adopted by the NCRP in 1959. Because these interpretations lead to policy rather than purely scientific positions, and because their misinterpretation and rigid application could lead, in the extreme, to the elimination of many radiation uses, it is essential to better judge each situation and prescribe for it those usages that are justifiable. At the same time it has clearly become necessary to simplify, wherever possible, the standards of practice which, in some instances, have become detailed and complex far out of proportion to our knowledge of the subject; with the exception of fetal exposure (see paragraph 240), any numerical changes in the dose-limiting recommendations of this report reflect the urge for simplification rather than biomedical necessity.

Examination of the "benefits" and of appropriate balances between benefit and risk will be conducted independently of the NAS–NCRP risk evaluation. The exact mode of examination has not yet been developed. It is clear that any determination of "benefit" will be necessarily qualitative in nature, and based not only on the needs and general welfare of the public, but on economic, political, and social factors as well.

In considering the risk–benefit balance, it may be important to make separate assessments based on the source of radiation. For example, with current medical uses resulting in an estimated genetically significant per capita dose (GSD)* in the range of 100 mrem per year, the related benefit must be weighed on a different balance than the projected less than 0.01 mrem per capita-year from nuclear power plants (Rogers and Gamertsfelder, 1970). Indeed, the identification of the greatest public good, resulting from more stringent application of both controls and technology in the radiation area, is clearly indicated in this comparison; i.e., a reduction in average exposure from medical uses of only a few percent would reduce whatever genetic effect—if any—such exposures entail, more than the total genetic effects of all the nuclear facilities projected for years ahead.

The present procedures for setting radiation protection standards for public health and safety have been questioned by some members of the public who advocate the broadening of the "weighting or balancing groups" beyond the scientific and engineering community. While current procedures do allow for open public comments in the writing of proposed radiation protection standards, the decision-making is not preceded by public hearings. The question of the adequacy of radiation standards has frequently become a key point of issue at the licensing hearings for nuclear facility projects. In order to eliminate detailed discussion of radiation protection standards for each facility license, the present rule-making procedure could be modified to require public hearings on this issue. Such hearings might significantly help through providing for open public debate on radiation protection standard issues, thereby instilling more public confidence in present standards and their application to nuclear facility operations. In regard to the relative national radiation dose, there are several areas in which significant reductions in dose could be accomplished. These are:

1. If it is considered important to control dose contributions to large populations of less than 1 mrem per year, R&D funds could achieve vastly greater returns if applied other than to nuclear power plants. Better con-

---

* The genetically significant dose (GSD) is defined as the gonad dose which, if received by every member of the population, would be expected to produce the same total genetic effect on the population as the sum of the individual doses actually received.

trol of diagnostic x-ray dose with no sacrifice to medical requirements would seem the most productive. Proper columnation of the beam alone would reduce the gonadal dose of the entire population by several tens of mrems per year.

2. Restriction on the use of building materials of high natural radioisotope content would reduce the radiation dose to the entire population on the order of 5 to 10 mrems per year.

3. Restriction on occupancy on those portions of the nation above the 4,000-foot elevation would reduce the average cosmic ray dose to the entire population by about one mrem per year, and for the individuals currently at high elevations would reduce their dose by several tens of mrems per year.

4. Due to the additional cosmic ray dose received in commercial airplane flights, it is estimated that the elimination of air traffic delays would shorten time aloft to the extent that the incremental radiation dose per flight would be reduced about 0.1 mrem.

2   RADIOLOGICAL CONSIDERATIONS OF NUCLEAR FACILITY SITING

Nuclear power plants have many advantages with respect to environmental contaminants, but the fission process inevitably creates radioactive products with a wide variety of chemical properties which decay with half-lives ranging from fractions of a second to about 30 years. The longer a nuclear facility operates on a given fuel loading, the larger the inventory of fission products becomes. Thus, when a nuclear power station first begins operation, the only radioactive materials present are the naturally radioactive uranium isotopes comprising the nuclear fuel. After initial operation, some of this fuel inventory is converted to fission products in the fuel elements. These fission products are highly radioactive. The radiological safety of both plant personnel and the public in the vicinity of the nuclear power plant depends on containing these fission products within the fuel elements.

In normal operation, very small quantities of these fission products may be released from the fuel through small defects in the cladding. Management of these materials does not usually provide a constraint in siting nuclear power facilities. The consideration which usually limits siting of these facilities is the potential for the release of large quantities of these radioactive materials under abnormal or accident conditions.

FISSION PRODUCT CONTAINMENT

To assure that no course of events can lead to a significant release of radioactive material, the fission products are contained within three barriers. First, all fission products are held within the fuel-element cladding, which is designed to withstand the high temperature and irradiation conditions within the reactor core (AEC, 1967). Should a leak or loss of the fuel-element cladding integrity occur, the reactor coolant system provides a second containment boundary. This containment boundary is also designed to withstand high-temperature, high-pressure, and high-irradiation levels. The coolant within this system is constantly monitored to provide prompt detection of significant fuel-element failures. In the event that this second boundary layer should be ruptured, a third independent, overall containment capability is provided for all nuclear power plants. This containment capability is designed to function under conditions imposed by the worst hypothetical accident that can be reasonably postulated and such natural environmental stresses as seismic or tornado loads and to hold essentially all radioactive material resulting from any reasonable combination of accident forces. This containment shell and associated safeguard systems are tested before operation commences and periodically throughout operating life.

Reactors are designed with two safety objectives: to prevent accidents and to limit the consequences of accidents. Energy-producing accidents evaluated are those resulting from addition of reactivity by one of several means—such as the sudden introduction of cold water into a hot coolant system or the rapid ejection of a control rod assembly. Power transients resulting from this type of accident are largely limited by the inherent safety characteristics of light water reactors—the negative temperature coefficient, the negative void coefficient, and the Doppler effect. In addition, redundant and diverse instrumentation systems assure that the reactor power level does not exceed prescribed limits in order to maintain the integrity of both fuel and coolant boundaries.

The reactor coolant system provides the second containment boundary for the fission products. Maintaining the integrity of this boundary is a prime safety objective in the design, fabrication, and operation of nuclear power plants. Retention of the reactor coolant within this boundary in many designs also permits limited core heat removal (cooling of the fuel) even in the event of substantial reduction in coolant flow. A wide variety of operating conditions must be considered in the design of this system, including thermal stresses, plant pressure cycling, irradiation, embrittlement, water chemistry, and operational load transients. Stress analyses are required of major components such as the pressure vessels, and a major materials program has been conducted for several years (e.g., the heavy section steel program).

Despite the assurance that a high-integrity coolant boundary is provided for these plants, a further line of defense is incorporated to protect the public in the unlikely event of a failure of this boundary. If the reactor coolant boundary were to break, the fuel would overheat unless coolant were reintroduced in a very short time. The subject of emergency core cooling was extensively reviewed in 1967 [European Nuclear Energy Agency (ENEA), 1970]. This report stressed the importance of providing emergency cooling systems that would be effective in preventing significant fuel meltdown. To achieve this capability, nuclear plants are furnished with a sophisticated emergency core cooling system which can accommodate a wide range of potential break sizes (including a double-ended rupture of a primary coolant line). Water reactor safety research and development work has included some scale testing of emergency cooling systems.

For the most part, prediction of core cooling performance is based on calculational models. There have been only limited, small-scale experiments for confirmation of flow and heat transfer under specific conditions. Recent tests conducted on a small-scale mockup of the LOFT reactor pressure vessel at the National Reactor Testing Station, Idaho, failed to confirm events predicted by calculational techniques. Steam pressures prevented the emergency cooling water from entering as predicted. Although designers believe their calculation models provide a reasonably accurate representation of heat transfer conditions under these circumstances, this Working Group believes it would be highly desirable to have the additional confirmation that can be provided only by larger-scale engineering experiments. Increased public

acceptance of large nuclear plants may be enhanced by the ability of the industry to provide such assurance.

The AEC adopted interim acceptance criteria on June 19, 1971, relative to emergency core-cooling systems that impose considerable conservatism in the evaluation models used to apply the criteria to specific reactors. They apply to the required performance of the emergency core-cooling system following an unlikely loss of coolant accident. The general criteria applicable to all plants are as follows:

1. The calculated maximum fuel element cladding temperature does not exceed 2300°F. This limit has been chosen on the basis of available data on embrittlement and possible subsequent shattering of the cladding. The results of further detailed experiments could be the basis for future revision of this limit.

2. The amount of fuel element cladding that acts chemically with water or steam does not exceed one percent of the total amount of cladding in the reactor.

3. The clad temperature transient is terminated at a time when the core geometry is still amenable to cooling, and before the cladding is so embrittled as to fail during or after quenching.

4. The core temperature is reduced and decay heat is removed for an extended period of time, as required by the long-lived radioactivity remaining in the core.

In its policy statement, the AEC explained its position by stating that "ideally one would have available analytical methods capable of detailed realistic prediction of all phenomena known or suspected to occur during a loss-of-coolant accident, supported in every aspect of definitive experiments directly applicable to the accident. In the absence of such perfection, adequate assurance of safety can be obtained from an appropriately conservative analysis based on available experimental information. In areas of incomplete knowledge, conservative assumptions or procedures must be applied. When further experimental information or improved calculational techniques become available, the conservatisms presently imposed will be reevaluated and a more realistic approach will be taken."

In accordance with the philosophy of doing everything possible and practical to insure the public safety, additional systems are provided against the release of fission products. These generally fall into three categories: containment and confinement, pressure reduction, and air-cleaning systems.

By far the most familiar safeguard is the containment system, which is in addition to the built-in containment factors previously mentioned. This is basically a pressure shell, designed to retain the energy stored in the reactor coolant or generated by chemical radiolytic reactions, and the radioactive materials which might be released as a consequence of a loss-of-coolant accident. This barrier can be fabricated of steel or reinforced or prestressed concrete; it may be a single or a double barrier; and it may be maintained at normal atmospheric pressure or at subatmospheric pressure.

Various accessories and auxiliaries are incorporated into the containment: access openings and closures including air locks for personnel, ventilating air inlets and isolation valves, penetration seals for pipes and electrical leads, and valves for closing pipes which penetrate the containment. Shields are included for protection against missiles which accidents can create. These safety systems are designed with special features for protection against earthquakes and high winds.

Due to the stored energy in water which might be released by a coolant-line break, the containment pressure would build up. If the pressure remained high for an extended period after the accident, any leakage of fission products would be enhanced, because of the higher pressure differential across the containment shell. One means of pressure reduction used for the boiling-water reactor system is the pressure-suppression system. This system utilizes a two-region containment, consisting of a dry well and a reservoir, called a "suppression pool." This pool contains water; if there were a break in the lines or any components in this part of the system, the resulting vapor would go into the suppression pool and be condensed, reducing the pressure.

In the BWR, direct access is provided to the external turbine. This raises another important point regarding containment. Consider cases where one assumes that isolation valves do not close. For example, on a line break within the *primary* containment area, the external steam loops of course are not broken and have lots of valves besides the isolation valves which would close anyway just for process reasons, so there are no escape paths. If a steam line breaks outside the isolation valves (two on each line, one within and one just outside the primary containment barrier), analyses show that backup core cooling systems operation would prevent any perforation of the core cladding, so release to the turbine building would be limited to the relatively minor portion of fission products in that portion of the reactor water which flashed to steam before boiling in the reactor vessel stopped. In spite of these facts, the redundant isolation valves are designed to close on either initiating event. Consequently, these valves become key items in any design of this type.

A second method of pressure reduction uses spray systems within a containment structure; these are included in both PWR's and BWR's to help condense the vapor. A third method involves vapor-space heat exchangers, which remove heat and thereby help reduce pressure in the system. A recent innovation has been the use of ice beds in the containment. The beds act as a heat sink, through which air–steam mixtures are directed in the event of a loss-of-coolant accident.

The last general category of safeguard systems incorporates those intended to reduce the inventory of radioactive materials in the containment or confinement atmosphere, and hence the inventory available for leakage. This group contains such devices as high-efficiency particulate filters, charcoal adsorption beds, and chemical spray systems. The last type combines both iodine-removal capability and pressure reduction. The former systems also usually incorporate heat-removal capability to accommodate the temperature limitations of system components.

RELIABILITY

As containment and other engineered safeguard systems become increasingly important in the design and siting of nuclear power plants, greater attention is being directed to the reliability of these systems. A system which appears highly effective under carefully controlled conditions is worse than useless if it cannot be counted upon to function when needed. In designing engineered safeguards for maximum reliability, several factors are considered. Static systems and devices are used whenever possible rather than those which require electrical or mechanical action. Mechanical devices are designed to fail in the safe position, so protection will not be lost during power failure or other malfunctions. They are also designed to operate after periods of inactivity. And, an obvious point which should not be overlooked, safeguard systems need to be designed to operate under accident conditions—not just under idealized design and test conditions.

For some critical subsystems upon which the entire operation of the containment system depends, multiple or redundant systems are required. This is particularly true for instrumentation which senses that an accident has occurred, provides an alarm, or initiates protective action. A highly reliable mechanical device is of no use if the sensors, circuitry, and/or power supply which cause it to function are unreliable. Spurious signals which cause a safety system to actuate in error could cause severe operating difficulties. For these reasons, several independent sensors, sometimes operating in coincidence, may be used to automatically actuate a safeguard device or provide an alarm. Two complete and independent containment-closure devices (and associated control systems) may be used in certain critical areas, such as for isolating an open containment ventilation system. Ideally, all safeguard systems should use components which are normally in operation, thus continuously used and their functioning noted. However, containment systems and several other components of engineered safeguard systems are not normally called upon to function; hence, their action is not checked by normal operation. It is, therefore, particularly important that careful inspection, testing, and maintenance of these safeguard systems be carried out to assure their continued operability and effectiveness. This implies that adequate administrative and technically sound procedures for performing these functions must be available, and the control necessary to assure that these procedures are followed is exercised.

SITING

For nuclear power plant siting, the radiological criteria of greatest significance are those related to potential exposure of the public under abnormal plant operating conditions. These exposure criteria determine the suitability of a given site to a much greater extent than do the dose criteria governing normal radioactive releases. A site with unfavorable meteorological or demographic characteristics is much less likely to be acceptable on the basis of accident conditions. The cost penalty associated with engineered safety compensation is usually much greater for accident conditions than in the case of normal radioactive release control. Accident considerations, therefore, play a very significant role in the siting and engineering of nuclear power plants. Unfortunately, accident considerations currently in use have frequently been misinterpreted.

The present Atomic Energy Commission regulations governing power plant siting do not contain accidental exposure criteria applicable to members of the public. The present 10CFR100 (1968), "Reactor Site Criteria," defines parameters to be used in judging the suitability of sites. It includes numerical values of whole-body and thyroid dose to be used as reference values for comparison of sites under defined, hypothetical conditions of releases of radioactivity. These dose values are not considered to be accidental exposure limits, however, but performance criteria under highly unrealistic, conservatively hypothesized conditions. Part 100 includes the following reactor site criteria:

1. An exclusion area of such size that an individual located at any point on its boundary for two hours immediately following onset of the postulated fission product release would not receive a total radiation dose to the whole body in excess of 25 rem or a total radiation dose in excess of 300 rem to the thyroid from iodine exposure.

2. A low population zone of such size that an individual located at any point on its outer boundary who is exposed to the radioactive cloud resulting from the postulated fission product release (during the entire period of its passage) would not receive a total radiation dose to the whole body in excess of 25 rem or a total radiation dose in excess of 300 rem to the thyroid from iodine exposure.

3. A population center distance of at least one and one-third times the distance from the reactor to the outer boundary of the low population zone. Population center distance means the distance from the reactor to the nearest boundary of a densely populated center containing more than about 25,000 residents. In applying this guide, due consideration should be given to the population distribution within the population center. Where very large cities are involved, a great distance may be necessary because of total integrated population dose consideration.

The limiting, or hypothetical, accident model used by the AEC is defined in 10CFR100, "Reactor Site Criteria." The fission product release fractions are defined in a related document referenced in that regulation, "Calculation of Distance Factors for Power and Test Reactors," TID-14844 (1962). Dispersion and fission parameters are given in AEC "Safety Guides" 3 and 4. This model, which deals with a loss-of-coolant accident, is referred to variously as the "maximum hypothetical accident" or "design basis accident." It assumes a spectrum of leak size and location up to the instantaneous severance of the largest pipe in the coolant system, postulates a melt of a gross amount of the fuel, and neglects the effect of emergency core-cooling systems and any population control measures. To deal with the release, a combination of consequence-limiting safeguard systems is required, which must be sufficient to reduce the calculated doses to those listed above.

The fission product releases into the primary containment barrier are specified to be 100% of the equilibrium core inventory of noble gases, 50% of the halogens, and 1% of the solids in the fission product inventory. Of the released halogens, 50% are assumed to deposit on the surfaces inside containment, leaving only 25% of the core inventory available for leakage. In actuality, such a release could not occur unless all the redundant emergency core-cooling systems malfunction. The combined probability of both these failures and the pipe rupture is so small as to be numerically indefinable. It is certainly very much smaller than similar failure probabilities in other engineered structures such as dams, aircraft, bridges, and chemical process equipment, whose failure would have major safety implications.

Meteorology evaluated for accidents usually covers a variety of circumstances, including those least favorable at the site under consideration for transport and diffusion of released gases and vapors—usually a moderately severe inversion coupled with light winds, which are assumed to be invariant in direction during the first phase of release.

Protection is assumed to be afforded only by the so-called "passive" systems; i.e., those for which no system action is required, such as a containment system. Limited or partial protection is credited to such active consequence-limiting systems as sprays and charcoal filters, in which pumps, blowers, valves, etc., are required to operate. No credit is granted for evacuation or other population protection measures which may be executed in a reasonable time period.

With these extremely stringent ground rules for evaluation, nuclear plant sites and the associated engineering safeguard systems must meet the dose criteria cited above. Obviously, in any untoward incident, actual exposures to members of the public would be very much less, if only because of the functioning of duplicate plant safeguard systems (which are provided in replicate and required to be tested periodically throughout the plant life). Other emergency measures would be brought into play, depending upon the condition of the plant and the actual dispersion conditions which existed at the time.

The only other dose criteria applicable for emergency exposure of the American public are those promulgated by the FRC (FRC functions have been transferred to EPA), as guidance for protective actions relating to the normal production, processing, distribution, and use of food products contaminated as a result of an acute, localized event—such as an inadvertent release at a reactor or nuclear fuel reprocessing plant.

Internationally, standards for emergency exposure of local populations have not been formally adopted by ICRP, although the group has stated that the British Medical Research Council (Brit. Med. J., 1959, 1961) and FRC (FRC, 1964, 1965) approaches constitute "a sound approach." A comparison of the BMRC and FRC basic guides (presented in Table 2) indicates their strong similarity.

Since these latter guides or standards do take into account appropriate and available protective measures applied within the plant or by control of local populations, the AEC's site criteria seem reasonable, although they may be difficult to understand by someone not familiar with their application.

TABLE 2 Radiation Protection Guides

Radionuclide	Organ	Individual Dose, rads	
		BMRC	FRC
^{131}I	Thyroid	25	30
^{89}Sr	Bone marrow	15 (total)	—
^{90}Sr	Bone marrow	1.5 (per year)	15 (total)
^{137}Cs	Whole body	10	15 (total)

The philosophy of design and safety assessment used by the nuclear power industry, which postulates a design basis accident to define boundary conditions, provides a built-in mechanism for public misunderstanding. If safety systems are required, and then assumed not to function so that subsequent consequences can be examined, the hypothesis can easily be taken as a vote of no confidence in reliability and availability. This has, in fact, been the case. Since this approach to safety assessment, which has provided a high degree of public safety, is not found in other similar industrial activities, a problem is created which must be faced by the nuclear engineering community: to explain to the public the nature and benefits of these assessments. Unless such explanation can be made by the broad non-nuclear-affiliated engineering community and accepted by the public, the difficulties that result from this rigorous self-appraisal technique will continue to exist.

FAST BREEDER REACTORS

In his June 4, 1971, Message on Energy to the Congress of the United States, President Nixon presented a program which included a commitment to complete the successful demonstration of the Liquid Metal Fast Breeder Reactor (LMFBR) by 1980. The development of the breeder reactor is considered essential to meet the demand for an economical clean energy source because of its highly efficient use of nuclear fuel. Concomitantly, the breeder reactor could extend the life of the natural uranium fuel supply from decades to centuries with far less impact on the environment than the light water-cooled reactors which are currently in operation. In order to successfully achieve the LMFBR and bring the breeder into large-scale commercial usage, many difficult and complex technological and engineering studies would have to be successfully completed. The initiation of the LMFBR demonstration plant program has been authorized, but detailed information (such as the design, owner, location) has not been established. However, it has been indicated that the plant will be a 300 to 500 megawatt electrical sodium-cooled, fast neutron reactor, fueled with a mixture of plutonium and uranium oxide (PuO_2–UO_2).

The demonstration plant will be designed so that no routine releases of radioactive effluents to the environment will occur during normal plant operation. However, it is recognized that some small gaseous releases may occur over the lifetime of the plant through phenomena such as diffusion and leakages through or around seals. The technology exists for restricting these releases, and the resulting radiological effects, to a very low level during normal operation, so that they should not have a significant adverse effect on the environment and will probably not be measurable at the site boundary. All other radioactive material produced in the course of the plant operation, such as fission products in fuel elements, will, as in the case of light water reactors, be transported to another location and processed to recover the unspent nuclear fuel.

The demonstration plant will contain substantial quantities of plutonium (900–1,600 kg), both in its initial fuel charge and as a result of breeding. The existence of this material involves safety and environmental considerations but are not any greater than those already considered for light water reactors. The toxic nature and other potentially hazardous aspects of plutonium are well recognized. The technology exists to assure the safe containment and control of plutonium in all phases of fuel handling, processing, storage, and transportation, as well as in reactor operation, just as it does for other radioactive materials. The demonstration plant will use sodium as a reactor coolant. This selection was based on an evaluation of a number of potential coolants which could be used in fast neutron breeder reactors. Although liquid sodium reacts readily with oxygen and water, the consequences of sodium fires can be precluded or inhibited by special systems or design features.

It is planned that for the demonstration plant, no routine releases of radioactive effluents to the environment will occur during normal plant operation, with the exception of small leakages through seals. The light water reactors have been permitted to release to the environment very small quantities of radioactive effluents, within established guidelines and under well-controlled and monitored conditions. The technology now exists for the demonstration plant to substantially improve on this situation through in-plant collection and storage of various radioactive materials produced during reactor operation. Exposures to the public at the site boundary of the demonstration plant due to plant operation are expected to be less than 1 percent of that due to natural background radiation. At further distances from the plant, exposures will be correspondingly smaller.

The nation needs a fast breeder reactor program to meet its future energy demands, but the program must

be developed and the reactors designed and operated to insure public safety. To this end the National Academy of Engineering could assist by establishing a Nuclear Power Safety Committee to conduct studies relating to potential reactor accidents, identify priorities for research, and recommend national policy on siting.

WASH-740

Opponents of nuclear power frequently cite the conclusions of AEC report WASH-740, published in 1957 and entitled "Theoretical Possibilities and Consequences of Major Accidents in Large Nuclear Plants." A brief review of the findings and significance of this report is appropriate.

This study was performed for the Atomic Energy Commission by the Brookhaven National Laboratory in early 1957 at a time when the technology of central station power reactors was in its infancy, and even before the Shippingport reactor (the first central station power reactor) went into operation. WASH-740 reports the results of a study in which the completely hypothetical assumption was made that large amounts of fission products were released from a reactor in some unspecified manner into the atmosphere in a highly dispersible form.

Although a variety of assumptions were made, there were, in effect, two boundary conditions. On the one hand, if integrity of containment structure were maintained, there would be no significant exposure to the surrounding population. At the other extreme, it was assumed that a major breach of containment occurred with release of 50% of the core in aerosol form. These assumptions are not relevant to contemporary reactor technology with respect to several major points, including the absence of engineered safeguards such as core cooling mechanisms, washdowns, or filters. The assumption that 50% of the core was dispersed as an aerosol is highly unreasonable—experiments performed since the report was published indicate that no more than 1% of the nonvolatile constituents would be released to the containment structure.

This study was made at the Commission's request in order to establish an upper limit on potential consequences in connection with consideration of the Price-Anderson indemnity legislation. Since that report was written, many power reactors have been designed, built, and operated. In the process, the technology of reactor design and nuclear power plant construction and provisions of engineered safety features have made very great advances.

In 1965, the AEC staff, in consultation with some

of the Brookhaven staff members, reconsidered WASH-740. (At this time, the extension of Price-Anderson indemnification was being considered.) With complete disregard for the greatly improved safety precaution system, it was concluded that, since current reactors were substantially larger than those considered in the Brookhaven study—and assuming the same type of hypothetical releases of fission products as those in the 1957 study—the theoretical damages would not be less and under some circumstances would be more than those assumed in the earlier study. However, it was further concluded that the positive safety factors that had developed during the ensuing years supported the firm conviction that the likelihood of major accidents (such as those posed in the Brookhaven study) was extremely low—in fact, even lower than the remote probability that had been estimated in the 1957 study. These factors included the favorable safe operating experience accumulated by power reactors as shown by the safety record since 1957, the substantial advances in reactor technology since that time, the safety incorporated in the design of each vital component and system, and the successive defenses built into safety features designed both to prevent accidents and to limit the consequences in the highly unlikely event that they should occur.

There has never been an "accident or near-accident" at any licensed power reactor where the consequences even remotely approached the situation postulated in WASH-740. The significant factor from the viewpoint of public safety is that a "defense-in-depth" concept has been built into licensed nuclear power plants. This assures that, should a system or component which is important to safety malfunction, systems or components will meet the safety requirements, and backup containment barriers are provided. As a result of this "defense-in-depth" approach, no malfunctions, misoperations, or equipment failures at licensed facilities have constituted a public safety problem in the United States. We know of no instance where their operations have resulted in exposure of any member of the public to radiation exceeding annual limits specified in nationally and internationally recognized safety standards. WASH-740 served the useful purpose that was intended: It identified the kinds of safeguards that should be provided and the kind of research needed to better inform about the physical consequences of a reactor accident.

In considering the overall philosophy of design for the abnormal situation which establishes the acceptability of a particular nuclear plant at a particular site, a number of observations can be made: Safety in any human activity is not an absolute. There is no "all or none" in the risk of accident. The engineering profession has customarily and historically employed safety

factors in the design of structures whose failure would significantly affect health or safety. These factors are usually based on judgments and modified by experience.

Safety assessments and design margins incorporated in nuclear power plants also involve large areas of judgment. Particularly for this technological development, the industry and its regulators are dealing in large measure with precautions against consequences of highly improbable accidents for which there is no experience base. Confidence that serious accidents will be prevented rests heavily upon proper design, construction, and operation of these plants. No incident of public exposure due to accidental releases from nuclear power reactors in the United States has ever occurred. It is, perhaps, due largely to the absence of this failure experience that the safety margins imposed on nuclear facilities substantially exceed those in common use in other engineered structures. However, learning from accidents in the nuclear power field is an alternative to be avoided. *Improved confidence can be achieved by experiments of sufficient scale to normalize and verify calculational models. In a number of important areas, these experiments or simulations remain to be done.* Recognizing that such simulations are expensive, it is felt that their cost is more than justified by the reassurance required to instill greater public confidence alone. Their cost may be economically justified by eventual relaxation of present siting restrictions and engineered safeguard requirements, or the discovery of things which should be done to ease siting restraints.

A final comment relates to public misunderstanding created by the stringent examination of improbable hypothetical accidents typically performed for nuclear plants. Efforts by the nuclear community to clarify the intent of this examination cannot be expected to be fully successful, in view of questions raised about "credibility." It is, rather, for the broader engineering community to undertake this effort, explaining to the public the implications of such self-examination, possibly in comparison to other engineered structures with significant safety implications.

The conditions within which reactor plants need to be designed and operated to assure public safety are set forth largely in general criteria published by the USAEC. They are being detailed further through the development of supplemental criteria by the nuclear community through their standards-writing organizations. The criteria embody much of the philosophy of safety design discussed above. *A rigorous effort on the part of the broader engineering community to thoroughly examine the existing basis for design and operation of these plants, to enforce and change these as needed, and to push for such other criteria and standards as may be required in the public interest, could contribute significantly to resolution of the siting problem.*

A particularly pressing need, for example, is examination of earthquake design criteria, especially those criteria relating to the possibilities of earth displacement at the plant site. Some areas of our nation that might benefit most from nuclear power also involve zones of seismic activity. Siting of plants in such regions will depend greatly on an ability to assure public safety should earthquakes occur. Better analytical tools to assess the possibility of common mode failures and of design guidance to preclude them are still needed.

An allied situation comes about through protective-systems testing requirements. Circuits under test frequently are out of service during the test period. Designs must assure that protection provided through redundancy is not nullified during the testing period. While emphasis on safety has resulted in multiple standby systems, the question has been raised whether this proliferation of systems is leading to a complexity that can compromise safety rather than enhance it.

In examining the potential role of the engineering community at large in improving the siting of nuclear power stations, this Working Group feels that a major contribution could result from a thorough review of the current nuclear power plant-safety design criteria, emergency core-cooling systems, standards, and practices by an independent, knowledgeable organization, such as the NAE or a "national board" created for such purpose. This review—conducted by senior members of the profession with substantial experience in design and operation of projects of similar complexity and safety significance—might do much not only to provide a fresh perspective of currently accepted approaches in the nuclear industry, but also to instill added public confidence in the safety aspects of plant design.

3 RADIOLOGICAL WASTE GENERATION AND MANAGEMENT AT NUCLEAR POWER STATIONS

Radioactive waste at a nuclear power station is produced as a by-product of the fission process within the fuel element and from neutron activation of structural materials, corrosion products, and coolant impurities within the reactor vessel. The characteristics of the radioactive waste from a particular power plant are highly dependent on the type of reactor system used. Naturally, the characteristics of wastes strongly influence the design of a particular waste-treatment system. In all cases, however, the transfer primary-loop activity via the waste processing systems to the gaseous and liquid effluents discharged from the plant is the primary mechanism for local contamination of the environment.

CORE INVENTORY

The quantity of fission products within the reactor fuel elements depends upon (1) the average operating power level of the reactor, (2) the fuel residence time in the core, and (3) the time transpired for radioactive decay. Typically, about 1.8×10^6 curies of radioactive fission products would be in the reactor core per megawatt of reactor power one hour after shutdown following a two-year operating cycle. At one day after shutdown, the activity would decay to about 10^6 Ci/megawatt; the inventory of the most biologically significant radionuclides for this operating history is shown in Table 3. Shorter irradiation times result in lower inventories; however, after a few weeks of operation, the fission-product inventory is up to about 40%

of the two-year inventory. A relatively small inventory of tritium is produced within the reactor fuel element by ternary fission.

ACTIVATION PRODUCTS

In addition to the fission products generated within the fuel elements, smaller quantities of radioisotopes—

TABLE 3 Inventory of Selected Radionuclides Following Two-Year Operation with 1-Day Decay [a]

Selected Isotopes	Half-Life	Activity in Fuel [kCi/MW(t)]
^{3}H	12.3 yr	0.0043
^{85}Kr	10.7 yr	0.25
^{89}Sr	51 days	24
^{90}Sr	28.9 yr	1.8
^{90}Y	64 hr	1.8
^{91}Y	58.8 days	32
^{99}Mo	66.6 hr	40
^{131}I	8.06 days	28
^{133}Xe	5.3 days	54
^{134}Cs	2.06 yr	0.61
^{132}Te	78 hr	34
^{133}I	20.8 hr	22
^{136}Cs	13 days	0.74
^{137}Cs	30.2 yr	2.4
^{140}Ba	13 days	46
^{140}La	40.2 hr	49
^{144}Ce	284.4 days	35

[a] Statistics from C. M. Unruh, Manager—Radiation Protection, Battelle Memorial Institute.

171

such as those of iron, chromium, cobalt, and manganese —are produced from neutron activation of corrosion products which normally form during reactor operation; these isotopes are entrained in the circulating coolant. Both soluble and insoluble compounds of these nuclides are formed. These can be removed from the coolant by appropriate use of filtration, ion exchange, and evaporation. The reactor coolant water itself is subject to neutron activation, producing small quantities of ^{16}N. Radiolytic decomposition also occurs but is not significant in pressurized water reactors because of the rapid rate of hydrogen/oxygen recombination at high system pressure. The introduction of boron and lithium in the coolant of pressurized water reactors to control core reactivity has the simultaneous effect of generating tritium by neutron absorption by the boron.

RADIOACTIVE WASTE MANAGEMENT

The plant design is to process and recycle waste streams so as to minimize both volume and radioactivity of effluent wherever practical. Releases to the environs are controlled by batch processing and/or continuous monitoring before discharge to assure that planned releases are within established short-term and annual average permissible emission rates.

The radionuclides in gaseous radioactive wastes tend to have shorter half-lives than those in the corresponding liquid wastes, and the atmosphere is a larger medium in which to achieve dose reduction than the typically available waterway. Current gaseous waste management techniques are basically retention to permit radioactive decay before release of the airstream, or removal of radioactive species from the airstream for storage. Retention refers to storing wastes long enough to decrease the associated treatment problem or hazard by permitting some radioactive decay to occur before release. The usefulness or efficacy of these techniques as a means for reducing activity levels in gaseous wastes depends on the particular isotopes present. In a typical BWR, an overall decontamination factor (DF) of 48 may be expected from a 30-minute delay for a representative gas mixture.

Using the 30-minute delay time as a reference point, DF's on the order of 10^3 could be attained for most krypton nuclides (except krypton-85) with a time delay of just under three days. A similar DF could be attained for all xenon nuclides with about 35 to 40 days of delay time, assuming the typical BWR gas mixture. After a three-day holdup, the xenon activity is largely determined by the ^{133}Xe decay constant. By then, the only krypton isotope of significant activity is long-lived ^{85}Kr. Thus, the efficiency of increased time delay as a means of reducing gaseous activity release is limited by the xenon decay constant.

Reservoirs intended to achieve gas holdup times of several days or more need to be designed in such a manner as to assure that the gas is not released prematurely as a result of mixing gases generated at different times. The design should, therefore, utilize either very long, narrow passages to minimize mixing effects or, alternatively, a series of separate chambers which are filled, left undisturbed for a predetermined time, and then discharged sequentially. In either event, the volumetric storage capacity of such a system should be roughly proportional to the delay time selected.

Filters collect radioactive solid particles formed when a gaseous parent nuclide decays to a particulate radioactive daughter. The performance of High-Efficiency Particulate Air (HEPA) filters is well documented. Tests run at the Oak Ridge National Laboratory (ORNL) in 1968 indicated removal efficiencies of 99.97% as a minimum for such HEPA filters installed there [International Atomic Energy Agency (IAEA), 1968c]. For maximum effectiveness, HEPA filters should be placed where the particulate concentration is highest. For off-gas systems in water-cooled reactors, this location is just below the point in the stack from which the gas is released, thus allowing a maximum of transport time for gaseous radionuclides to decay to particulate daughter products before filtration.

Liquid waste management systems currently employ four basic treatment techniques to reduce levels of radioactivity. These techniques are: (1) holdup and decay, (2) filtration, (3) evaporation, and (4) demineralization. A final reduction in liquid radionuclide concentrations is achieved by dilution of the wastes in the condenser cooling water to insure that radionuclide concentrations are at the lowest level possible before reaching the site boundary.

Holdup and decay for liquid waste is identical in principle to that for gaseous waste, although little reduction in liquid radioactivity levels is accomplished by this method. Radionuclides found in liquid radioactive waste have relatively long half-lives, since short half-life radionuclides decay away in the time it takes for their transfer through the plant. A relatively long holdup time would be needed to achieve any appreciable reduction in liquid-waste radioactivity levels—it is estimated that approximately 40 days would be required to reduce typical liquid waste radioactivity levels by about a factor of 5.

Filtration is usually utilized as the sole means of radioactive waste treatment for waste streams containing primarily insoluble or particulate contaminants. Filtration is also frequently employed with other types of waste treatment as a pre- or post-treatment step for the process liquid. Thus, the potential of the principal treatment process (e.g., ion-exchange) can be fully realized. The objective of prefilters is removal of suspended solids to prevent interference of particulates in subsequent processes and possible clogging of process-liquid flow channels, such as in ion exchangers. In post-treatment application, filters perform such tasks as collection of resin "fines" escaping from ion exchangers. Filter types used include natural filtration (using sand or other media), activated carbon, vacuum and pressure precoat type filtration, and fibrous and knife-edge filtration.

Evaporation separates water from nonvolatile dissolved and insoluble radioactive wastes by boiling. This results in a concentration of the wastes, permitting easier ultimate disposal. The efficiency of evaporation for radioactive waste treatment can vary widely, depending on the radioactive materials present. Overall DF's of 10^2 to 10^5 (between feed and condensate) are experienced, depending on the mass velocity of the vapor in the evaporator and decontamination efficiency for nonvolatile radioactive contaminants. If volatile radioactive materials such as tritiated water, iodine, or ruthenium are present, the overall DF may be substantially reduced, due to carryover of these materials (IAEA, 1968b). Evaporation is a common method of liquid-waste treatment, because streams of relatively high dissolved-solids content can be accommodated. It is, consequently, a suitable process for use with subsequent ion-exchange treatment. Care needs to be taken, however, to provide pretreatment of feed streams, which contain organic agents such as laundry wastes.

The efficiency of ion-exchange treatment of waste streams depends on the type, composition, and concentration of waste liquid, the type of exchanger, regeneration methods, radionuclides present, and operating procedures. Decontamination factors as low as 2 and as high as 10^5 are reported (IAEA, 1967). Only low total dissolved and suspended solids waste can be processed efficiently by ion exchange, because bed exhaustion occurs rapidly for liquids with a high total dissolved-solids content. Also, suspended solids will clog an ion exchanger and prevent its efficient operation. Thus, the use of ion-exchange treatment needs to be restricted to radioactive wastes with low total dissolved solids and low suspended solids.

Dilution of liquid wastes has been used to bring the concentration of radionuclides within acceptable limits for discharge. This approach is based on the premise that concentration needs to be maintained below specified values to control both direct ingestion and indirect ingestion of aquatic or marine species whose radioactivity uptake is directly proportional to the discharge concentration. If the receiving body of water is restricted, as in the case of a cooling pond or lake, dilution becomes less useful. The total quantity of radioactive material released in this case needs to be decreased due to accumulation and recycling of the effluent stream.

BWR WASTE MANAGEMENT SYSTEM

For boiling water reactors using nominal 30-minute decay on gases removed by the condenser air ejector, about 99% of the radiogas emission is via this route. The stream includes radiolytic hydrogen and oxygen and the normal condenser system air inleakage. Additional sources of gaseous waste are the gland seal condenser, plant ventilation exhausts, gases from the plant's radiochemical laboratory, radwaste facility,

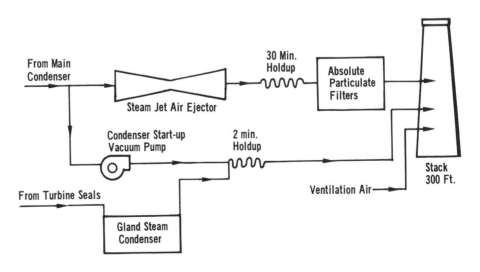

FIGURE 1   Typical BWR off-gas system.

laundry, decontamination operations, and various tank vents. The gaseous waste contains primarily the activation product ^{13}N, isotopes of the noble-gas fission products krypton and xenon. About 90 percent of the ^{13}N will decay to a nonradioactive isotope in the time it takes for the gaseous waste to be transferred through the plant to the stack and from there to the site boundary. Consequently, it is of minimal concern in offsite areas. The isotopic composition of the krypton and xenon depend on type of cladding defects, the radiation history of the reactor fuel, and on the age of the mixture at the time of release, because many isotopes of krypton and xenon have relatively short half-lives. Some radioactive particulates appear in the gaseous waste as a result of entrainment and decay of noble-gas precursors, and isotopes of some of the more volatile elements, such as iodines, will remain in the vapor state. Removal of iodine occurs by retention in reactor water, retention in condensate, and retention in the air ejector condenser condensate. The ^{131}I emission via the air ejector decay system is typically about 10^{-6} of the noble gas emission.

A typical operating BWR off-gas system is shown in Figure 1. Noncondensible gases are drawn from the main condenser through steam jet air ejectors and condensers into a delay line, where they are retained for 30 minutes. After this holdup time, the gases pass through an absolute particulate filter and are discharged through the stack. The stack is usually about 300 feet high, although its actual design height is influenced by site topography and heights of adjacent buildings. Since a major portion of the activity released from a BWR is composed of short-lived gases, an elevated release (opposed to a ground release) provides effective dispersal and decay. Off-gases from the turbine gland seals are processed similarly. Because these are mainly short-lived activation gases, only a two-minute delay is provided before they are discharged through the stack.

A typical BWR liquid radwaste processing system is shown in Figure 2. The unique feature of this system is the segregation of wastes by chemical and physical properties. Influent is collected and processed according to its classification as high-purity (equipment drains), low-purity (floor drains), chemical, or laundry wastes. Contents of equipment drains are filtered and demineralized, and can then be either reused in the plant or discharged. Plant floor drains, chemical wastes, and laundry wastes are filtered and discharged from the plant. Since the laundry wastes tend to foul filtering media, they are processed separately through their own filter.

The major sources of solid radioactive wastes at BWR's are filter sludges and demineralizer resins. This waste is first centrifuged to remove excess water, and then measured into 55-gallon steel drums by a vibrating hopper. Cement and make-up water, if necessary, are added; the resulting slurry is mixed by a mechanical mixer lowered into the drum from above. Enough water is added to assure that the final slurry will solidify, and mixing continues until the slurry is completely homogenized. The mixer is then withdrawn; the drum is capped and conveyed to a storage area. Drums normally

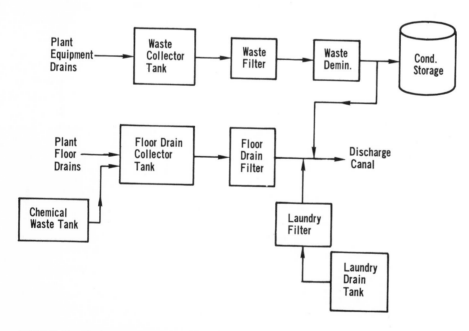

FIGURE 2 Typical BWR liquid radwaste system.

spend from three to six months in storage before being shipped offsite for permanent burial. Shipments per year per facility involve some 100–175 drums.

BWR OPERATING EXPERIENCE

A radiological surveillance study performed by the Radiological Engineering Laboratory of the Environmental Protection Agency (formerly the Division of Environmental Radiation of the Bureau of Radiological Health) at an operating BWR power station (Kahn et al., 1970) measured the characteristics of the gaseous effluent discharged to the environment. The average fission-product noble-gas release rate was 12,500 μCi/sec while the plant was operating, which was during 64 percent of the year. (The AEC license prescribes the stack-discharge limit as 700,000 μCi/sec. Therefore, the station discharged at slightly more than 1% of the stack-discharge limit for the year.) The principal radioactive noble gases found in the laboratory effluent analyses were ^{85m}Kr, ^{87}Kr, ^{88}Kr, ^{133}Xe, ^{135}Xe, and ^{138}Xe. One day after release, only ^{85}Kr, ^{133m}Xe, and ^{135}Xe were detected in the sample; after one month the only noble gases detectable were ^{133}Xe and ^{85}Kr. Tritium could also be detected in the laboratory sample. The principle non-noble gas fission product found was ^{131}I. Data for ten operating BWR's are shown in Tables 4 and 5.

The radioactive liquid waste constituents from a BWR are the activated corrosion products and fission products. The fission-product levels are attributable to tramp uranium and to leakage from fuel elements through cladding defects. The relative contribution of leaking fuel elements, of course, depends directly upon the number of leaking rods and the severity of the leaks.

The characteristics of the liquid waste effluent were also measured in the study mentioned. The average concentration of all detected soluble and insoluble radionuclides in the high-conductivity waste prior to dilution was approximately 2×10^{-3} μCi/ml. The radioactive constituents at highest concentrations were ^{3}H, ^{58}Co, ^{89}Sr, ^{90}Sr, ^{131}I, ^{134}Cs, ^{137}Cs, ^{140}Ba, and ^{144}Ce. The average contribution to the total unidentified activity in the water utilized for radioactive waste dilution was 0.189×10^{-7} μCi/ml over a one-year period. The permissible limit from Appendix B, Table II, of 10CFR20 is 1×10^{-7} μCi/ml for an unidentified mixture containing

TABLE 4 GE–BWR Calendar Year 1970, Experience on Radiogases [a] to Atmosphere Operating Data and Conformance to Technical Specification [b]

Plant	Date of Commercial Operation	Nominal Rating [MW (e)]	Gross Electrical Generation (10^6 MWh)	Nominal Off-gas Holdup (minutes) [c]	Annual Average Emission — Microcuries per Second — Reported Actual	Annual Average Emission — Microcuries per Second — Tech. Spec. Permissible	Rounded [d] Percent of Tech. Spec.	Total [e] Annual Release (curies)	Unit [f] Average Release [μCi/sec/MW(t)]
Dresden 1	6/60	200	1.50	20	30,000	560,000 [g, h]	5	950,000	50
Big Rock Point	3/63	72	0.38	18	9,000	1,000,000	<1	280,000	60
Humboldt Bay	5/63	70	0.43	18	16,000	50,000	30	500,000	100
Garigliano	5/64	150	0.74	30	16,000	1,000,000	2	500,000	60
KRB	4/67	237	1.84	[i]	1,000	68,000	1	31,000	1
Tarapur (2 units)	10/69	380	2.17	30	14,000	580,000	2	440,000	20
Oyster Creek	12/69	640	3.56	30	3,500	300,000	1	110,000	3
Nine Mile Point	2/70	600	1.63	30	<1,000	800,000	<1	<30,000	1
Tsuruga	3/70	342	1.89	[j]	1,800	50,000 [k]	4	57,000	3
Dresden 2	(1971)	809	1.25	30	8,600	700,000 [h]	1	270,000	20

[a] Includes noble and activation radiogases only.
[b] Statistics from J. M. Smith, General Electric Company, Atomic Power Equipment Division.
[c] Original design basis holdup time. In practice, is usually longer due to power level dependency and extent of air inleakage.
[d] Rounded off to one significant figure, recognizing accuracy of data.
[e] As stated in typical AEC summaries. Does not account for radioactive decay. Rounded to two significant figures.
[f] Shown to remove power level dependency. Rounded to one significant figure. MW(t) is annual average.
[g] Dresden limit was 700,000 μCi/sec for past decade. Recently reduced to 560,000 μCi/sec by AEC during development of overall Dresden site limit.
[h] Dresden site limit for combined D-1 and D-2 operation is: $\dfrac{\text{D-1 emission}}{560,000} + \dfrac{\text{D-2 emission}}{700,000} = $ not more than unity.
[i] Design includes recombiners, 30-minute delay, and charcoal adsorption. Effective holdup during 1970 based on influent and effluent values probably approached 1 day for kryptons, and two weeks for xenons.
[j] Design includes recombiners and storage tanks which provide about a day holdup when operating pressurized. During 1970 most operation was at atmospheric pressure with holdup probably in the 1–2-hour range.
[k] Limit based on 0.15 MeV gamma at 24 hours decay; for shorter holdup, limit is adjusted inversely as the gamma MeV of radiogas mixture.

TABLE 5   GE—BWR Experience on Radiogases to Atmosphere Estimation of Environs Radiation Dose [a]

Plant	Annual Average Emission (microcuries per second)		Calendar Year 1970			Four Years, 1967–1970			
	Reported Actual	Best [b] Estimate Permissible	Worst Fence-post dose (mrem/ yr)	Nearest [c] Neighbor Dose (mrem/ yr)	Large Popula-tion Group Dose [d] (mrem/ yr)	Average Emission (μCi/sec)	Worst Fence-post dose (mrem/ yr)	Nearest [c] Neighbor Dose (mrem/ yr)	Large Population Group Dose [d] (mrem/yr)
Dresden 1	30,000	700,000	22	5	0.04	18,000	13	3	0.03
Big Rock Point	9,000	2,600,000	2	0.4	0.004	8,000	2	0.3	0.003
Humboldt Bay	16,000	125,000	64	13	0.13	22,000	88	18	0.18
Garigliano	16,000	1,000,000	8	2	0.02	9,000	5	1	0.01
KRB	1,000	1,500,000	0.3	0.07	0.0007	<1,000	0.2	0.04	0.0004
Tarapur (2 units)	14,000	580,000	12	3	0.02	—	—	—	—
Oyster Creek	3,500	600,000	3	0.6	0.006	—	—	—	—
Nine Mile Point	<1,000	800,000	0.5	0.1	0.001	—	—	—	—
Tsuruga	1,800	500,000	2	0.4	0.004	—	—	—	—
Dresden 2	8,600	1,000,000	4	0.9	0.009	—	—	—	—

[a] Statistics from J. M. Smith, General Electric Company, Atomic Power Division.
[b] GE–APED calculation, or estimate, of continuous emission rate that would deliver a dose of 500 mrem/year to the most exposed location (fencepost) at or beyond the plant property. In this regard, adjacent waterways are considered to be available as effective exclusion area, as annual occupancy factor for any individual would be very low.
[c] Actual dose to any individual is estimated to be about one-fifth of "fencepost" dose, considering distance and direction of normally occupied locations, occupancy factor, and shielding by structures, etc.
[d] Estimated average dose to entire population within a 50-mile radius. Based on IAEA Paper SM 146/8, Rogers and Gamertsfelder, USAEC, wherein, based on meterorology and population distribution for eleven operating power reactor sites, average dose in 50-mile radius was estimated as 1/500 of worst fencepost dose. This factor has been applied to all sites and thus its accuracy is probably not better than order of magnitude.

no ^{129}I, ^{226}Ra, or ^{228}Ra; and 3×10^{-7} μCi/ml for soluble ^{90}Sr or ^{131}I. Therefore, the station discharged at approximately 19% of the limit based on a gross beta analysis. The unidentified limit is usually conservative by a factor of about 100. Historically, BWR's have discharged liquid radioactive wastes at levels only a small fraction of the allowable limits.

PWR WASTE MANAGEMENT SYSTEM

The primary coolant in pressurized water reactors does not boil. Thus, most of the gases are contained within the boundaries of the primary coolant system. Gases which leave the primary coolant system are collected and routed to storage tanks. The general composition of the gaseous waste produced in a PWR is somewhat different than that produced in a BWR. The difference in the composition of the discharged gaseous waste is that caused by the increased in-plant residence time of the PWR gases before discharge. The shorter half-life isotopes are much less abundant in the gaseous waste of a PWR than in that of a BWR. Occasionally, primary coolant will leak into the secondary system through defective steam-generator tubes. When this

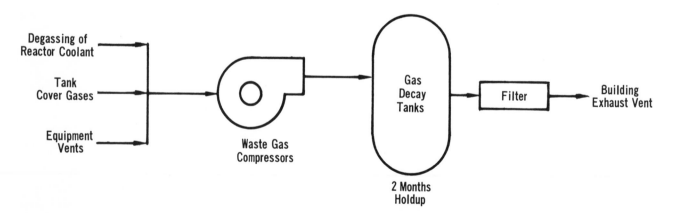

FIGURE 3   Typical PWR gaseous radwaste system.

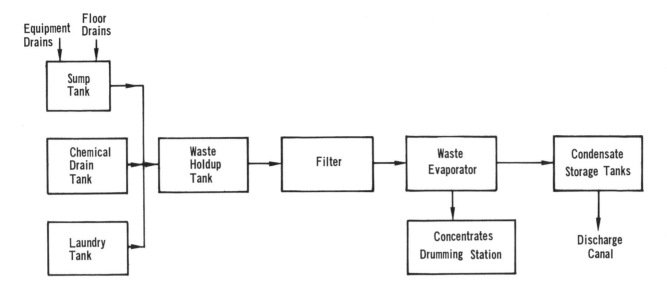

FIGURE 4 Typical PWR liquid radwaste system—dirty wastes.

occurs, short-lived gaseous radioactive wastes may be released through the main condenser air ejector.

A typical PWR gaseous radwaste system is shown in Figure 3. The waste gases from various sources are collected in a vent header and discharged by a waste-gas compressor into one of several decay tanks. When the tank reaches a set pressure and activity level, it is isolated, and a second tank placed in service. Gas is held for decay from one to two months before being discharged through a filter to the environment from a building vent. Since these gaseous wastes are small in quantity and contain minimal activity, dispersal from an elevated stack is not necessary.

A typical PWR liquid radwaste system is shown in Figure 4 and Figure 5. The dirty wastes from various sources (Figure 4) are collected in separate tanks, and, when ready for processing, are discharged to the waste-holdup tank. This tank serves primarily as a batching tank for the waste evaporator. The distillate from the evaporation process is condensed and stored in a condensate storage tank before discharge from the plant. The concentrates are collected and stored for processing through the solid-radwaste system.

Clean wastes (Figure 5), which consist primarily of reactor coolant, are collected in holdup tanks. After filtration and demineralization of the wastes, the boric-acid

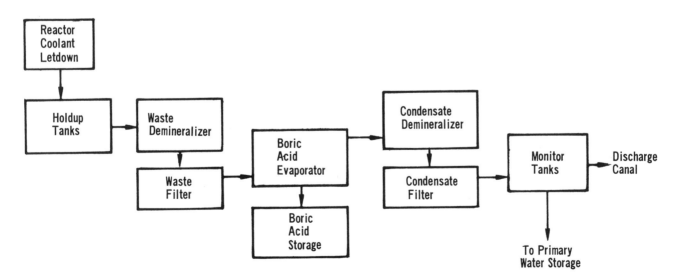

FIGURE 5 Typical PWR liquid radwaste system—clean wastes.

evaporator serves primarily to recover boric acid and primary-grade water. The boric-acid evaporator condensate, after being filtered and demineralized, can be recycled to the primary coolant make-up tank or released to the discharge canal.

Solid radwastes from a PWR may be processed in a different manner, depending upon the source, the activity, and the utility's choice. Evaporator concentrates are measured into 55-gallon steel drums prefilled with a cement or cement–vermiculite mixture. The rate of concentrate addition is controlled to allow complete mixing of the resultant slurry. The vermiculite acts as a dewatering agent, and removes water not needed for solidification. Capping can be done after concentrate addition, but before complete solidification is achieved.

PWR's use cartridge-type filters that consist of paper or fiber elements held within a steel cage. Usually, only the elements need changing, but occasionally the entire cage assembly needs to be replaced. In either case, the component to be disposed of is placed inside a 55-gallon steel drum lined with cement for shielding purposes. More cement is added to completely encapsulate the component. The drum is then capped. All processed solid radwaste is stored on-site before being shipped to a burial ground for final disposal. These shipments range from 100 to 175 drums per year.

PWR OPERATING EXPERIENCE

A recent radiological surveillance study performed at an operating PWR power station (Kahn, 1971) measured the characteristics of the gaseous wastes at several points in the waste-management systems. The maximum an-

nual gross beta–gamma level of activity discharged by gaseous effluents was approximately 22 curies for the year 1962 and averaged 5.3 Ci/yr between 1962 and 1968. This maximum annual release of gross beta–gamma gaseous activity represents 0.32% of the plant's annual average stack-discharge limit for noble gases. The radionuclides found in the gaseous waste were ^3H, ^{14}C, ^{85}Kr, ^{133}Xe, and ^{135}Xe. No gaseous ^{131}I was found at the minimal detectable level. Based on measured concentrations in the surge drum and subsequent dilution factors, four radionuclides were estimated to be found at the site boundary at concentrations corresponding to 0.002%, 1.0%, 0.03%, and 0.017%, respectively, of their individual 10CFR20 (Kahn, 1971) concentration limits in air. The concentrations of particulate radioactivity in air at the site boundary were conservatively estimated to be 0.01% of the limit for the most abundant radionuclide, ^{90}Sr. Secondary sources of gaseous waste such as the condenser air ejector and secondary-system liquid waste-tank vents appear to be the major contributors to effluent radioactivity. For example, all the recorded gaseous ^3H discharges (except in one case) occurred during months when the vapor container was vented or gas was released under special circumstances.

Liquid wastes from PWR's are similar to those from BWR's except that tritium is a much more significant constituent. This abundance of tritium arises mainly from diffusion of the fission-produced nuclide through stainless clad fuel from extensive use of boron and lithium in PWR coolants. The boron undergoes a neutron-capture reaction with the fission neutrons; this reaction generates tritium, which has a relatively long half-life of 12.3 years, and which combines with oxygen to form water. Tritiated water is chemically identical to

TABLE 6   W–PWR Operating Experience [a]

Plant	Design Power [MW(t)]	Dilution Parameters			Total Liquid Released (Curies)	Total Gases Released (Curies)	Total Tritium Released (Curies)	Fuel Clad Material
		Circulation Water Flow (gpm)	Site Boundary Meteorology X/Q (sec/m³)	Year				
Yankee	625	138,000	5.0×10^{-5}	1967	0.055	2.3	1,690	Stainless
				1968	0.008	0.68	1,170	Stainless
				1969	0.019	4.14	1,225	Stainless
				1970	0.036	16.5	1,375	Stainless
Conn. Yankee	825	372,000	3.3×10^{-5}	1968	3.96	3.74	1,740	Stainless
				1969	12.2	190.0	5,100	Stainless
				1970	29.5	876.0	7,376	Stainless
San Onofre	1,345	346,000	7.0×10^{-6}	1969	8.90	251.0	3,500	Stainless
				1970 [b]	3.80	1,606.0	4,769 [b]	Stainless
RG&E	1,520	334,000	1.5×10^{-6}	1970	9.35	9,974.0	107	Zirconium

[a] Statistics from J. H. Wright, Director of Environmental Systems Department, Power Systems, Westinghouse Electric Company.
[b] Does not include the month of August.

ordinary water, making separation and removal extremely difficult and, to date, impractical.

Reactor coolant contributes to the PWR liquid waste through expansion overflow, reactor letdown flow, component leakage, and sampling. Other sources of liquid waste include floor drains, decontamination, and laundry wastes.

The characteristics of PWR liquid radioactive wastes were also reviewed in the study mentioned (IAEA, 1967). The major constituent was tritium. Other major radioisotopes were ^{14}C, ^{51}Cr, and ^{58}Co. Lesser amounts of ^{32}P, ^{54}Mn, ^{55}Fe, ^{59}Fe, ^{60}Co, ^{63}Ni, ^{90}Sr, ^{95}Zr, ^{95}Nb, ^{110m}Ag, ^{124}Sb, ^{131}I, and ^{137}Cs were also discharged. All radionuclides were released in quantities well below AEC limits. Data on the discharge of radioactive wastes from four operating PWR's are shown in Table 6.

4   RADIATION EXPOSURE IN RELATION TO MAN

The radiation which may be expected to reach people living nearby is a prime consideration in locating a nuclear facility. In other sections of this report, the anticipated releases from normal operations as well as accidental releases are presented. As these radioactive materials are released, they may constitute widely different radiation hazards to man. For example, a highly insoluble radionuclide which may be adsorbed on soil materials may never come in contact with people directly or enter human food chains. If this material emits radiation of low penetration, it would constitute no hazard to the population; the ionizing radiation would not have an opportunity to affect human tissue. But radioactive material which is discharged into the air and immediately comes in contact with large human populations may have a maximum impact in terms of human exposure. In this section, the ways in which people can be exposed to ionizing radiation from power plant operations will be discussed and estimates of the relative radiation exposure from radioactive releases that may occur from the power program will be given.

Environmental surveillance at power plants or nuclear facilities to evaluate critical exposure pathways is very difficult. A common approach involves monitoring gases and airborne particles prior to release from a stack into the atmosphere and monitoring liquid waste before discharge into the coolant water canal (prior to dilution with the receiving water). In addition, certain environmental samples are routinely analyzed for the presence of radionuclides which may have been released from the reactor site. The best estimates of radiation dose to man are obtained from actual environmental measurements, but it is often more practical to estimate exposures from release rates and to assume the processes by which radiation may reach persons near the facility.

CRITICAL EXPOSURE PATHWAYS OF RADIOACTIVE MATERIALS TO MAN

The most important potential pathways by which radionuclides from a facility may reach the immediate population are considered to be the following (in order of significance):

For atmospheric discharges:

1. Atmospheric discharge→whole body external exposure
2. Atmospheric discharge→inhalation exposure
3. Atmospheric discharge→deposition on grass→cattle→milk→man
4. Atmospheric discharge→deposition on leafy vegetables→man
5. Atmospheric discharge→deposition on grass→cattle→beef→man
6. Atmospheric discharge→deposition on soil→plants→man

For liquid discharges:

1. Aqueous discharge→waterway→drinking water supply→man
2. Aqueous discharge→waterway→seafood→man
3. Aqueous discharge→waterway→aquatic plants→animals→man

4. Aqueous discharge→waterway→external exposure

5. Aqueous discharge→waterway→sediments→external exposure

Each pathway is considered individually for specific radionuclides; the exposure resulting from each radionuclide in the particular pathway can be estimated. For the nuclides not considered by the FRC, data of the International Commission of Radiation Protection (ICRP) can be used, taking $\frac{1}{10}$ of the (MPC)$_W$ value * for a 168-hour week and assuming an average water consumption of 2.2 liters per day for adults. Dose estimates for populations should include fetuses, infants, and children as well as adults. Special dietary intakes may become important as the number of nuclear facilities increases.

A useful concept for evaluating radiation dose to human population is that of "man-rem." This is defined as the calculated product of the exposure to man-made radiation of individuals in the population group times the number of people exposed. Interest in this integrated number stems from the genetic considerations discussed elsewhere in this report. It will be discussed in the following paragraphs in the context of a tool that may be useful in assessing the impact (or lack thereof) of the nuclear industry. However, it should be kept in mind that the nuclear industry is likely to contribute but a very small fraction of man's exposure to radiation. Variations in man's exposure due to background are greater than the exposure he is likely to receive from this man-made source. This concept of integrated dose can be useful in assessing the health impact of future doses from radionuclides introduced as the result of contaminating events. It could apply to a single intake episode (e.g., a dose of a thorium compound or other radiopharmaceutical) or describe a continuous radiation exposure, in which each person's exposure is considered to start at the time of conception. Although the integrated dose concept has mainly been applied to world population exposure from nuclear tests, the concept is, in principle, applicable to smaller groups and cases where releases have already occurred or can be predicted. To estimate health significance, it may be necessary to consider specific radiation properties affecting the distribution of dose among members of the population. Thus, the highest dose to an individual will be greater than the average to a group. Calculating the integrated dose, then, becomes mainly a problem of predicting changes in the amounts and distribution of relevant radionuclides in the environment and in exposed individuals. This may involve considering the metabolism of the nuclides in various organs and tissues, and environmental vectors which contribute to both internal and external radiation dose.

The internal radiation dose to the whole body or to a specific organ or tissue of an individual in a population group is the product of (1) the concentration of the radionuclide in the intake media, (2) the weight or volume intake of the media containing the radionuclide per individual per unit time, (3) the factor of converting activity intake of the specific radionuclide to organ burden, (4) the length of time the radionuclide remains in the organ or tissue, and (5) the conversion of organ burden and residence time to dose.

The particular radionuclides that can be identified as being important in relation to man depend upon the specific nuclear facility and the type of operation. Table 7 shows these radionuclides and their relationship to specific sources of environmental radiation. In evaluating this relationship, the radionuclides have been classified into groups, depending upon their relationship to man as a source of exposure. A description of each group and its biological effect follows:

Group I includes radionuclides that make up basic organic constituents. These would be dispersed through-

* Same as 10CFR20 values.

TABLE 7 Radionuclides and Relationship to Specific Sources

Group	Source			
	Nuclear Reactor	Fuel Re-processing	Uranium Mines, Mills, and Fuel Fabrication Facilities	Aerospace
Basic Organic Constituents				
^{3}H	X	X		
^{14}C	X			
Noble Gases				
^{133}Xe, ^{135}Xe, ^{138}Xe	X	X		
^{87}Kr, ^{88}Kr	X	X		
^{85}Kr	X	X		
Halogens				
^{129}I		X		
^{131}I	X	X		
Bone Seekers				
^{238}Pu, ^{239}Pu	X	X		X
U–Th		X	X	
^{89}Sr, ^{90}Sr	X	X		
^{226}Ra, ^{210}Po			X	
Trace Constituents				
^{95}Zr–Nb	X	X		
^{58}Co, ^{60}Co	X	X		
^{134}Cs, ^{137}Cs	X	X		
^{147}Pm, ^{144}Ce, ^{106}Ru	X	X		
^{65}Zn	X			

out the body, irradiating all organs. They would thus be of concern for both genetic and somatic effects. These nuclides have short biological half-lives.

Group II includes radioactive noble gases. These would not be incorporated into body tissues in significant amounts. They would be primarily of concern from external exposure, and would involve both genetic and somatic effects. Exposure would occur while the person was present in or near the plume containing the radioactive gases.

Group III includes halogens which consist primarily of isotopes of iodine. Radioiodines are produced in the fission process in large quantities compared to other radionuclides. They may be dispersed to either the liquid or gaseous environment, and have several potential exposure pathways to man. Iodine isotopes are primarily an internal-exposure somatic hazard concentrating in the thyroid gland.

Group IV includes the radionuclides which tend to concentrate in bone—generally the heavy metals, plus strontium, which substitutes for calcium in the bone. These radionuclides tend to have long biological half-lives and may produce somatic effects of the bone, bone marrow, or blood. They may also be of concern for lung exposure when inhaled in particulate form.

Group V includes other radionuclides which are present in smaller quantities but which may be concentrated through environmental media and then further concentrated in a particular body organ, resulting in somatic effects to that organ.

One step in calculating doses to populations is the identification of dietary intake of certain foodstuffs. For example, because of ethnic or cultural differences, some individuals may select diets which other groups would not. There may also be economic factors; for example, among fishermen the consumption of fish or seafoods may be very much higher than would be found in the normal population. Other variables in considering population exposures are metabolic differences in the absorption, storage, and excretion of radionuclides. These metabolically determined processes may vary in populations by as much as a factor of 10. This raises the question of whether the dose calculation methods presented in the ICRP publications should be modified to take into account the small proportion of the population which may have markedly different transfer processes.

Tritium resulting from the operation of pressurized water reactors and fuel reprocessing plants is released in substantial quantities to the environment. Although tritium is considered to be among the least hazardous of all radionuclides, its continued production, difficulty of retention in the plant, and relatively long radioactive

half-life will lead toward increased levels in the environment as the number of power plants increases. Tritium is usually present in the form of tritiated water. It is not removed from the reactor coolant by either filtration or ion exchange, but is reduced by use of zirconium clad fuel. Purified water containing tritium may be transferred back to the reactor coolant system, stored for future use, or transferred to the liquid radioactive waste processing system (Jacobs, 1968).

Since tritium may be incorporated in genetic material, DNA, it might lead to unusual biological effectiveness of tritium disintegrations. Most experimental evidence for chronic exposure to date, however, has not shown an unusual biological effectiveness. The possibility that more sophisticated experimentation may reveal that the biological effectiveness is somewhat greater does exist, nevertheless.

The maximum permissible concentration calculations for tritium are based on the turnover time of tritiated water in the human body (twelve days) and the assumption that 100% of the ingested tritium is taken into the body water. For tritium in forms other than water, the 12-day turnover time may not be valid. However, the extreme assumption of complete random distribution of the tritium in organic molecules of adults also would not be true. In chronic exposure, the distribution of tritium will be low for that portion of tritium which occurs in materials with a slow turnover rate [Joint Committee on Atomic Energy (JCAE), 1969]. However, tritium would accumulate in that part of the system with a long turnover time, assuming uniform distribution in the body.

In the developing fetus, random distribution of tritium would be expected regardless of the form in which it is ingested. However, there is no evidence that the tritium-to-hydrogen ratio will not be above that of the ingested material. In the case of the fetus, the slowly exchanging rate of tritium could cause retention of tritium for long periods of time. The dose rate produced from ingestion of either tritiated water or tritiated organic compounds depends primarily on the ratio of tritium to ordinary hydrogen in these materials. The dose rate can be anticipated if the specific activity (ratio of tritium to hydrogen) is known.

The projected environmental levels of tritium released from light water power reactors in the year 2000 are estimated to be 10^8 curies, or approximately 5% of the maximum tritium present in the year 1963. At present, tritium releases from operating power reactors are only a small fraction of the radiation protection guide, and there is no indication that these levels will result in a public health problem. However, because of its biological consequences and behavior in the environment, a continuing research and surveillance effort will be required to obtain data upon which to make future judgments. If it is noted that the levels in the environment increase at a constant rate over a period of time, an action program should proceed forthwith to reduce the levels.

RADIOACTIVITY IN THE MARINE ENVIRONMENT

The National Academy of Sciences has published (in the fall of 1971) a report entitled "Radioactivity in the Marine Environment." This report was prepared by the Panel on Radioactivity in the Marine Environment of the National Research Council Committee on Oceanography, now the Ocean Affairs Boards. The report discusses subjects such as distribution of ^{137}Cs and other fallout radionuclides, physical processes of water movement, accumulation and distribution of radionuclides by marine organisms, radiological interactions and considerations, and the effects of radionuclides on marine organisms and man (NAS, 1971).

Movement of radioactivity in the marine environment has been studied by several groups, including the Bureau of Commercial Fisheries in Beaufort, North Carolina (Rice and Baptist, 1970). The radionuclides in the marine environment can remain in solution or in suspension, precipitate and settle on the bottom, or be taken up by plants and animals. Certain factors interact to dilute and disperse these materials; other factors simultaneously tend to concentrate them, as shown in Figure 6. Currents, turbulent diffusion, isotopic dilution, and biologic transport dilute and disperse radionuclides. Concentrating processes may be biological, chemical, or physical. Radionuclides are concentrated by the biota, by uptake directly from the water, and by passage through food chains; they are concentrated chemically and physically by adsorption, ion exchange, coprecipitation, flocculation, and sedimentation. Radionuclides are exchanged between the water, the sediments, and the biota. Each radionuclide tends to take a characteristic route, and has its own rate of movement through the many components (reservoirs) of the marine environment.

Radionuclides accumulated by organisms may become new compounds. These compounds may be incorporated into the tissues of the organism and later transmitted to other organisms through the food web, or excreted and subsequently recycled. Even after death, the biota are involved in the cycling process. Dead tissues are attacked by bacteria and other organisms which decompose them to their component elements. Organisms not only cycle radionuclides in a local environment; some animals transport the radionuclides great distances during migration.

The availability of radionuclides to marine organisms depends to a large extent upon the physical state in which these nuclides occur in seawater.

Radionuclides may find their way into the marine environment in the particulate or dissolved state. Isotopes introduced as particles may dissolve or remain particulate; those introduced in the dissolved state may be sorbed onto particulate matter, form colloids, precipitate as particles, or remain in solution. Both particulate and dissolved radionuclides can enter into the marine environment and be taken up by organisms.

The concentration factor of a radionuclide or an element in an aquatic organism is usually expressed as the ratio of the amount of element per unit weight of organism to that of an equal weight of the water. For a concentration factor to be valid, the element in the organism and in the water must have reached equilibrium; that is, the rates of uptake and loss of element by the

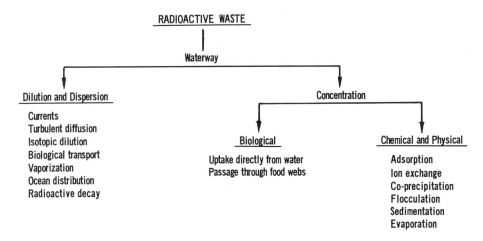

FIGURE 6 A schematic diagram showing the processes tending to dilute or concentrate radioactive materials added to a marine environment.

organism have become equal. Many published concentration factors, however, were determined from short-term experiments in which equilibrium was not attained. A method of assessing effects in the marine environment is based on the use of the maximum permissible specific activity * for radionuclides in critical organs and tissues of man. This approach relates a given radionuclide to the corresponding stable element and may be used for hazards prediction by determining (1) the distribution of the stable elements in the biogeochemical system, (2) physical and isotope dilution rates for the radionuclides in the environment, and (3) biological half-lives in food organisms of man. The approach provides a method for the step-by-step evaluation of isotope dilution of an introduced radionuclide as it passes through the hydrosphere, geosphere, and biosphere to man. The method does not require the determination of environmental and biological compartments for each radionuclide or detailed transfer routes and rates in food webs. Values for elemental compartments in both the environment and the organisms vary greatly under natural conditions, and minor errors in the measurement of compartment values may introduce serious errors in prediction. In most marine areas the total biomass usually accounts for 10^{-6} or less of the total mass of the biogeochemical system actively associated with biologically important elements. Because of this, the organisms exert an insignificant influence upon the distribution patterns of added contaminants. They may, however, provide transport of radionuclides to man through his food (Lowman, 1969).

Concentration factors of stable elements in various organisms are often used as a basis for predicting the eventual radionuclide concentrations which will be reached in organisms near nuclear power plant effluent outlets. Such predictions estimate maximum levels and tend to be too high, because stable elements are accumulated during the organism's entire life and are usually in equilibrium with the elements in the water. On the other hand, low levels of radioactive wastes are released intermittently, and it is often difficult to detect radioactivity in water and aquatic organisms. In these areas, therefore, a concentration factor has little significance, but actual radioactivity measurements have practical value. Some typical food chain concentration factors are shown in Table 8 (Freke, 1967).

* Maximum permissible specific activity used in this report refers to the radionuclide in μCi per gram of the corresponding stable element allowed in the critical organ of man. The specific activity which is allowed is dependent on the annual radiation dose levels recommended by the ICRP for the general population. It is equal to 1/10 the continuous exposure allowed to occupational workers.

TABLE 8 Food Chain Concentration Factors—Marine Organisms [a]

Isotopes	Fish	Mollusks	Crustacea
^{3}H	1	1	1
^{32}P	10^4	10^4	10^4
^{54}Mn	3×10^3	5×10^4	10^4
^{55}Fe	10^3	2×10^4	4×10^3
^{58}Co, ^{60}Co	10^2	3×10^2	10^4
^{65}Zn	5×10^3	5×10^4	5×10^3
^{89}Sr, ^{90}Sr	1	1	1
^{90}Y, ^{91}Y	30	10^2	10^2
^{99}Mo	10	10^2	10^2
^{131}I, ^{133}I	20	10^2	10^2
^{132}Te	10	10	10^2
^{134}Cs, ^{137}Cs	30	10	50
^{140}Ba	3	3	3
^{140}La	30	10^2	10^2
^{144}Ce	30	10^2	10^2

[a] Statistics from Freke (1967).

TRANSFER OF RADIOACTIVITY THROUGH FOOD CHAINS *

Radionuclides can enter the food chain at any trophic level, but their transfer to high trophic levels varies considerably, depending upon the radionuclides and species involved. Fission products in a particulate state which are concentrated by primary producers do not usually progress beyond the second trophic level—that is, the primary consumer. Some induced radionuclides such as ^{65}Zn, ^{58}Co, ^{60}Co, ^{55}Fe, ^{59}Fe, and ^{54}Mn are found in the third and fourth trophic levels. Transfer of radionuclides was traced through sample food chains under controlled conditions in the laboratory. How well the results can be extrapolated to field conditions is not known. Specific studies for tracing the radionuclides in the environment may be merited as part of ecological studies around selected proposed nuclear facilities. With the emphasis placed on keeping radioactive releases very low, the radionuclides actually released from present reactors can be found only with considerable difficulty. When advanced waste treatment systems are routinely used, tracing may be virtually impossible.

Radionuclide concentrations do not ordinarily increase as they move through the food chain in the environment. Radionuclides are metabolized in the same way as their stable isotopes. With the present levels of artificial radionuclides in the environment, concentrations reaching the higher trophic levels are very small

* Note: Both "concentration factors" and "critical pathways" are considered in setting emission limits for plants. See Chapter 1.

—often less than that of naturally occurring radionuclides such as ^{40}K. Even with a continuous source of radionuclides, experiments with species tested thus far have indicated that radioactivity concentrations decrease rather than increase in the higher trophic levels, compared with concentrations in the lower trophic levels.

Concern is expressed that detrimental effects to species other than man may occur when these species are exposed directly to the effluents from nuclear facilities. The issue involved here is whether the process of dilution, which may occur in human exposure as a result of multiple sources of food or water, would not apply in the case of organisms living in the vicinity of nuclear facilities and deriving all food from the material which could contain radionuclides. For example, one might infer that the radioactive material from the outfall of a nuclear reactor might contain a thousand times the concentration any human population might come in contact with, since the sources to which man would be exposed would be diluted by other nonradioactive sources. An example might be shellfish grown in a waterway exposed to effluents. Thus, the question remains whether this nonhuman population of either animals or plants may be adversely affected by the radioactivity, even though the conditions may be safe for man. Basic to this issue is the radiosensitivity of various forms of marine life. Lethal effects are considered particularly, although genetic effects cannot be ignored. One operating assumption is that all nonhuman forms of life are more resistant to radiation, or less sensitive, than man. This assumption has some theoretical support in the fact that the most critical part of the cell appears to be the cell nucleus and the nucleoprotein contained within. The radiosensitivity of a cell is proportional to the amount of DNA in the nucleus of a cell. This generalization has actually proved to be a reasonable one, on the basis of a wide variety of comparative studies of radiosensitivity. Nonetheless, the possibility remains that the exposure of the lower forms of marine life to as much as a thousand times or more of that reaching man could have harmful effects on them, even though these effects would not be observed in man.

On this basis, it is conceivable that a proportion of the population of plants or marine animals exposed directly to the radioactive material in the effluents from nuclear facilities might be affected to some degree. It should be borne in mind, however, that at the most only a very small proportion of the organisms in any particular waterway would be affected. Therefore, the prospect of obliteration of a species is extremely unlikely under any conceivable conditions. Effects would be sharply limited to the region in the vicinity of the plant, if they occurred at all. In this regard, it should be noted, however, that no such effects have been observed in the vicinity of such AEC installations as Hanford, Oak Ridge, and Savannah River, where levels are and have been much higher than those expected from commercial nuclear power facilities.

MULTIPLE PATHWAYS CONSIDERATIONS

Greater use of nuclear power will result in increased numbers of plants which will share common resources. Utility companies need to control reactor discharges on a regional basis. Discharges of gaseous and liquid waste to the environment from multiple sites could require the Atomic Energy Commission to impose stricter discharge limits than for single sites.

These factors need to be considered in establishing environmental surveillance programs, so that the impact of radioactivity which might be released from nuclear power sources can be evaluated over relatively large geographical areas. Such evaluations must consider the long-term buildup of radioactivity in the aquatic environment, including the reconcentration phenomena in biological media, which might cause population exposure. Multiple reactor sites may also present regulatory problems relative to the establishment of radioactive-effluent discharge limits. These limits need to be carefully developed, utilizing the best available information. The many environmental aspects which could influence population exposure should be considered to assure that the total dose to the population from multiple reactor sources is within acceptable limits.

DOSE ESTIMATES FROM RADIONUCLIDE RELEASES FROM NUCLEAR POWER PLANTS

Elsewhere in this section, average values for releases of radionuclides from operating power plants are given. From this information, one can calculate the maximum human dose expected under usual dilution conditions for airborne and liquid wastes. Only the highest expected concentrations are used for calculating the internal or external doses from these elements.

From a BWR

Blanchard et al. (1968) calculated radiation dose rates to a hypothetical individual from the principal exposure pathways at Dresden-1 Nuclear Power Station at Morris, Illinois [210 MW(e) BWR]. These results are presented to illustrate a technique for estimating the dose from a nuclear power plant by using liquid and airborne particulate effluent concentrations and minimal data from field measurements of the gaseous discharge. The technique is applied to the Dresden nuclear plant, where the seven pathways that were considered are listed in Table 9 in order of dose contribution. The critical radionuclide or radionuclides, which in every case are estimated to contribute almost the entire dose, are given in the second column of the table. The critical pathway was external whole-body radiation from the plume emitted at the stack; the second pathway of consideration was milk consumption. In all pathways except beef consumption, ^{131}I also was the estimated critical radionuclide. Facilities in other environments may require the consideration of different pathways in a different order of estimated dose.

The whole-body external dose was a measured quantity which agreed with estimated values within a factor of 2 (Kahn *et al.*, 1970). The dose rates calculated for the radionuclides from the other pathways are based on discharge rates at the reactor. This necessitated the use of a number of assumptions and parameters whose values in some cases were highly uncertain. Hence, calculated dose rates for these pathways of lesser criticality are probably uncertain by a factor of 10, and possibly more.

The total dose rates to each critical organ, estimated for the more significant radionuclides by summing the doses from each pathway, are listed in Table 10. These values represent maxima, in that no one person would be exposed to all pathways. Shown in the last column of Table 10 is the fraction of the Radiation Protection Guides (RPG) for a "suitable sample" of the exposed population as recommended by the FRC (1961). In all cases, the dose to the critical organ is a very small fraction of the RPG.

TABLE 9 The Estimated Dose Contributed by the Major Radionuclides in Each Pathway [a]

Pathway	Critical Radionuclide	Critical Organ	Estimated Dose Rate (mrem/ yr)
Atmosphere—external	Gaseous fission products	Whole body	14.0
Milk	^{131}I	Thyroid	0.28
Drinking water	^{131}I	Thyroid	0.35
Leafy vegetables	^{131}I	Thyroid	0.063
Fish	^{131}I	Thyroid	0.043
Beef	^{89}Sr	Bone	3.7×10^{-4}
Atmosphere—inhalation	^{131}I	Thyroid	2.0×10^{-4}

[a] Statistics from Blanchard *et al.* (1970).

TABLE 10 The Estimated Total Dose Rate to the Critical Organs and Its Fraction of the Maximum Permissible Dose-Rate (RPG) [a]

Critical Organ	Nuclides	Dose (mrem/ yr)	Fraction of RPG
External whole body [b]	Mixed gaseous fission products	14.0	8×10^{-2}
Thyroid	^{131}I	0.74	1×10^{-3}
Bone	^{89}Sr, ^{90}Sr	0.026	2×10^{-4}
GI(LLI)	^{58}Co, ^{60}Co, ^{140}Ba	0.0027	5×10^{-6}
Internal whole body	^{134}Cs, ^{137}Cs, ^{3}H	0.011	6×10^{-5}

[a] Statistics from Blanchard *et al.* (1970).
[b] Based on highest field measurements, at TLD Station 10, 1.2 km NE of stack, and the dose-to-exposure ratio of this device.

From a PWR

A radiological surveillance study was conducted during the period October 1968 and February 1971 by the Radiochemistry and Nuclear Engineering Branch, Division of Research of the Environmental Protection Agency, at the Yankee Nuclear Power Station, a pressurized water reactor.* Yankee is located at Rowe, Massachusetts, and operates at a maximum power of 185 megawatts electrical [MW(e)] and 600 thermal megawatts [MW(t)]. It has produced more than 1×10^7 megawatt-hours between 1960 and 1969 and has passed through seven fuel cycles. Liquid and gaseous wastes obtained from the site visits were analyzed for their radionuclide content. Annual totals were estimated by averaging the measured concentrations—usually for two sets of samples—by isotope, and multiplying these averages by the annual effluent volumes reported by the station operator. The estimated amounts of radionuclides in the gaseous and liquid effluents discharged annually are summarized in Tables 11 and 12 only to indicate the magnitude of the releases. For accurate annual values, sampling and analysis would have to be performed more frequently, in view of the observed variations in the radionuclide concentrations of the wastes.

External radiation measurements with survey meters yielded an exposure rate above background of 1 to 3 μR/hr at the 0.3-km perimeter at Yankee, 0.7 ± 0.3 μR/hr at the nearest habitation (0.4 km distant on the west side of Sherman Reservoir), and 0.3 ± 0.3 μR/hr at Monroe Bridge (1.1 km distant). The natural radiation background at somewhat greater distances ranged from 5.7 to 8.5 μR/hr, depending on the time of year. The radiation values above background are attributed to

* Report to be published: "Radiological Surveillance Studies at a Pressurized Water Nuclear Power Reactor."

TABLE 11 Estimated Amounts of Radionuclides Discharged Annually into Effluent Circulating Coolant Water [a]

Radionuclide	Radionuclides in Liquid Effluent (Ci/yr) [b]	
	Reactor Plant	Secondary Plant
^{3}H	6×10^2	2×10^2
^{14}C	1×10^{-2}	ND
^{32}P	8×10^{-5}	ND
^{51}Cr	ND	2×10^{-2}
^{54}Mn	2×10^{-4}	8×10^{-3}
^{55}Fe	1×10^{-2}	4×10^{-4}
^{59}Fe	6×10^{-4}	8×10^{-3}
^{58}Co	4×10^{-4}	2×10^{-2}
^{60}Co	2×10^{-4}	4×10^{-3}
^{63}Ni	ND	2×10^{-3}
^{90}Sr	2×10^{-5}	4×10^{-5}
^{95}Zr	ND	6×10^{-3}
^{95}Nb	ND	4×10^{-3}
^{110m}Ag	3×10^{-4}	2×10^{-3}
^{124}Sb	8×10^{-5}	8×10^{-4}
^{131}I	2×10^{-4}	4×10^{-3}
^{137}Cs	1×10^{-4}	8×10^{-5}

[a] Statistics from "Radiological Surveillance Studies at a Pressurized Water Nuclear Power Reactor," work done at Yankee Nuclear Power Station, Rowe, Mass.
[b] ND, not detected.

direct radiation from radioactive waste stored at the Yankee Nuclear Power Station.

The radiation exposure from Yankee to persons living approximately 1 km distant was 3 ± 3 mR/yr due to direct radiation. Radiation exposure from this source to persons living at greater distances would be essentially zero because of the terrain and distance. The radiation dose from Yankee to avid fishermen and fish eaters through ingesting fish caught at the southern end of

TABLE 12 Radionuclides in Gaseous Effluent (Ci/yr) [a]

Radio-nuclide	Main Coolant Sampling	Air Ejector [b]	Surge Drum [b]	Vapor Con-tainer [b]
^{3}H	5×10^{-4}	4×10^{-3}	1×10^{-2}	13.0
^{14}C	2×10^{-3}	1×10^{-2}	1×10^{-1}	2×10^{-1}
^{41}Ar	4×10^{-1}	$<1 \times 10^{-1}$	NA	NA
^{85m}Kr	2×10^{-2}	NA	NA	NA
^{85}Kr	6×10^{-4}	$<4 \times 10^{-4}$	1×10^{-2}	3.0
^{87}Kr	2×10^{-2}	NA	NA	NA
^{88}Kr	3×10^{-2}	NA	NA	NA
^{133m}Xe	2×10^{-3}	ND	$<2 \times 10^{-4}$	NA
^{133}Xe	1×10^{-1}	4×10^{-2}	8×10^{-3}	ND
^{135}Xe	7×10^{-2}	1×10^{-1}	NA	NA

[a] Statistics from "Radiological Surveillance Studies at a Pressurized Water Nuclear Power Reactor," work done at Yankee Nuclear Power Station, Rowe, Mass.
[b] NA, not analyzed; ND, not detected.

Sherman Reservoir was 0.1 mrem/yr as inferred from effluent radioactivity data, and was considerably less on the basis of direct radionuclide analyses of fish muscle. The radiation dose from stack effluent was estimated to be 0.1 mrem/yr at the Yankee exclusion boundary. Thus, operation of the Yankee nuclear power station under the observed conditions had an extremely small impact on the radiation dose in the environment. The direct exposure rate was so far below the natural radiation background that it could not be measured with certainty, while inferred radiation doses by two other pathways were each only a fraction of 1 mrem/yr. No other exposure pathway was observed.

It could be concluded from this study that

1. No radioactivity attributed to Yankee could be observed in suspended solids, including plankton, from Sherman Reservoir. These samples were of relatively small volume, however, because the water was low in suspended solids.

2. No radioactivity attributed to Yankee was found in snow within the station perimeter at Yankee, in vegetation and soil just beyond the Yankee perimeter, in milk from a dairy at Rowe, or in deer that had died accidentally within 3 km of Yankee.

3. Computations based on measured effluent concentrations and a simple model of dispersion in air indicated that radionuclide concentrations in air and on the ground near Yankee were so low that they could not be detected with the available sample volumes and analytical procedures.

The estimated radiation doses to individuals in the vicinity of the Yankee Nuclear Power Plant are in the same magnitude and comparable to the calculated radiation for rates to a hypothetical individual in the vicinity of the Calvert Cliffs Nuclear Power Station, as discussed by Goldman, Calley, et al. (1971). Data from the Preliminary Safety Analysis Report (PSAR) have been used to calculate radiation dose rates to a hypothetical individual from the principal exposure pathways at the Calvert Cliffs Nuclear Power Station [825-MW(e) PWR—2 units]. As the design of the plant changes, the radiation dose estimates may also change. Both the gaseous emissions and the liquid discharges were evaluated in terms of whole-body dose—which is also the genetically significant dose—and thus these doses can be added directly. The totals are shown in Table 13. Under the most pessimistic conditions, the maximum population whole-body dose within a 50-mile radius resulting from both gaseous and liquid emissions is 453 man-rads per year, and the average annual per capita dose within a 50-mile radius from both sources is 0.000154 rads per year, less than one-thousandth of the natural background radiation level.

TABLE 13 Total Radiation Dose from Gaseous and Liquid Releases [a, b]

Dose	Maximum Dose [c]	Highest Expected Dose [d]
Individual Dose		
Dose rate (rads/year) at site boundary	0.0135	0.000337
Average dose rate (rads/year) per person eating 120 lbs of seafood per year	0.000989	0.000148
Natural background radiation dose (rads/year)	~0.125	~0.125
Population Dose		
Total man-rads/year within 50-mile radius	453.0	44.7
Total man-rads/year from natural background	300,000– 594,000 [e]	300,000– 594,000 [e]
Annual average per capita dose within 50-mile radius (rads/year) [e]	0.000154	0.0000182

[a] Statistics from Goldstein, Sun, and Gonzalez (1971).
[b] Based on 80% load factor for two units, or 11,200,000 MW-hr per year.
[c] Maximum dose assumes no dilution in Bay water, no decay, and 1% fuel defects in both reactors.
[d] Highest expected dose; same as "maximum," except for dilution in summer or drought freshwater flow of 35,000 cfs.
[e] Based on 1965 population of 2,372,000 to maximize per capita dose.

The maximum individual dose would occur to a hypothetical individual who lived on the site boundary and ate seafood grown in the cooling water discharge canal. Using the least favorable operating experience as a basis for predicting Calvert Cliffs gas release performance and the freshwater flow dilution factor for liquids under drought conditions, this individual could receive a total whole-body dose of 0.000337 rads per year, about 3/1,000 of natural background radiation, and 7/10,000 of the criteria specified by national and international organizations.

The total radiation doses to which individuals and the general population within a 50-mile radius of the plant would be exposed are also listed in Table 13. The table has two columns labeled "maximum" and "highest expected." All of these are annual body doses in addition to natural background radiation, which is also included to provide a basis for comparison.

RADIOACTIVITY FROM FOSSIL-FUEL AND NUCLEAR POWER PLANTS *

Trace quantities of uranium and thorium and their products of radioactive decay are released in fly ash from large, fossil-fuel steam electric stations, raising the question of the significance of these releases compared with those from nuclear power plants. A study to evaluate the radioactivity discharge from fossil-fuel plants compared to nuclear plants was conducted in 1967–1968 by the Eastern Environmental Radiation Laboratory of the Environmental Protection Agency (formerly the Bureau of Radiological Health's Southeastern Radiological Health Laboratory). This study showed that the radioactive material discharged to the environment from a fossil-fuel plant is not a form that is readily transferable to man. However, the radioactive material discharged from a nuclear power plant is in a form that can result in an external exposure and is transferable to man via the food chain. Comparisons were also made between coal-fired and nuclear power plants of current design to show the effects modern technology has on the relative radiological significance of each. The long-range effects of power plant fuel use are considered relative to the buildup of ^{85}Kr in the atmosphere and the release of carbon (except ^{14}C) from fossil-fuel plants. It was concluded that nuclear power reactors over the long term represent a greater overall radiological burden on the environment than fossil-fuel plants, although all are well below radiation protection guides established by the Federal Radiation Council.

* Martin *et al.* (1969); Goldstein, Sun, and Gonzalez (1971).

5 FUEL CYCLE CONSIDERATIONS

During the past few years, considerable attention has been focused on the environmental effects of electric-power production. The releases of even low levels of radioactivity from commercial nuclear power stations are criticized by a number of environmentalists and conservationists. While these effluents are discharged at only small fractions of radiation protection standards, it is correctly noted that the nuclear power plant is only one step in an overall nuclear fuel cycle. To fully evaluate the environmental impact of nuclear-power production, one should examine the total fuel cycle, beginning with the mining and milling of uranium ore and proceeding to the final disposal of various residues from different parts of the cycle.

The Working Group has explored in detail only that part of the fuel cycle related to the transportation and disposal of high-level wastes which are considered more important from a radiological standpoint.

PLUTONIUM RECYCLE

Recycling of plutonium presents two potentially important considerations—the safeguarding of plutonium from unauthorized diversion and the toxic nature of this radioactive element. At every step of plutonium handling, automatic protection systems and administrative procedures are employed to preclude the loss of plutonium to the environment. Although pure plutonium metal is subject to rapid oxidation at elevated temperature, the oxides and most other chemical compounds of plutonium found in the fuel cycle are relatively unreactive.

Because of its potential use in explosive devices, plutonium safeguard controls have been implemented through AEC regulations (10CFR50, 1968; Wischow, 1971) published in Title 10 of the Code of Federal Regulations. These regulations require that special nuclear material activity be limited to the quantity, purpose, and locations authorized in the license issued by the AEC.

The AEC, from time to time, conducts audits of the plutonium inventories to verify that the facility's safeguard program is effective.

TRANSPORTATION

The nuclear fuel cycle involves the transportation of wide varieties of radioactive substances in many different forms, shapes, and quantities. The radiation safety associated with shipment of these materials is provided for in regulations of the AEC and the Department of Transportation. Transportation requirements are detailed in Appendix A of this report.

DISPOSAL OF HIGH-LEVEL WASTES

The radioactive fission products contained in the spent fuel, after separation of the uranium and plutonium, are prepared for transportation and storage in a permanent repository. Several methods for containing and storing these high-level wastes are presented in Appendix B of this report.

189

FUEL REPROCESSING

The fuel in reactors becomes progressively depleted of fissionable material and increasingly contaminated with fission products. It therefore needs to be recycled to recover the unused fissionable material produced in the reactor and to remove the fission products. In the course of this reprocessing, highly radioactive fission products are separated chemically; these must be stored safely for long periods of time. Operational experience at a privately owned plant is limited to one licensed facility (Nuclear Fuel Services, Inc., West Valley,

N.Y.). The degree of containment of radionuclides and distribution in the environment is not presently as well established as in the case of nuclear power plants. The location of the fuel reprocessing plant, in terms of its radioactive effluents, may therefore be significant in relation to other sources, such as nuclear power reactors or the ultimate disposal site.

Figure 7 (representing a 1-year fuel cycle) shows the estimated skin surface dose from unrestricted discharge of ^{85}Kr to the year 2060.* Present regulations

* Assumption made that there would be no change from the present in krypton-85 waste treatment to the year 2060.

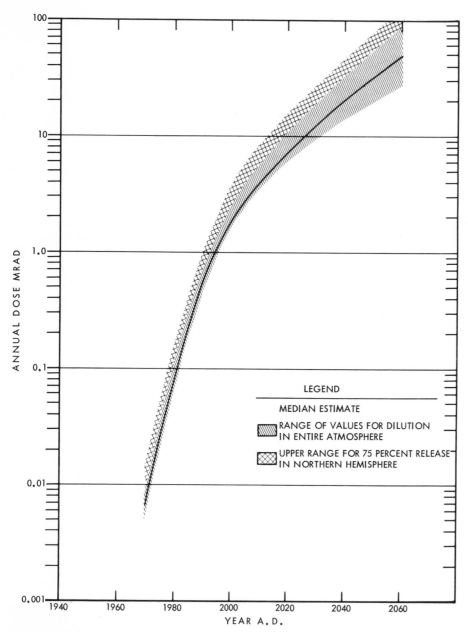

FIGURE 7 Estimated skin surface dose from unrestricted release of krypton-85, 1970–2060.

permit discharge of ^{3}H and ^{85}Kr into the atmosphere, although processes are available for trapping of the ^{85}Kr (Coleman and Liberace, 1966).

Spent fuel reprocessing, at the present time, represents the largest potential source of radioisotope emissions to the environment. While current and near-future releases from fuel reprocessing plants do not appear to constitute a public health problem, significant reductions of releases of certain long-lived isotopes, especially krypton, may be necessary. Processes such as cryogenic distillation or freon absorption permit such reductions, if they become necessary.

Presently operating fuel reprocessing plants make use of organic–aqueous solvent extraction processes to separate valuable fissionable and fertile materials (uranium, plutonium, and thorium) from mixtures of fission products and inert materials. Figure 8 provides a simplified flow diagram of an idealized waste-handling operation for gaseous, liquid, and solid wastes resulting from the reprocessing of spent reactor fuels. These wastes, and radioactive particulates from the process, are removed from the plant off-gas as required, before discharge by sorption, chemical interactions, and filtration. Liquid waste streams, which contain over 99% of the radioactivity, are combined and solidified by any of several processes which have been developed for this purpose. The water and acid from these processes will be further purified by conventional treatment until they can be either recycled or safely released to the environment. On-site storage as liquids, solids, or both permits decay of short- and intermediate-lived nuclides and reduces the heat dissipation problem. Solidified high-level wastes will eventually be shipped to a national repository for permanent storage.

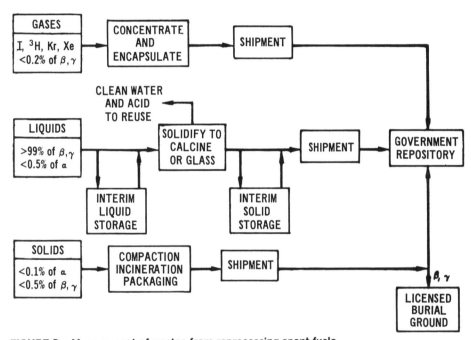

FIGURE 8   Management of wastes from reprocessing spent fuels.

6

NEAR-FUTURE DEVELOPMENTS IN TREATMENT OF RADIOACTIVE WASTES

Many technical advances are being made to improve waste-management procedures for nuclear power stations. Many of these techniques have been developed to the stage where prototypes are being installed in new nuclear power stations, or have been included in the design of proposed stations. Other techniques are still undergoing research and development.

TECHNIQUES FOR TREATING GASEOUS WASTES

An activated charcoal component is a special case, a device which incorporates the functions of both filtration and holdup. Commercially available charcoal filters have a very high (greater than 99%) removal efficiency for the iodine radioisotopes. Charcoal also adsorbs noble gases and retains them for varying lengths of time, thereby providing holdup time to allow radioactive decay to particulate daughters which are retained in the charcoal element which acts as a filter. If there is enough filter material of proper configuration, the efficiency of noble-gas removal can be as high as desired (except for ^{85}Kr, which has a half-life of 10.4 years).

For maximum adsorption effectiveness, humidity control must be accomplished in the gas stream to the adsorber unit. Although the trapping efficiency for noble gases increases as the temperature of the adsorber is reduced, a reduced temperature system must provide adequate safety provisions in the event of loss of refrigeration. This would include means for stopping the flow of gas and provision of sufficient storage capacity to hold the gas as the adsorber warms to room temperature.

Permselective membranes have been investigated for the removal of krypton and xenon from other gases. The permeabilities of a dimethyl silicone rubber membrane have been measured (ORNL, 1967), and a laboratory-scale unit has operated satisfactorily. Design calculations for a full-size unit plant, assuming the full-size unit will have the same operating characteristics as the small-scale unit, indicate a decontamination factor (DF) of 2×10^4. The development has been closed out because of marginal feasibility and doubtful economics compared to other alternatives.

Cryogenic enrichment is a simple concept for increasing the efficiency and decreasing the size required of a solid adsorber bed. A cryogenic stripping column is used to increase the concentration of noble gases from a few ppm to about 80 ppm. This reduces the adsorber flow rate by a factor of approximately 10, with a corresponding decrease in the volume of the adsorber bed.

The off-gas is compressed to about 3 atmospheres, cooled to $+40°F$ to remove most of the water vapor, and then cooled to $-290°F$. Entering the bottom of a packed column, the cooled gas passes upward countercurrent to a descending stream of liquid nitrogen. This condenses the relatively high boiling-temperature krypton and xenon, and concentrates them in a comparatively small oxygen-rich fraction in the bottom of the column. This liquid is vaporized and passed through the adsorber bed, where the noble gases are adsorbed. Tests with the cryogenic enrichment process show a decontamination factor of 714 (ORNL, 1963).

Much operating experience and design information exists for commercial processes of cryogenic distillation,

192

such as the distillation of krypton and xenon from air. The Rare Gas Recovery Facility at the Idaho Chemical Processing Plant uses the cryogenic distillation process to recover xenon and krypton from spent fuel elements (Offutt and Bendixsen, 1969). The gaseous effluent from this plant is compressed to 30 psig, precooled to $+40°F$, the water condensed and removed by a mechanical separator, then cooled further until the remaining water vapor and carbon dioxide are frozen out. The air is further cooled to $-290°F$ before it enters the distillation column. Upon leaving the distillation column, the liquid enriched in the noble gases is vaporized, compressed, and stored in cylinders.

The selective absorption of krypton and xenon in a fluorocarbon solvent is a direct method for stripping krypton and xenon from contaminated off-gas streams. This method is possible because krypton and xenon are much more soluble than argon, oxygen, and nitrogen in the fluorocarbon solvent. In this continuous absorption process, krypton and xenon are preferentially removed from an incoming gas stream by selective dissolution in a fluorocarbon solvent at relatively low temperature and high pressure. This is followed by one or more fractionating and stripping operations which yield a product gas stream rich in the noble gases and an essentially pure solvent stream which can be recycled.

The overall process is composed of three basic steps: (1) absorption, (2) fractionation, and (3) stripping. The absorption system consists of a single packed column, with heat exchangers, solvent pump, and gas compressor. The initial primary separation of gas constituents takes place in the absorber section. The contaminated gas is compressed and cooled to the desired working pressure (300 psia to 500 psia) and temperature ($-30°F$ to $+30°F$) of the absorber column. The cooled, compressed gas enters the bottom of the absorber column where it comes in contact with the downward-flowing solvent. The xenon and krypton, plus small amounts of oxygen, nitrogen, and argon, are dissolved by the solvent. The liquid solvent, enriched in the contaminants, leaves from the bottom of the absorber and is fed to the fractionator flash unit.

The fractionator system includes a packed column, a reboiler, an overhead condenser system and a flash drum. The fractionator's function is to enrich the gases (xenon and krypton) dissolved in the liquid solvent by removing undesired contaminants, such as oxygen, nitrogen, and argon. The fractionator is operated at a lower pressure (50 psia to 100 psia) and higher temperature (140°F to 190°F) than the absorber. Thus, part of the solvent is vaporized as the solvent enters the flash unit. The resulting solvent vapor and gas mixture passes into the overhead condensers for recycling, while the rest of the solvent—still liquid—is fed into the top of the fractionator packed column. The upflowing solvent vapor generated in the reboiler comes in contact with the downflowing solvent liquid. The solvent vapor and gas mixture leaving the top of the column is condensed and drained back into the column.

Most of the undesired contaminants are removed in the fractionation step. The solvent, now enriched in krypton and xenon, is pumped to the stripping section for noble gas-product concentration and solvent purification. The stripping section contains the same types of components as the fractionating section. It differs in size and operating temperature range (90°F to 120°F) and pressure range (20 psia to 30 psia). The stripper section operation is identical to that previously described for the fractionator. However, because of the lower pressure and temperature of the stripper section operation, the remaining absorbed gases are given off by the solvent. The pure solvent flows from the stripper into a liquid storage tank to be reused as needed. The product, enriched in krypton and xenon, is collected at the top of the stripper. Four years of testing and pilot plant studies have been conducted at the Oak Ridge Gaseous Diffusion Plant on the selective absorption of xenon and krypton in a fluorocarbon solvent. Decontamination factors as high as 1,000 have been measured for streams simulating fuel recovery plant off-gases, using stable krypton and xenon (Stephenson, Merriman, and Dunthorn, 1970). Tracer studies are currently being conducted to optimize this process between economics and performance.

The results of the studies and tests indicate that this system could be used directly in BWR condenser off-gas systems, although tests are currently under way to verify this on a simulated BWR off-gas system. A simple scale-up of the pilot plant is required. The application of this system to a PWR is not as direct, since removal of krypton and xenon from off-gas streams might involve separation from hydrogen or nitrogen.

Catalytic recombination of radiolytic oxygen and hydrogen in BWR off-gas streams appears to be useful for reducing the discharge of radioactive gases. As generated, the off-gas from a BWR is an explosive mixture of oxygen and hydrogen. The percent of hydrogen in the total gas volume must be reduced to less than 4% in order to avoid possible explosion. Since hydrogen makes up about 60% of the total volume of the off-gas, the volume of diluent gas is of necessity large. Steam can be used as a diluent, and is easily condensed out after the recombination takes place, thus removing the diluent as required.

The recombiners convert the free hydrogen and oxygen in water, in the presence of a catalyst. The water is condensed from the gas stream along with the diluent steam, with a consequent reduction in the typical BWR off-gas volume of about 90%. The waste-treatment system for the remaining gas may thus be considerably simplified by reduction in the component sizes.

TECHNIQUES FOR TREATING LIQUID WASTES

Several chemical separation processes have been developed for treating liquid wastes. Decontamination factors of about 10 to as high as 200 have been reported for specific isotopes and processes (Pottier, 1968). The effluents from nuclear power stations tend to be mixed fission and corrosion products with varying chemical properties. Thus, various treatment techniques will be required. Attention must be directed to those nuclides which need the highest DF's to reduce their activity to an acceptable level. Because of varying biological import, an isotope of significantly lower concentration may require a high DF, further complicating the task of radionuclide treatment.

Chemical separation is accomplished by adding chemicals which react with specific elements in the waste to form a product which may be separated by physical methods (e.g., filtration). Since no single chemical separation technique can remove all isotopes with the same efficiency, chemical processes may have to be used in conjunction with other treatment methods, such as ion exchange or foam fractionation, to increase the overall DF.

Gas stripping of liquid-waste streams is a standard chemical process which has been used for many years; however, no specific data are available concerning the effectiveness of the operation. Design data on gas stripping units to be installed are generally not available, because of the proprietary nature of the information.

ADVANCED WASTE TREATMENT SYSTEM PROPOSED FOR PWR'S

The philosophy used in the design of this system is to concentrate the liquid and gaseous radioactive wastes into manageable quantities, contain them for extended periods within closed systems, and ultimately ship them from the plant for permanent offsite storage instead of diluting and dispersing them to the environment. The system is further based upon segregation of waste drains and recycling of processed liquids to the Reactor Coolant System (RCS). Liquid radioactive wastes are kept separate from nonradioactive liquid wastes. The latter are collected, filtered, monitored, and then released to the condenser circulating-waste discharge system. By recycling the processed liquids to the RCS, the re-

sulting small volumes of solid and gaseous wastes can be put in containers and disposed of offsite. One should point out, however, that commerical burial grounds that handle solid wastes do not accept gaseous wastes for disposal.

The major factor in the growth of the system has been the technological development of a method of boron removal from the reactor coolant. By using an ion-exchange process in place of a gas stripper/evaporator train to achieve the RCS boron concentration, making changes required for daily load-follow operation, radioactive and other gases remain in solution in the RCS. The RCS is continually degassed as a low flow rate (~1 scfm *) to control the buildup of reactor coolant system noble gases. The system is degassed to remove 90% of the operating gas levels and then depressurized prior to refueling or major maintenance. Consequently, the volume of radioactive gases that must be discharged during refueling is significantly reduced. It is claimed that the bottling of the stripped radioactive gases for onsite storage or offsite disposal is feasible, although the method of offsite disposal to be used has not yet been specified.

The system consists of three subsystems: (1) the boron thermal-regeneration subsystem, (2) the waste-liquid handling subsystem, and (3) the waste-gas handling subsystem. The boron thermal-regeneration subsystem uses ion-exchange resins to provide the necessary boron concentration changes. It is based upon the fundamental principles that ion-exchange resins will increase their capacity for boron retention with decreasing fluid temperature, and that the process is reversible. By regulating the temperature of reactor coolant flowing through the ion-exchanger, a rapid means of changing boron concentration is provided.

The waste-liquid handling subsystem is arranged in two loops. The first reprocesses and retains waste liquid containing potential sources of high radioactivity. The second loop collects, processes, monitors, and discharges wastes expected to be nonradioactive. Radioactive liquids are evaporated and the condensate returned to the RCS. The evaporator concentrates are either reused as boric acid or drummed for offsite disposal. Thus, no radioactive liquid wastes from this source are intentionally discharged.

The waste-gas handling subsystem is designed to separate all radioactive particulates or vapors from the radioactive noble gases, and to retain the resulting noble gases within closed plant systems. It is also designed to reduce the total gas volume handled to a minimum, and to store the resulting radioactive noble gases onsite until only the long-lived ^{85}Kr remains. The subsystem receives gases from the recycle evaporator, equipment

---

* Standard cubic foot per minute.

vents, and from the volume-control tank during degassing of the RCS prior to refueling or a major maintenance shutdown.

It is claimed that this system will reduce the gaseous discharge from an 1,100-MW(e) PWR plant during normal power operation from 4,150 Ci/yr for a standard plant to essentially zero. Liquid releases supposedly will be reduced from 40 to 1 Ci/yr, and tritium from 385 Ci/yr to essentially zero.

However, it should be noted that achieving this performance depends on three very important factors: (1) tritium releases from the fuel must be very low to prevent in-plant hazards to personnel; (2) there must be no steam generator-tube leaks, a situation which has yet to be achieved in operating PWR's; and (3) gaseous wastes from containment purge and ventilation must be controlled.

This system will cause an increase in the quantity of solid radioactive waste shipped offsite for disposal by a total of approximately *twenty-five* 55-gallon drums per year. For plants not using the system, the usual solid-waste shipments are *100 to 175 55-gallon drums per year*.

In general, this liquid handling system appears to accomplish little more than current systems—if they were used to the fullest extent possible—except that the boron ion exchanger permits more economic recycling of the waste water back into the facility. Although discharges of tritium are nearly zero from the facility site if primary coolant system leaks do not materialize, a long-range problem which has not been solved is the ultimate disposal of tritium-laden water.

ADVANCED WASTE TREATMENT SYSTEM PROPOSED FOR BWR'S

A system has been developed to reduce the amount of gaseous radioactivity discharged from BWR's. In this system, noncondensible gas from the main condenser is diluted with steam, superheated, and then passed through a catalytic recombiner to convert the hydrogen and oxygen into water, in order to reduce the volume of gases to be handled. The bulk moisture is removed from this effluent in a condenser cooled by reactor feedwater. After a 30-minute holdup, particulate decay daughters condense on the walls of the pipe and are subsequently removed by filtration. The offgas is further processed by a cooler–condenser to remove additional moisture, a de-entrainer and reheater to reduce the relative humidity, and a high-efficiency filter prior to entering charcoal adsorbers. The charcoal adsorbers provide further delay of the offgas, to the extent that discharged radioactivity, except for ^{85}Kr, may be reduced to a specified level. Heat is removed from the vault housing the adsorbers in order to maintain the charcoal beds at an approximately constant temperature. The offgas effluent from the adsorbers is passed through another high-efficiency filter prior to release or treatment by the Kr–Xe bottling system.

A Kr–Xe bottling system provides a means for concentrating and bottling xenon and krypton that otherwise would be released to the atmosphere, making the amounts discharged to the atmosphere negligible. In this system, the gases are adsorbed on a silica bed; the kryptons and xenons are separated by cryogenic distillation. A compressor is then used to compress these radioactive gases into high-pressure gas-storage bottles.

An alternative system has been proposed for several BWR nuclear power stations. In this system, the effluent gas, after passing through the recombiner and condenser, is held in storage tanks long enough for much of the radioactivity to decay. Times on the order of several days to weeks are being considered. These systems are estimated by the designers to result in DF's of the discharged gases of 25 to 600, depending on the amount of holdup time, compared with the discharges from the 30-minute delay system used in current BWR designs.

The need for the packaging of the final gaseous effluents from the advanced gaseous treatment systems of both BWR's and PWR's is debatable. After the long delay provided by these systems, there remain only about 500 curies of ^{85}Kr annually from each reactor. This amount is negligible in terms of radiation exposures resulting from its release. However, there is a need to reduce the quantities of ^{85}Kr from fuel reprocessing plants, since these facilities release several thousand curies per day.

7   CONCLUSIONS AND RECOMMENDATIONS

Using nuclear power to generate electricity has meant development of a whole new technology. This new technology may be accompanied by some form of risk—either of damage to property, injury to people, or possible genetic consequences from radiation exposure.

Two radiological considerations apply to nuclear facility siting: the first dealing with the radiological implications of normal operation, the second with the consequences of accidental radioactivity releases. In reviewing the conditions governing site selection for nuclear power plants, the Working Group feels that the primary constraint is exercised by the latter.

Experience has demonstrated that present technology is adequate to maintain radiation doses in the vicinity of nuclear power plants—on the average—to small percentages of natural background radiation. Releases from operating reactors have, in fact, been very low: for the most part, less than 1% of maximum allowable under radiation protection standards.

Technology for both preventing accidents and mitigating their consequences, in terms of control of radioactivity release under the assumptions used by the AEC for site evaluation, has been more difficult to demonstrate. Such accidents as have occurred have been well within the capability of plants to safely sustain abnormalities. It is the accident considerations that lead to the primary constraints on the siting of nuclear power plants. From the technological standpoint, greater flexibility in nuclear plant siting could be achieved only by some substantial change in either the present approach to evaluation used by the AEC or in present engineering concepts for prevention or mitigation of accidents. The

Working Group feels that such changes are sufficiently unlikely in the near future that major modifications in present U.S. siting practices cannot be reasonably anticipated.

In the area of waste-control technology, large-scale demonstration of the safety and feasibility of long-term storage of high-activity fuel-recovery wastes is by far the most crucial aspect of continued development of nuclear power. Because of the long-term environmental commitment, the public requires reassurance as to the feasibility and safety of burial in salt formations prior to accepting large-scale operations.

Oak Ridge studies (ORNL-4451) indicate that ^{85}Kr dose rates may increase man's exposure by the year 2000 by approximately 1 mrem/yr. Public Health Service studies (as reported in Chapter 6 of this report) indicate that the estimated skin surface dose due to ^{85}Kr may be as much as 10 mrem/yr by the year 2020. These projections of ^{85}Kr discharges from fuel reprocessing plants indicate that, for the long term, significant exposure may result in the northern hemisphere. Because of international implications, the ^{85}Kr will be present throughout the earth's atmosphere; therefore, it will probably be desirable to incorporate capture processes before the atmospheric inventory of this isotope has been substantially increased.

No methods for tritium control are currently available for incorporation into fuel reprocessing plants. Although projections of population dose from this isotope indicate generally small contributions (fractions of an mrem/yr) in comparison with natural radiation exposure, the Working Group feels the ubiquitous nature of hydrogen merits continued surveillance of tritium releases to the biosphere.

The present procedures for setting radiation protection standards for public health and safety have been questioned by some members of the public, who advocate the broadening of the "weighting or balancing groups" beyond the scientific and engineering community. While current procedures do allow for open public comments in writing on proposed radiation protection standards, the decision making is not preceded by public hearings. Rather, the question of the adequacy of radiation standards has frequently become a key point of issue at the construction-permit licensing hearing for nuclear facility projects. As a means, perhaps, of eliminating detailed discussion of radiation protection standards at the construction-permit stage, the present rule-making procedure could be modified to require public hearings on the rules themselves. Such hearings might significantly help through providing for open public debate on radiation protection standard issues, and thereby instill more public confidence in present standards and their application to nuclear facility operations.

The engineering outlook for environmental systems is such that the curies of radioactive material discharged in the liquid and gaseous waste can be reduced, and result in a minimum activity release system. These systems, designed on the basis of the lowest practicable level concepts, can be expected to be effective in minimizing population exposure.

Mounting public awareness of the projected growth of power and the potential health risks entailed, however small, makes public acceptance as crucial an aspect in siting as technology itself. In this connection, the Working Group believes that the engineering community at large could make a significant contribution to improve the siting of nuclear power stations by establishing a Nuclear Power Safety Committee to conduct a thorough review of current nuclear power plant safety-design criteria, standards, potential reactor accidents, emergency core-cooling systems, and operating practices. An independent engineering review and analysis—conducted under the direction of the National Academy of Engineering by senior members of the profession with substantial experience in design and operation of similar complexity and safety significance —might do much not only to provide a fresh perspective of currently accepted approaches in the nuclear industry but also to instill added public confidence in the safety aspects of plant design.

The critical pathway for a possible exposure of the population from a BWR is via the atmosphere through discharges of noble gases; and from a PWR, via the discharges of liquid radioactive waste.

Medical uses are currently estimated to result in a genetically significant per capita dose in the range of 100 mrem per year. A reduction in per capita exposure from medical uses of only 1% or 1 mrem would more than compensate for potential nuclear facility exposure (less than 0.01 mrem per capita year), while total elimination of nuclear facilities would not materially change the per capita gonadal exposure and resulting genetic burden in the United States. Thus, stringent application of technology, education, and control in the area of medical radiation uses is indicated.

APPENDIX A: TRANSPORTATION REQUIREMENTS

TYPES AND QUANTITIES

The nuclear industry involves the transport of a wide variety of radioactive substances in many different forms, shapes, and quantities. Regulations govern such shipments (AEC, 1966; 10CFR71). Shipments are reported (Beck, 1969) to be now in the neighborhood of 250,000 to 300,000 per year. By far the largest portion of such shipments involve small-quantity, low-level radioactive materials of activation (artificially radioactive) products, or separated fission-product radionuclides, radioactive sources, and assorted other materials. These materials are being used in thousands of institutions for medical diagnostic and therapeutic research, and for agricultural and industrial purposes. Of most interest to this study is the shipment of the enriched uranium or plutonium raw material to fuel fabrication plants for manufacture into fuel elements, the shipment of the spent fuel, and the transport of wastes from fuel reprocessing.

SPECIAL HAZARDS INVOLVED IN FUEL, WASTE, AND BY-PRODUCT SHIPMENTS

Now about the characteristics of these materials of significance to transportation:

1. Fuel or fuel elements or enriched raw materials for fabrication into fuel elements.

a. These materials are radioactive, but outside the body, there is no significant radiation or safety problem from this characteristic. They usually can be handled without radiation shielding.

b. Radiation exposure problems could result from transportation accidents or other situations which would disperse such materials under conditions which would permit them to be ingested into the body. Leakage of such materials in liquid or powder form (which could contaminate other materials) could lead to the possibility of ingestion and, hence, could lead to radiation exposure problems.

c. These fuel materials, either raw or in processed fuel element form, can become critical. That is, a sustained fissioning chain reaction can occur if the fuel in sufficient quantity is brought together in an appropriate configuration. Within a reactor, such a configuration is purposely contrived, and, with the aid of control rods, protective shielding, and other devices, the fissioning reactor is controlled at the desired rate with the release of radiation, heat, and fission products. This can also happen, accidentally and inadvertently, outside a reactor assembly. In this case, radiation, heat, and fission products would be generated, but without the benefit of control rods and protective shielding; a real danger to people nearby could result.

No accidental criticality has occurred in transportation activities. Criticality would occur if too much fuel were assembled into one mass, or if a sufficient number of individually safe masses were brought too close together. It is prevented by maintaining the quantity of fuel in any one container below specified limits and maintaining the required separation between individual quantities. If one of these limits is not satisfied and inadvertent criticality does occur, the chain reaction would cause its own termination. Some steam pressure could build up if liquids were involved, but there could be no nuclear explosion like that of a bomb—that would be a physical impossibility. Further, there is little likelihood of enough fission products being formed to present much of a residual radiation hazard.

The enrichment of the nuclear fuel—that is, the amount of the fissionable ^{235}U isotope in the fuel over that contained in natural uranium—is also an important factor in determining the possibility of criticality. Most nuclear fuels are of low enrichment, 2–3% of the total fuel; thus large quantities would be required to produce accidental criticality, on the order of hundreds of pounds or more. High-enrichment fuels can become critical with a mass of only a pound or so. Therefore, highly enriched fuels are handled strictly in small individual quantities, and need to be kept properly separated from each other.

2. Spent fuels containing fission products. These would normally only be transported from a nuclear reactor to a chemical reprocessing plant.

a. Such fuels are highly radioactive and give off much heat due to radioactivity. They require massive shielding to absorb the radiation and cooling system to prevent the fuel from overheating. Because these materials need to be shipped inside massively shielded containers (called casks) and for economy, as much fuel as possible is included in each shipment [Shappert *et al.,* 1970; Southern Interstate Nuclear Board (SINB), 1970]. A load of spent fuel casks may weigh about 3 tons.

b. In shipment of such materials, the only substantial hazard arises from the possibility that some of the fuel elements or the radioactivity contained in the fuel elements within the cask could be released accidentally outside the container. Radiation exposure, radioactive contamination, and radioactive material ingestion could follow, depending on how much is released and in what form. Such release could conceivably come from accidents that damage the container, impair the shielding, inactivate the cooling system, or cause leakage of radioactive liquids or gases. Accident probabilities have been extensively studied (Heinisch, n.d.; Leimkuhler, 1961, 1963; Stewart, 1963; Kelly, in press). Container design requirements and shipping regulations reflect transport accident considerations.

3. Miscellaneous radioactive materials in general, including the fresh and spent fuels mentioned above, are divided into several different categories for transportation purposes.

a. Exempt quantities of low-activity radioactive materials: Materials in this category, though radioactive, present such low hazard from radioactivity that no significant danger to health and safety could arise therefrom. Such materials may be transported without any special requirements as to packaging, procedures, or labeling. They are "exempt."

b. Small quantities of radioactive or fissile materials: In this category, Type A, the quantity and nature of the radioactivity needs to be so low that no danger arises under the normal conditions of transport, and little would arise even if the package were damaged and the contents dispersed. For such materials, the packages must withstand the "normal transport" test conditions. Specific limits are placed on the quantity and activity of the contents, and the package must carry appropriate labels as to the contents.

c. Intermediate quantity fissile or radioactive materials: In this category, Type B, the quantity of radioactivity can range up to 20 curies for more dangerous isotopes, 200 curies for less dangerous isotopes, and up to 5,000 curies of "special form," e.g., encapsulated sources. Transportation packages in this category need to withstand "hypothetical accident" conditions, carry labels that state the identity, quantity, and nature of the contents, and must carry clearly marked radiation unit values if required.

d. Large quantity shipments are all those above the intermediate range. Here the quantity or intensity of radioactivity, the nature of the materials involved, or the quantity or arrangement of the fissile materials require special handling or unique nonstandard arrangements.

Thus, to summarize, for transportation purposes, materials are divided into several categories with respect to the degree and hazards involved. The nuclear fuel, waste, and by-product materials each, to one degree or another, give rise to potential hazards of the following types:

1. Radiation from radioactive materials which could lead to external exposure of people.

2. Internal radiation exposure from radioactive materials, and poisoning from their chemical toxicity if ingested.

3. Radioactive contamination if materials are released and dispersed.

4. For the large quantity shipments, thermal effects resulting from radioactivity. These would require cooling precautions.

5. For fissile materials, inadvertent criticality would impose limitation on net quantities and separation between quantities.

CONCEPTS AND PRINCIPLES RELATING TO SAFETY IN TRANSPORTATION

In the transportation of all types of hazardous materials, there is a vast difference between potential hazard and realized damage. For such materials a system of protection is designed to prevent the potential danger from becoming reality. To protect health and safety during the transportation of special nuclear and nuclear by-product materials, a highly developed system of protection has evolved. Major features of this system include the following:

1. Primary reliance for safety is placed on the container in which the material is shipped. That is, clear definition is made of the characteristics a container needs to possess and of the contents that may be shipped therein. For example, if in normal (undamaged) condition of the container and its contents, hazard might arise from the contents (e.g., radiation, heat), the container design should protect against these hazards (e.g., shielding, cooling). If hazards would result from accidental damage to packages in transit, the packages should be built and tested to standards that would prevent release of the contents under "hypothetical accident" conditions.

These principles perhaps are already explained sufficiently. For exempt quantities of materials, where the degree of hazard is so low as to be of no concern, there are no special requirements as to packaging.

For shipment of Type A intermediate level materials, where little hazard would result if the materials were accidentally released, containers must be built to prevent release of the contents under conditions of "normal transport." These conditions are temperatures of 130°F to −40°F, reduced pressure to ½ atmosphere; vibrations normally incident to transport; a water spray heavy enough to keep the entire exposed surface except the bottom wet for 30 minutes. Other conditions are specified, including a free drop through a distance of 4 feet onto a flat, unyielding horizontal surface 1½–2½ hours after the water test; a corner drop onto each corner of the package in succession, or, for a cylindrical package, onto each quarter of each rim from a height of 1 foot; impact of the hemispherical end of a vertical steel cylinder 1¼ inch in diameter and weighing 13 pounds dropped from a height of 40 inches; and for packages less than 10,000 pounds in weight, a compressive load equal to either 5 times the weight of the package or 2 psi multiplied by the maximum horizontal cross section of the package applied during a period of 24 hours uniformly against the top and bottom of the package. Packaging for liquids must also meet additional provisions of either a 30-foot free fall drop

test without leakage, or be surrounded with enough absorbent material to absorb twice the volume of liquid.

For Type B quantities (fissile materials and large quantities where accidental release of the contents could be hazardous) the containers must be designed, constructed, and tested to prevent release of the contents and loss of shielding beyond specified limits, under "hypothetical accident" conditions. These conditions, which should be considered in sequence, are: 30-foot drop of the package in its most vulnerable position onto a hard flat surface; 40-inch drop onto a 6-inch diameter steel "spike"; a fire at 1475°F for 30 minutes, followed by immersion in water.

2. Where safety depends on restricting the contents of a container to defined limits—which is frequently the case for radioactive materials and almost always the case with nuclear fuel materials generally—the dimensions or capacity of the container are so designed that it is difficult or impossible to load beyond the limits. Further, where safety in transit or storage (i.e., against criticality) depends on maintaining a certain spacing between individual packages, the packages are built with rugged extended perimeters, commonly called "bird cages," which insure that this spacing will be maintained.

3. Explicit and definitive specifications are published in the regulations concerning the required characteristics of containers and the quantities and characteristics of materials that may be shipped in them. When containers of a given design are built, tested, and found to be adequate for transportation of certain materials, additional containers of the same design are acceptable for transport of the specified materials. Any authorized shipper may then use such containers for transport of the specified materials. A nonspecification container must be separately evaluated and approved for each type of proposed use by the Department of Transportation.

When "specification containers" are used or approval of a container is obtained, the shipper may prepare and load the container as specified with the materials authorized, and deliver the package to any public carrier for transport. The responsibility rests fully on the shipper to adhere to all published regulations and requirements in the preparation of the shipment. All shippers (and receivers) are AEC licensees (or Agreement States, i.e., the twenty-three states which have an agreement with the AEC to regulate radioactive materials within their boundaries). All common and contract carriers are exempt from AEC regulations but are subject to the transport regulations of DOT.

Substantial safety functions are accomplished by the labels on the containers. Hence, explicit specifications on container labeling are included in the regulatory requirements. Labels must be attached (on all packages other than those in the exempt category) giving notice that hazardous radioactive materials are contained, designating the class or category of the shipment, specifying the identity and quantity of the radioisotopes in the container, and assigning to the container its prescribed transport index number.

· The transport index relates to (a) the level of radiation outside the package (which in any case cannot exceed defined limits, but may be lower); (b) the fraction of a composite shipment of fissile containers allowed together in transit from a criticality-prevention standpoint; or (c) both (a) and (b). In arriving at the transport index for an individual package, the values assigned to the radiation and/or the criticality potential of each package are such that the number of packages giving a total transport index of 50 are all that may be grouped together in transit without exceeding the total radiation allowed for each shipment in transit or exceeding the criticality limits for a group of packages in transit. That is, only those packages may be grouped together in one location in transit whose combined transport indexes do not exceed 50. Such groups must be kept at least 20 feet from other shipments which would be sensitive to the composite (still very low) level of radiation (photographic film, etc.), and 20 feet from other packages which might add to the degree of criticality.

This leads to the next safety principle: carriers must be well-informed, efficient, and responsive to DOT regulatory requirements relative to transportation and related matters. Aside from the usual matters of good practice in transportation procedure, the carrier is required to perform, in behalf of safety, only one specific and unique function as a matter of regulatory requirement: he must assure that the cumulative transport index does not exceed 50 for any group of packages, and he needs to assure that such groups are separated from radiation-sensitive shipments or from other transport-index labeled packages by prescribed distances.

There is, however, one further principle which is observed in behalf of the nation's safeguards program designed to prevent unauthorized diversion of nuclear fuel materials to nonauthorized uses. Packages containing more than 5 kilograms of uranium-235 (of more than 20% enrichment), uranium-233, and/or plutonium, with some minor exceptions in defined circumstances, may move from the custody of the shipper to that of the carrier and to the custody of each successive carrier and finally to the receiver, only when there is written, jointly signed, hand-to-hand transfer of custody receipt between the two parties involved in transfer.

REGULATORY AGENCY RESPONSIBILITIES

Two federal agencies, the Atomic Energy Commission and the Department of Transportation are responsible for protecting the health and safety of the public against possible radiological hazards arising from radioactive materials in transit over public carriers' systems. The DOT has a general obligation to exercise regulatory surveillance over the transportation of all dangerous goods. The AEC is responsible for protecting against radiologic hazards of source, special nuclear, and by-product materials in all activities involved in possession, processing, and use of such materials.

Traditionally, the agencies in DOT have focused on the regulatory aspects of transportation and carrier systems, but have had little technical expertise in radiological matters. The AEC, on the other hand, possesses considerable radiological expertise.

In 1966, AEC and the Interstate Commerce Commission (ICC) signed a Memorandum of Understanding designed to minimize unnecessary duplication of effort and maximize the use of the special expertise of the two agencies. This function of the ICC has now been transferred to DOT.

Under the agreement, DOT is now responsible for authorizing all shipments to be delivered to carriers and maintains the normal, customary surveillance over carriers during transport. DOT regulations identify the package specifications for materials in the various categories, the design details, and authorized contents of approved "Specification Containers," labeling requirements. The regulations also define requirements to be observed by carriers.

To facilitate actual shipping approval, DOT divides radioactive shipments into three categories and follows particular procedural patterns for each category:

1. Exempt and small quantity shipments below Type B quantities (discussed earlier) and less than 15 grams of fissile material. The DOT regulations contain requirements which the shipper must meet for shipping these quantities, but no specific approval of DOT or AEC is required.

2. Type B quantities and above, including large quantities and fissile materials exceeding 15 grams. Specification containers are described in the regulations, and a shipper may use any of these containers with no prior approval from DOT or from AEC.

3. Type B quantities and above, including large quantities and fissile materials in excess of 15 grams. If the shipper uses a container other than a specification container, the shipment must be authorized by DOT, based on a demonstration that the type of container to be used satisfies the published container standards for

the type and quantity of material to be transported. To obtain approval, the prospective shipper must submit an application to the Department of Transportation in which the container and intended contents are described; the results of applicable tests are given; the structural, shielding, and criticality safety characteristics of the container are evaluated; and any special procedures related to use and transportation of the container are discussed.

Under the agreement, the AEC

a. Assists DOT in developing and revising its regulations relating to radioactive materials

b. Joins with DOT in evaluating and approving proposed general-use specification containers before they are formally authorized in DOT regulations

c. Furnishes technical advice to DOT on the adequacy of or special conditions that apply to proposed large quantity shipments

d. May join with DOT in investigating transportation accidents

Each agency makes its reports and experience freely available to the other and collaborates on matters of mutual interest. For international shipments, the DOT is the "competent authority," i.e., "the responsible transportation authority." The AEC provides technical staff expertise for purposes of dealing with foreign agencies on transportation matters, and in developing international standards for transportation.

THE RECORD OF SAFETY IN TRANSPORTATION OF NUCLEAR AND RADIOACTIVE MATERIALS

It is estimated that 250,000 to 300,000 packages of radioactive material are now shipped each year in the United States. This includes AEC shipments by AEC vehicles and by public carriers, and non-AEC (licensee) shipments by public carriers. In earlier years of the atomic energy program, almost all shipments were of AEC materials, and most were in AEC vehicles. Since then, the AEC shipments have been transported more and more by public carriers. Altogether, AEC accounts for about one quarter of the total radioactive and fissile shipments in the country. To date, there have been no injuries due to radiation from fissile or radioactive materials in transportation.

For AEC shipments—whose records exist over the past 19 years—119 transportation incidents were recorded. In 84 of these, no radioactive material was released from the packages. None of the 35 cases in which material was released resulted in any serious exposure to radiation. Only one case resulted in dis-

persal into the air, and only one case involved costly cleanup.

For non-AEC shipments (all by public carriers) over the past 2 years, 43 transportation incidents have come to the attention of the AEC. In 21 of these, no radioactive material was released. Eleven incidents involved release of material; two produced increases in radiation levels outside the packages. Yet, to our knowledge, none of these cases resulted in serious radiation exposures, and only one case resulted in widespread contamination and costly cleanup. The remaining nine incidents involved millicurie amounts of ^{63}Ni, ^{131}I, ^{147}Pm, ^{14}C, ^{203}Hg, ^{60}Co, ^{144}Ce, ^{210}Pb, ^{65}Zn, and tritium; one larger amount (5.8 curies) of tritium was lost or stolen and not recovered.

FUTURE OUTLOOK

Currently, the number of shipments of nuclear fuel and associated materials is relatively small. With the rapidly growing nuclear power program, such shipments will increase greatly. The Nuclear Assurance Corporation (Heinisch, n.d.) has listed shipment statistics for U.S. reactors (Table A-1).

TABLE A-1 Shipment Statistics for Reactors in the United States [a]

	Number of Shipments		Expected Number of Accidents	
	1971	1974	1971	1974
(Mill → Converter) U₃O₈ → UF₆ Converters (Converter → Enrichment Plant)	704	1,071	2.59	3.94
UF₆ → USAEC Enrichment (Enrichment Plant → Pellet Mfgrs.)	938	1,438	1.57	2.59
Enriched UF₆ → Powder and Pellet Mfgrs.	781	1,347	1.20	2.01
Powder and Pellet Mfgrs. → Fabricators	136	223	0.38	0.70
Fuel Fabricators → Reactors	272	360	0.74	1.22
Reactors → Spent Fuel Reprocessing	76	1,417	0.17	3.14

[a] Statistics from Nuclear Assurance Corporation.

Projecting the spent-fuel shipping requirements, it is estimated (SINB, 1970; Walchli, 1970) that by 1980 approximately 4,000 shipments per year will be required. By 1987, the number would be close to 25,000 (assuming about 1 metric ton of fuel or the equivalent of two PWR-type fuel assemblies/load).

While the chance of accidents involving transport of these materials must be faced squarely, one should not equate transport accident with public exposure to radiation. It appears that in this area, engineering design of transport devices can do much to assure public safety. Perhaps an examination of the transport problem might well be a topic for separate examination by NAE.*

With the expected commercial advent of fast breeder reactors in the 1980's, some comment should be made about the spent fuel shipment of fast-reactor fuels. The plutonium fuel elements (ORNL, 1970) to be used in fast reactors are designed to be about 17 ft long, with most of the heat produced in the center third of the elements. Because of the high plutonium inventory, there is a strong economic incentive to ship the spent fuel to reprocessing plants after cooling times of only

---

* It is understood that this matter is under study by NAS Committee on Waste Management.

30 days or less. For shipments of one or two tons of fuel, the high heat output of the fuel at these short decay times precludes the use of lead-shielded casks, since the operating temperature of the shield material under the loss-of-coolant condition is above the melting point of lead. In order to ship 18 fuel elements per cask, a steel cask of approximately 120 tons will be required, and the mode of shipment will be limited to rail.

Because of heat-transfer problems, fast breeder reactor fuel may require a sodium coolant. This implies that each element needs to be separately encapsulated in sodium (constituting a "special form," as defined in the regulations). Such control over this fuel may make the problems of contamination and potential leakage less severe than they are for thermal reactor fuels, although the transfer of heat becomes somewhat more difficult.

APPENDIX B: WASTE DISPOSAL CONSIDERATIONS

ORIGIN AND NATURE OF RADIOACTIVE WASTES

Despite the extensive discussions and published information on this subject, "radioactive wastes" continue to be considered by many people as a single uncategorized entity. The word "radioactive" is so strongly impressed that it has become an all-inclusive term, to the point where wastes from nuclear reactors, from laboratory research, from medical use, from chemical reprocessing of irradiated fuel elements, etc., are all considered one and the same thing. Important characteristics such as the quantity and concentration of radioactive material involved and its detailed chemical and physical nature are often completely ignored. However, these are paramount to a meaningful understanding and essential to any discussion of radioactive waste operations.

Radioactive wastes are generated in all areas of the nuclear fuel cycle; they accumulate as liquids, solids, or gases at varying radiation levels. The liquid radioactive wastes are generally classified as high-, intermediate-, or low-level, based on the concentration of radioactivity in specific waste streams (e.g., curies per liter or a concentration relative to a particular radiation protection guide). These classifications are important to the plant operator primarily as an approximate indication of the degree of confinement and control needed for the processing or interim storage of each type of waste.

High-level liquid wastes are those which, by virtue of their radionuclide concentration, half-life, and biological significance, require perpetual isolation from the biosphere. The chemical reprocessing of irradiated fuels is the primary source of all high-level wastes. Intermediate-level liquid wastes refers only to radioactive liquids in a processing status which must eventually be treated to produce a low-level liquid waste (which can be released) and a high-level waste concentrate (which needs to be isolated from the biosphere). Low-level liquid wastes are those wastes which, after suitable treatment, can be discharged to the biosphere without exposing people to concentrations in excess of those permitted by AEC regulations. Wastes generated in the cold or preirradiation phase of the fuel cycle (from the mine to the reactor), as well as wastes resulting from research laboratories, from medical and industrial applications of radioisotopes, and from all types of nuclear reactors, are generally considered low-level or low-hazard potential wastes.

ESTIMATED FUEL REPROCESSING WASTES FROM CIVILIAN NUCLEAR POWER ECONOMY *

Before proceeding further with the management of wastes from fuel reprocessing, let us look briefly at the estimated magnitude of this operation during the coming years. The extent of this potential problem with an expanding nuclear power industry is under continuing assessment as an integral part of the AEC's radioactive effluent control R&D program. Table B-1 summarizes the estimated quantities of high-level wastes from the

---

* Belter (1970).

TABLE B-1 Estimated High Level Waste from the Civilian Nuclear Power Industry [a]

	Calendar Year			
	1970	1980	1990	2000
Installed capacity, 10^3 MW(e)	6	150	450	940
Spent fuel processed, metric tons/yr	55	3,000	9,000	19,000
Volume of high-level liquid waste				
Annual production, 10^6 gal/yr [b]	0.017	0.97	3.3	5.8
Accumulated, 10^6 gal (if not solidified)	0.4 [c]	4.4	29	77
Volume of high- level waste, if solidified				
Annual production, $10^3 ft^3$/yr [d]	—	9.7	33	58
Accumulated, 10^3 ft^3	—	44	290	770

[a] Statistics from Belter (1970).
[b] Assumes wastes concentrated to 100 gal per 10,000 MWd (thermal).
[c] Assumes 1 ft^3 of solidified waste per 10,000 MWd (thermal).
[d] Actual at Nuclear Fuel Services' West Valley Plant.

civilian nuclear power program. One might note that 77 million gallons of high-level liquid waste will have been generated by the year 2000. This amount will undoubtedly not be on hand in liquid form. Much of this total will have been solidified and shipped to a repository. [Note: Table B-2, taken from ORNL-4451, (ORNL, 1970) was done on a somewhat different basis, but provides a more detailed breakdown by principal radionuclides.]

Table B-3 estimates the volumes of fuel element-cladding wastes through the year 2000. These projections of cladding wastes assume that the cladding and associated fuel "hardware" are compacted to about 70% of theoretical density. Currently, the AEC buries about 1.8 million cubic feet of radioactive solid waste annually. This compares with the 4 million ft^3/year of total solid wastes (excluding solidified high-level wastes) projected for the year 2000.

HIGH-LEVEL LIQUID-WASTE TANK STORAGE

High-level liquid wastes originate mainly from the first cycle of solvent extraction, and contain greater than 99.9% of the nonvolatile fission products. Present practice is to concentrate and store these wastes on an interim basis in specially designed, underground, carbon and stainless-steel tanks equipped with devices for removing the decay heat if necessary. More than 20 years' experience with this method of handling highly radioactive liquid waste has shown it to be a practicable

means of interim handling. As noted, more than 80 million gallons of radioactive solutions and sludges are stored in nearly 200 underground tanks throughout the Atomic Energy Commission complex. Most of these wastes have concentrations in the range of 1–100 curies/gal; however, some of the high-level wastes have an initial concentration of 10,000 curies/gal.

Although corrosion data indicate expected tank lifetimes of 40 to 50 years, there are reported instances of tank failure, all in carbon-steel systems at Hanford and Savannah River. Stress-corrosion cracking and/or thermal stress of the reinforced concrete structures were established as causes of these tank leakages; this was taken into account in improved design and construction of new interim tank storage. While current tank storage practices successfully prevent significant quantities of radioactive materials from escaping to the environment, such operations require continual surveillance and tank replacement. This need for surveillance—as well as the necessity to transfer liquid waste from tank to tank over periods of hundreds of years—is a compelling factor for the extensive research and development programs directed at the conversion of high-level liquid waste to a solid form, and the ultimate storage of the solid-waste materials in selected geologic environments.

HIGH-LEVEL WASTE SOLIDIFICATION

The conversion of high-level liquid wastes to solids as a pretreatment for long-term storage is being developed in several countries with large-scale nuclear energy programs. Extensive R&D was conducted during the past 15 years on the development of high-temperature solidification systems for these wastes. Four processes for solidification were developed in the United States to the point of radioactive demonstration on an engineering scale. These processes are fluidized bed calcination, pot calcination, spray solidification, and phosphate-glass solidification.

The pot, spray, and phosphate-glass processes were developed and are being demonstrated at processing rates of 10 to 20 liters/hr of liquid waste, or waste from the processing of one ton of nuclear fuel per day. The fluidized bed process was demonstrated at high rates of 300 liters/hr of liquid wastes containing large amounts of aluminum and zirconium salts. In all processes, heat is applied to raise the temperature of the wastes from 400° to 1200°C. At these temperatures, essentially all the volatile constituents (primarily water and oxides of nitrogen) are driven off, leaving a solid or a melt that will cool to a solid. The resulting solids are relatively stable chemically, especially at temperatures lower than those used during

TABLE B-2 Projected Fuel Processing Requirements and High-Level Waste Conditions for the Civilian Nuclear Power Program [a]

	Calendar Year			
	1970	1980	1990	2000
Installed capacity, MW(e) [b]	14,000	153,000	368,000	735,000
Electricity generated, 10^9 kWhr/year [b]	71	1,000	2,410	4,420
Spent fuel shipping				
Number of casks shipped annually	30	1,200	6,800	9,500
Number of loaded casks in transit	1	14	60	85
Spent fuel processed, metric tons/year [b]	94	3,500	13,500	15,000
Volume of high-level liquid waste generated [c, d]				
Annually, 10^6 gal/year	0.017	0.97	2.69	4.60
Accumulated, 10^6 gal	0.017	4.40	23.8	60.1
Volume of high-level waste, if solidified [c, e]				
Annually, 10^3 ft³/year	0.17	9.73	26.9	46.0
Accumulated, 10^3 ft³	0.17	44.0	238	601
Solidified waste shipping [f]				
Number of casks shipped annually	0	3	172	477
Number of loaded casks in transit [g]	0	1	4	10
Significant radioisotopes in waste [h, i]				
Total accumulated weight, metric tons	1.8	450	2,400	6,200
Total accumulated beta activity, megacuries	210	18,900	85,000	209,000
Total heat-generation rate, megawatts	0.9	80	340	810
^{90}Sr generated annually, megacuries	4.0	230	560	770
^{90}Sr accumulated, megacuries	4.0	960	4,600	10,000
^{137}Cs generated annually, megacuries	5.6	320	880	1,500
^{137}Cs accumulated, megacuries	5.6	1,300	6,500	15,600
^{129}I generated annually, curies	2.0	110	440	670
^{129}I accumulated, curies	2.0	480	2,700	7,600
^{85}Kr generated annually, megacuries	0.6	33	90	150
^{85}Kr accumulated, megacuries	0.6	124	570	1,200
^{3}H generated annually, megacuries	0.04	2.1	6.2	12
^{3}H accumulated, megacuries	0.04	7.3	36	90
^{238}Pu generated annually, megacuries	0.0007	0.041	0.2	0.6
^{238}Pu accumulated, megacuries	0.0007	1.20	8.3	31
^{239}Pu generated annually, megacuries	0.00009	0.005	0.05	0.2
^{239}Pu accumulated, megacuries	0.00009	0.02	0.24	1.3
^{240}Pu generated annually, megacuries	0.00012	0.007	0.06	0.21
^{240}Pu accumulated, megacuries	0.00012	0.04	0.4	1.9
^{241}Am generated annually, megacuries	0.009	0.5	4.4	15
^{241}Am accumulated, megacuries	0.009	2.3	23	120
^{243}Am generated annually, megacuries	0.00021	0.01	0.1	0.5
^{243}Am accumulated, megacuries	0.00021	0.23	1.5	5.2
^{244}Cm generated annually, megacuries	0.13	7.4	18	23
^{244}Cm accumulated, megacuries	0.13	30	140	260
Volume of cladding hulls generated [j]				
Annually, 10^3 ft³	0.3	8	40	90
Accumulated, 10^3 ft³	0.3	40	320	1,030

[a] From ORNL (1970).
[b] Data from Phase 3, Case 42 Systems Analysis Task Force (Apr. 11, 1968).
[c] Based on an average fuel exposure of 33,000 MWd/ton, and a delay of 2 years between power generation and fuel processing.
[d] Assumes wastes concentrated to 100 gal per 10,000 MWd (thermal).
[e] Assumes 1 ft³ of solidified waste per 10,000 MWd (thermal).
[f] Assumes 10-year-old wastes, shipped in thirty-six 6-in.-diam. cylinders per shipment cask.
[g] One-way transit time is 7 days.
[h] Assumes LWR fuel continuously irradiated at 30 MW/ton to 33,000 MWd/ton, and fuel processing 90 days after discharge from reactor; LMFBR core continuously irradiated to 80,000 MWd/ton at 148 MW/ton, axial blanket to 2,500 MWd/ton at 4.6 MW/ton, radial blanket to 8,100 MWd/ton at 8.4 MW/ton, and fuel processing 30 days after discharge.
[i] Assumes 0.5% of Pu in spent fuel is lost to waste.
[j] Based on 2.1 ft³ of cladding hulls per ton of LWR fuel processed, and 8.7 ft³ of cladding hardware per ton of LMFBR mixed core and blankets processed.

TABLE B-3 Solid Wastes from Spent Fuel Processing [a]

	Calendar Year		
	1970	1980	2000
Volume of cladding waste [b]			
Annual, 10^3 ft³	0.1	8.3	110
Accumulated, 10^3 ft³	0.1	37	1,300
Total volume of solid waste [c]			
Annual, 10^6 ft³	0.01	0.8	4
Accumulated, 10^6 ft³	0.1	4	62
Burial ground area [d]			
Annual, acres	0.2	16	81
Accumulated, acres	1.5	70	1,230

[a] Source: Oak Ridge National Laboratory-DWG 70-1506.
[b] Cladding hulls are compacted to 2.1 ft³/ton and 8.7 ft³/ton of LWR and LMFBR fuel, respectively.
[c] Based on an average volume of 200 ft³ of solid wastes per ton of fuel processed.
[d] Based on burial of 50,000 ft³ solid waste per acre of burial ground.

processing. The addition of glass-forming materials, such as phosphoric acid, lead oxide, sodium tetraborate, etc., for the purpose of providing a relatively non-leachable final product was also intensively studied. Table B-4 summarizes the present worldwide status of solidification technology for the various processes.

In the United States, the Waste Calcining Facility (WCF) at the National Reactor Testing Station (NRTS) has demonstrated the use of the fluidized bed process during the past 7½ years, principally on waste from the processing of test reactor fuels. Over 2.0 million gallons of aluminum nitrate waste and zirconium-type waste were calcined in the WCF through May 1970, resulting in about 27,500 cubic feet of waste solids. An average volume reduction of approximately 10 has been obtained since plant startup. Within the past year, cold-pilot-plant scale experimental work at the NRTS has indicated that this process may be applicable to a wide variety of nuclear fuel wastes, including those from Purex processing of power reactor fuels.

Similarly, the technology for solidification of high-level waste from power reactor fuel reprocessing (using the pot, spray, and phosphate-glass processes) has been undergoing an engineering-scale demonstration in the Waste Solidification Engineering Prototype at Hanford since November 1966. As of May, 1970, 33 solidification runs were completed in the pilot plant, including 9 pot calcination runs, 13 runs with the spray solidification process, and 11 phosphate glass solidification runs. Over 52 million curies of high-activity waste were solidified in the 33 runs, with a maximum of 3.6 million curies being converted in a spray-solidification run (209 watts/liter). Results of the U.S. R&D programs are being provided to industry on a continuing basis, for

use in the design and construction of solidification systems in commercial processing plants.

ULTIMATE DISPOSAL OF HIGH-ACTIVITY SOLID-WASTE MATERIALS

After high-level liquid wastes are converted to solids, there still exists the requirement for storage or ultimate disposal of these solid wastes. Several natural geologic formations were suggested as possible media for permanent storage of such wastes. The initial stimulus for this part of the AEC effluent control R&D program commenced in September 1955, when, at the request of the Atomic Energy Commission, a committee of geologists and geophysicists was established by the National Academy of Sciences–National Research Council to study the possibilities of disposing of high-level radioactive waste material on land and to indicate what research was needed to determine the feasibility of such a program. The results of this committee's study and its recommendations can be summarized as follows:

Disposal in salt is the most promising method for the near future. Research should be pushed immediately on the structural problem of stability vs. size of cavities at a given depth; on the thermal problem—getting rid of the heat or keeping it down to acceptable levels.*

In response to the NAS suggestion, a major part of the AEC ground disposal R&D program in recent years has been directed toward establishing the suitability of using underground salt formations for the disposal of high-level solidified radioactive waste. Bedded salt formations are unique in that they are practically always dry and not associated with usable sources of groundwater. Because of the plasticity of salt, fractures seal or close rapidly. Salt beds are generally located in areas of low seismic activity and widely separated from flowing aquifers, as evidenced by the very existence of such deposits. Salt has considerable compressive strength, being similar to concrete in this respect. Large spaces may be mined out, and, even at depths of 1,000 feet, two-thirds of the salt area may be removed with only slight deformation of the support pillars.

During the 1960's, salt-disposal technology was developed to the point where considerable confidence has developed in a system that appears—in concept—practical and promises assurance of perpetual isolation of high-level waste from man's biosphere. Project Salt Vault (a demonstration disposal of high-level radioactive solids in a Lyons, Kansas, bedded salt mine

---

* Report of the Committee on Waste Disposal of the Division of Earth Sciences, NAS–NRC Publication 519, April 1957.

using Engineering Test Reactor fuel assemblies in lieu of actual solidified waste) has successfully demonstrated waste-handling equipment and techniques similar to those required in an actual disposal operation. A total of about 4,000,000 curies of fission-product activity in 21 containers, each averaging about 200,000 curies, was transferred to the experimental area in the mine (three separate shipments of seven canisters each on a six-month schedule) and back to the NRTS in Idaho at the end of the test program. During the period of the field test program (from November 1965 through May 1967), the salt received a maximum radiation dose approaching 10^9 rads. Results of the field demonstration program can be summarized as follows: the feasibility and safety of handling high radioactive materials in an underground environment were demonstrated. The

TABLE B-4   Summary of Research and Development on Solidification of High-Level Waste

Process and Sites [a]	Time Span	Lab Scale Radioactivity [b]	Pilot Plant Radio-activity [b]	Capacity, liter/hr	Product	Chemical Additives	Status of Work
Pot Calcination							
ORNL [c]	1958–1965	None	None	25	Calcine	Calcium, sulfate	Completed
BNW	1959–1962	None	None	10	Calcine	Sulfate	Completed
	1962–1970	H	H	20	Calcine	Sulfate, calcium	In progress
Spray							
BNW	1959–1970	H	H	20	Ceramic, glass	Phosphate, boro-phosphate	In progress
USSR	~1961 to date	?	?	20	Calcine, glass	Borosilicate	In progress
Phosphate Glass							
BNL	1960–1970	None	None	20	Glass	Phosphate	Completed
BNW	1964–1970	H	H	20	Glass	Phosphate	In progress
Fluid Bed							
ANL	1955–1959	None	L	6	Granules	None	Completed
INC	1955 to date	L	I	300	Granules	None	In progress
BNW	1959–1961	No work	None	20	Granules	None	Completed
USSR	~1962 to date	?	?	30	Glass, granules	Borosilicate	In progress
Pot Glass							
AERE	1959–1966	None	H	6	Glass	Borosilicate	Completed
FAR	1962 to date	H	L	20	Glass	Boroalumino-silicate	In progress
CPP	1969 startup	No work	H	20	Glass	Phosphosilicate Boroalumino-silicate	In progress
ORNL	1961–1966	None	None	3	Semiglass	Phosphate, borophosphate	Completed
Rotary Kiln							
BNL	1955–1963	None	None	20	Powder	None	Completed
FAR	1960 to date	None	None	6	Glass	Phosphosilicate, borosilicate	In progress
Ceramic Sponge							
LASL	1959–1964	None	L	4	Ceramic balls	None	Completed

[a] Work is also being done in Canada, Germany, Denmark, India, Japan, and Czechoslovakia.
[b] Values are based upon:
   H is > Ci/kg of solid
   I is 0.07 to 70 Ci/kg of solid
   L is < 0.07 Ci/kg of solid
[c] Abbreviation summary:
   ORNL   Oak Ridge National Laboratory, Oak Ridge, Tennessee
   BNW   Battelle-Northwest, Richland, Washington
   USSR   Union of Soviet Socialist Republics
   BNL   Brookhaven National Laboratory, Upton, Long Island, New York
   ANL   Argonne National Laboratory, Argonne, Illinois
   INC   Idaho Nuclear Corporation, Idaho Falls, Idaho
   AERE   Atomic Energy Research Establishment, Harwell, Berks, England
   FAR   Center for Nuclear Studies, Fontenay-aux-Roses, France
   CPP   Center for Plutonium Production, Marcoule, France
   LASL   Los Alamos Scientific Laboratory, Los Alamos, New Mexico

stability of salt under the effects of heat and radiation was shown. Data were obtained on the creep and plastic-flow characteristics of salt which make possible the design of a safe, long-term disposal facility.

It is estimated that between 1,000 and 3,000 acres of salt area may be required for disposal purposes by the year 2000. This is only a small fraction of the 400,000-square-mile area that is underlaid by salt in the United States. Oak Ridge National Laboratory and Kaiser engineers are now completing a conceptional design of a proposed salt disposal facility at a site tentatively selected in the Lyons, Kansas, area. ORNL, in cooperation with the Kansas State Geological Survey, is also carrying out geohydrologic field studies to determine the environmental suitability of the proposed site.

Other disposal alternatives which were investigated include (1) disposal as solids in surface or near-surface vaults; (2) disposal as liquids or slurries in caverns in deep geologic formations; and (3) disposal as solids in caverns in deep geologic formations.

While disposal of wastes as solids in surface or near-surface vaults on commercial reprocessing plant sites may be economically attractive, the concept suffers from the aspect of proliferation of high-level waste disposal sites. From a long-term safety standpoint, surface disposal is also considered a greater inherent risk than disposal in selected deep geologic formations, because of a closer proximity to man's bioenvironment, and because it requires continued surveillance.

Since safety considerations militate against the bulk shipment of high-level waste liquids or slurries, their storage in deep geologic formations would require suitable geohydrologic conditions at the fuel reprocessing site. This limits the number and location of suitable reprocessing plant sites. The relatively small volumes of waste associated with a single commercial fuel-reprocessing plant and cost (probably several million dollars to determine if a selected geologic formation is suitable) would appear to limit this concept. While liquid or slurried waste disposal in deep underground formations cannot, at the present time, be conclusively eliminated on the basis of either safety or economics, it is not expected to be an important concept in the nuclear power economy. Further details on other ground disposal methods are provided in the attached bibliography.

TRANSPORTATION OF HIGH-LEVEL RADIOACTIVE WASTE FROM REPROCESSING PLANTS

Shipment of high-activity waste from the fuel reprocessing plant will be governed by the Code of Federal

Regulations (10CFR71) and the Department of Transportation regulations. Engineering studies which were carried out to date indicate that unacceptable safety and economic problems would be encountered in the design and construction of a bulk liquid high-level waste shipping system which would meet AEC and DOT shipping regulations. It was decided that bulk quantities of liquid high-activity waste should be converted to a solid form prior to shipment beyond the fuel reprocessing site. It is envisioned that solidified waste will likely be in the form of a calcined oxide or glasslike product encased in a steel container with a welded closure. From a safety standpoint, it is likely the solidified waste product will be doubly contained, first in its welded-steel can, and then in the shipping cask itself. The calcined or glasslike waste product is quite immobile. Impact and fire accident conditions, as specified in the shipping regulations, can apparently be met.

THE DISPOSAL OF RADIOACTIVE HULLS AND OTHER CONTAMINATED FUEL STRUCTURAL HARDWARE

One type of radioactive waste peculiar to fuel reprocessing plant operations is the fuel structural hardware and fuel hulls which, after hot nitric acid leaching processes, require disposal. This waste contains, typically, thousands of curies of induced activity consisting of comparatively short half-lived radioisotopes and small amounts of fission products and plutonium which survived the acid leaching processes and remain affixed to the fuel hulls. At the present time, one might anticipate that these wastes may be disposed of in the same manner as high-level solidified wastes (i.e., transferred to a federal repository), or they may be buried in a licensed waste-burial facility.

HIGH-LEVEL WASTE DISPOSAL POLICY

The waste treatment and disposal technology which was developed and demonstrated during the 1960's is now being used to establish new high-level waste management policy and regulatory procedures. The AEC issued a statement of policy in the Federal Register dated November 14, 1970, to become effective 90 days after publication, which it did on February 12, 1971. A number of key provisions follow:

a. Public health and safety considerations associated with commercial fuel reprocessing plants do not require that they be located on land owned and controlled by the Federal Government. The plants, including facilities for temporary storage of high-level liquid radioactive wastes,* may be located on privately owned property. This conclusion is based (1) on the availability of technology for solidifying high-level waste in forms suitable for safe transport and disposal at a Federal repository and (2) on a recognition that fuel reprocessing plants can be so designed that radiologically significant contaminants may be removed or otherwise satisfactorily disposed of when the plant is retired from service.

b. A fuel reprocessing plant's inventory of high-level liquid radioactive wastes will be limited to that produced in the prior five years. These wastes shall be converted to a dry solid as required to comply with this inventory limitation, and placed in a sealed container prior to transfer to a Federal repository in a shipping cask meeting the requirements of 10CFR, Part 71. The dry solid shall be chemically, thermally, and radiolytically stable to the extent that the equilibrium pressure in the sealed container will not exceed the safe operating pressure for that container during the period from canning through a minimum of 90 days after receipt (transfer of physical custody) at the Federal repository. The specifications have been developed to give licensees maximum flexibility in producing solid forms satisfying safety requirements associated with on-site interim storage, transportation, and Federal reposi-

---

* For the purpose of this statement of policy, "high-level liquid radioactive wastes" means those aqueous wastes resulting from the operation of the first cycle solvent extraction system, or equivalent, and the concentrated wastes from subsequent extraction cycles, or equivalent, in a facility for reprocessing irradiated fuels.

tory operations. All of these high-level radioactive wastes shall be transferred to a Federal repository no later than ten years following separation of fission products from the irradiated fuel. These inventory restrictions are being imposed to limit the mobility of wastes temporarily stored at the reprocessor's site and to assure timely shipments to a Federal repository. It is believed that the inventory limitations are reasonable from the standpoint of plant operations and waste management economics.

c. Upon arrival of the solid fuel wastes at the repository the Federal Government will assume permanent custody of the materials, but industry will pay all costs of disposal. A schedule of charges for permanent disposal at a designated repository will be made available as soon as possible.

d. Disposal of high-level fission product waste material will be permitted only on land owned and controlled by the Federal Government. This policy of Federal control of repository sites does not imply, however, either that state involvement in the development of a repository will not be encouraged or that adoption of the policy will preclude state participation in high-level waste management activities at some time in the future.

On June 17, 1970, the AEC announced tentative selection of a site near Lyons, Kansas, for location of a demonstration repository. As of the date of this publication, the environment impact statement developed in accordance with the National Environmental Policy Act of 1969 has not been completed to meet the requirements of all interested parties.

APPENDIX C: SUMMARY OF WORKING GROUP I(c) REPORT*

Nuclear energy as a source of electrical power is being increasingly used in the United States, and it is anticipated that about 90 nuclear power reactors will be in operation by the end of 1975, compared to 21 early in 1971. During the same period, the numbers of commercial fuel reprocessing plants will increase from 1 to 4.

From the beginnings of the atomic energy industry in World War II until the present time, the industry has been carefully regulated by the application of radiation protection standards proposed by national and international organizations, such as the National Council on Radiation Protection and Measurements, the International Commission on Radiological Protection, the National Academy of Sciences, and the Federal Radiation Council. The Federal Radiation Council was absorbed recently by the newly created Environmental Protection Agency, and its basic function of providing radiation protection guides to be used by all federal agencies will be continued by the Division of Criteria and Standards of the Office of Radiation Programs. Thus, although AEC has statutory responsibility, under the Atomic Energy Act of 1946, for regulating the safety of the atomic energy industry, the numerical guides that limit the amount of radiation a person is permitted to receive originate in the work of these other organizations.

The biological effects of exposure to nuclear radiations vary, depending on many factors, including the amount of radiation, whether it is delivered in a single dose or spread out over a long period of time, whether the radiation is administered externally (as from an x-ray machine) or internally (as when one ingests a radioactive substance), and whether the dose is delivered in the form of x-ray, gamma, alpha, beta protons or neutrons.

Exposure to all forms of radiation occurs naturally, with individuals in the United States annually receiving about 100 units † of radiation from both cosmic and terrestrial sources. The latter includes direct gamma radiation from members of the uranium and thorium families in the earth's crust, as well as an internal dose from specific radionuclides such as radium that enter the food chain directly from the soil. In addition, radioactive gases emanate naturally from the earth's crust and decay to a radioactive form of lead that eventually rains out to the surface of the earth. Although the dose from nature is about 100 units in most places, the dose is actually quite variable, depending on the altitude and the radioactive mineral content of the soils in any given geographical area.

The types of biological injury can be divided into those that affect the irradiated individual (somatic effects) and those that cause injury to the genetic mechanisms and are therefore not seen in the exposed individual but are passed on to future generations. There is a vast literature on this subject and it is generally accepted that more is known about the biological effects

---

* This summary has been prepared by Dr. Merril Eisenbud of the A. J. Lanza Laboratory, New York, a member of the COPPS Steering Group.

† In this summary the term "radiation units" will refer to millirems, a unit of ionizing radiation dose, which is basically a measure of the absorbed energy, modified by type of radiation.

of ionizing radiation than is known about the effects of any other contaminant of our environment. A considerable amount of information about the effects in humans exists from experience with the medical uses of x-ray and radium, as well as the misuses of the ionizing radiations in industry prior to World War II. The atom bombings of Hiroshima and Nagasaki provided additional information that has been exhaustively studied.

For most noxious substances it is generally assumed that the effects follow a dose-response curve that is S-shaped (Figure 1) with a "threshold" below which no effect takes place. There are theoretical, experimental, and epidemiological reasons to believe that certain of the effects of ionizing radiation exposure are directly proportional to dose and that there is no threshold. Because of the conservative approach that has been taken in establishing radiation standards, it has been assumed for many years that this in fact is the case, although there is considerable evidence that many of the more important effects do not follow this simple relationship. The assumption also implies that the effects are independent of the rate at which the dose is delivered. There is evidence with respect to both the genetic and somatic effects to show that this is not so and that the so-called "linear hypothesis" provides radiation effects estimates that are on the high side for many types of effects, particularly at low doses. It is not possible to demonstrate if effects do in fact occur at low doses because the effects would be so infrequent as to be undetectable against the frequency with which the effects normally occur. For example, most of the radiation effects that have been described in the literature were caused by doses in excess of 100,000 radiation units.

The permissible radiation levels may be best visualized by comparison with other well-known sources as shown in Table 1.

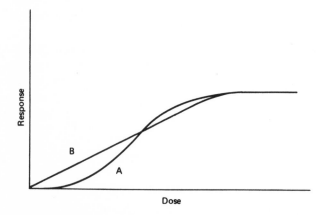

FIGURE 1 Types of dose curves: (A) Conventional sigmoid curve; (B) Linear response used to estimate radiation effects at low doses.

TABLE 1 Population Radiation Dose from All Sources

Source of Radiation	Radiation Units per Year
Natural radiation	100–125
Medical irradiation	50–100
Upper limit of permissible occupational exposure	5,000
Maximum permissible exposure to any member of the public from all sources other than natural and medical irradiation	500
Limit to average public exposure	170
Upper limit of exposure for the maximum exposed individual in the vicinity of power reactors	5–10

The apparent inconsistency between the maximum permissible dose permitted individuals in the vicinity of power reactors and the permissible per capita dose of 170 units per year, is explained by the general admonition contained in all radiation protection guidelines that the recommended upper numerical limits notwithstanding, the dose should be reduced as low as practicable. In actual practice, the doses to the populations in the vicinity of operating reactors is a small fraction of 1 radiation unit.

An operating nuclear reactor accumulates huge quantities of radioactive substances which must be kept under control during normal operation and also under accidental conditions. To assure that no course of events can lead to a significant release of radioactive material, the fission products (which are the waste materials produced by the reactor) are contained within three barriers. First, all fission products are held within the fuel element cladding which is designed to withstand the high temperature and irradiation conditions within the reactor core. Should a leak or loss of the fuel element cladding integrity occur, the reactor coolant system provides a secondary containment boundary, which is designed to withstand high temperature, high pressure, and high radiation levels. In the event this second layer of defense fails, a third independent system of containment is provided for all nuclear power plants. The third barrier is a containment structure designed to retain all of the radioactive products, under any credible circumstances. Associated with these three barriers are a variety of means by which the radioactive releases can be controlled in the event of a serious accident.

Reactor safety specialists have analyzed the various kinds of accidents that could occur and have laid down rules of design which are intended to prevent the accident from ever happening. However, having designed the reactor in this way, it is then assumed that despite all precautions, the accident does in fact occur and that safeguards must be provided to prevent the radioactive materials from escaping to the general environment.

There are fundamental physical reasons why a reactor cannot explode as a result of the release of excess nuclear energy. One of the most fundamental reasons is that the reactivity actually decreases as the temperature increases. Thus any accident that would cause the rate of fission to increase above that contemplated would cause the core temperature to increase, which in turn would cause the reactivity to decrease, thereby stabilizing the system. Other factors as well provide for inherent stability. Examples are the formation of steam bubbles in the water coolant, which has the effect of reducing the moderating effect of the water, and in extreme cases, slight mechanical disarrangements of the core structure which would have the effect of reducing reactivity.

The most serious type of accident that must be considered is the loss of coolant due to a massive rupture of a pipe. This would cause the water to flash to steam, following which the heat of radioactive decay would cause the temperature of the core to increase and the cladding to fail, releasing radioactive products to containment.

Means are provided to reduce the pressure of the steam released to containment. This can be done by providing sprays, or by allowing the steam to vent to a water-filled reservoir, in which it is condensed. A recent innovation has been the use of ice beds in the containment. These serve as a heat sink through which the air-steam mixtures are directed in the event of a loss-of-coolant accident.

It is assumed that the fission products released into the reactor building containment are 100% of the core inventory of noble gases, 50% of the radioactive halogens, and 1% of the solids in the fission product inventory. Such a release could not actually occur unless all of the redundant emergency core cooling systems malfunctioned. The combined probability of both these failures and the pipe rupture is so small as to be numerically indefinable, but it is certainly very much smaller than similar failure probabilities in other engineered structures such as dams, aircraft, bridges, and chemical process equipment whose failure would have major safety implications.

The AEC places stringent limitations on the dose to the surrounding population even in the event of a loss of coolant accident. The reactor designs, the safeguards provided, and the individual site must be such that the permissible doses will not be exceeded. In actuality, the reactor designs in many cases are such that even in the event of a loss-of-coolant accident, the dose to the surrounding population would be no greater than those permitted during normal operation.

The radioactive wastes generated during normal reactor operation may be liquid or gaseous and are of two basic types—fission products and activation products. As noted earlier, the fission products are produced in copious amounts, but are very largely retained within the fuel and escape to the environment in relatively small amounts. The activation products are caused by neutron irradiation of corrosion products or other impurities that pass through the reactor core. These are present in the liquid wastes as radioactive nuclides of iron, chromium, cobalt, manganese and other elements. In addition, the reactor coolant water itself is subject to neutron activation, producing small quantities of radioactive nitrogen. In pressurized water reactors, boron and lithium are frequently used in the coolant and result in the production of tritium, the radioactive isotope of hydrogen.

Nuclear reactors are designed to process and recycle waste streams so as to minimize both the volume and the radioactivity of the effluents to the lowest practicable extent. Releases to the environs are controlled by batch processing and/or continuous monitoring for discharge to assure that no release will exceed established limits.

The gaseous wastes consist mainly of very short-lived isotopes of the noble gases. The design of a pressurized water reactor is such that these are released in insignificant quantities, since the gases are retained within the primary system and rapidly decay. However, in the boiling water reactor, the radioactive gases pass from the reactor directly to the turbine and are removed from the condenser by the air ejector.

The basic waste management techniques currently used to treat radioactive gaseous wastes are: (1) holdup and decay; (2) filtration; (3) distillation; and (4) cryogenics. Holdup and decay refer to storing wastes long enough to permit substantial radioactive decay. As noted, this occurs automatically in a pressurized water reactor because of the manner in which a pressurized water reactor operates. In the boiling water reactor, an overall decontamination factor of about 48 is achieved by a 30-minute delay before release to the atmosphere. Reservoirs intended to achieve gas holdup times of several days of more are feasible, in which case the emissions are limited to the long-lived krypton-85, which is present in such a small concentration that it is not significant in the immediate environs. However, it has a 12-year half-life, and because of this can accumulate in the general atmosphere. However, the dose from this radioactive gas should not exceed 1 to 2 radiation units per year by the year 2000, so that there is adequate time to develop practical means to remove it from the waste gas streams. The preferred method of doing this is currently by freezing, but other systems have promise as well.

The main reason for filtering gases before release to the atmosphere is that many of the short-lived radioactive gases decay to radioactive particulates.

Cryogenic techniques are intended to freeze out the long-lived radioactive noble gases so that they may be removed from the premises and stored offsite until such time as essentially full decay has taken place.

Liquid wastes are also held for a period of time to permit decay of the shorter-lived nuclides. Following storage for this purpose, the wastes are treated by filtration, evaporation, and demineralization. A final reduction in liquid radionuclide concentrations is achieved by dilution of the wastes in the condenser cooling water to insure that the radionuclide concentrations are at acceptable levels before reaching the site boundary.

The amounts of the various radionuclides that can safely be discharged to the environment are limited by the resulting radiation dose to people, and this in turn is determined by the manner in which the radionuclides cause exposure to humans. The most important pathways by which radionuclides from a facility may reach the immediate population are considered to be the following, in order of significance:

For atmospheric discharges

1. Atmospheric discharge → whole body external exposure

2. Atmospheric discharge → inhalation exposure

3. Atmospheric discharge → deposition on grass → cattle → milk → man

4. Atmospheric discharge → deposition on leafy vegetables → man

5. Atmospheric discharge → deposition on grass → cattle → beef → man

6. Atmospheric discharge → deposition on soil → plants → man

For liquid discharges

1. Aqueous discharge → waterway → drinking water supply → man

2. Aqueous discharge → waterway → seafood → man

3. Aqueous discharge → waterway → aquatic plants → animals → man

4. Aqueous discharge → waterway → external exposure

5. Aqueous discharge → waterway → sediments → external exposure

An important step in calculating doses to people is the identification of dietary intake of certain foodstuffs. Because of ethnic or cultural differences, some individuals may select diets which others would not. There may also be economic factors, as for example among fishermen who would consume much more fish or other seafoods than the normal population.

One of the most difficult radionuclides to control is tritium, which is produced copiously in reactors, either by irradiation of the boron or lithium used in the pressurized water reactor coolants or, in both the PWR's and BWR's, as a fission product. Although tritium is considered to be among the least hazardous of all radionuclides, it has a 12-year half-life, and thus has the potential to accumulate in the environment. The projected environmental levels of tritium to be released from light-water power reactors by the year 2000 are estimated to be about 100,000,000 curies, but the projected dose from this quantity is expected to be very much less than 1 mr per year.

The standards set for limiting the emissions from civilian power reactors take into consideration the possibility that ecological concentration of radionuclides may take place in such a way as to contaminate food products. The operator of a nuclear power reactor is required to monitor the environment and to identify the critical pathways to provide assurance that such concentration is not taking place to a hazardous degree.

Conclusions

The radiological considerations that apply to nuclear facility siting are: (1) the exposure of people to the radioactive wastes from normal operations; and (2) the consequences of an accidental release.

Experience has demonstrated that present technology is adequate to maintain radiation doses in the vicinity of nuclear power plants on the average to small percentages of the natural radiation background. Releases from operating reactors have in fact been very low, for the most part less than 1% of the applicable standards.

It is the possibility of an accident that leads to the primary constraint on the siting of a nuclear power plant. It is not likely that major modifications in present U.S. siting practice will be made until some basic change takes place in the present engineering techniques for prevention or mitigation of accidents.

From the point of view of public exposure, the nuclear power industry contributes an insignificant dose compared to that contributed by the use of ionizing radiations in medical practice. Since the per capita dose from the medical uses of ionizing radiation are currently estimated to be about 100 radiation units per year, compared to a potential dose of less than 0.01 radiation unit per year from the nuclear reactor industry, it is clear that a reduction in per capita exposure from medical uses of only 1% would more than compensate for the projected exposure from the nuclear power industry. Greater attention to the ways in which the ionizing radiations are used in medical practice would seem to be indicated.

GLOSSARY

ATOMIC NUMBER   The number of protons in the nucleus of an atom. Each chemical element has its own atomic number, and together the atomic numbers form a complete series from 1 (hydrogen) to 103 (lawrencium) in order of increasing atomic weight (there are several exceptions). (*See also* atomic weight).

ATOMIC WEIGHT   The mass of an atom. The basis of a scale of atomic weights is the oxygen atom, and the commonest isotope of this element has arbitrarily been assigned an atomic weight of 16. Hence the unit of the scale is $\frac{1}{16}$ the weight of oxygen-16, or roughly the mass of the proton or neutron. The atomic weight of an element, therefore, is approximately equivalent to the total number of protons and neutrons in its nucleus.

BACKGROUND RADIATION   The radiation of man's natural environment, consisting of that which comes from cosmic rays and from the naturally radioactive elements of the earth, including that from within man's body. The term may also mean radiation extraneous to the source of radiation under discussion.

BODY BURDEN   The amount of radioactive material present in the body of man or animals.

BOILING WATER REACTOR   A nuclear reactor in which water, used as both coolant and moderator, is allowed to boil in the core. The resulting steam generally is used directly to drive a turbine.

BONE SEEKER   A radioisotope that tends to lodge in the bones when it is introduced into the body. Example: strontium-90, which behaves chemically like calcium.

BREEDER REACTOR   Usually a reactor that creates more fissionable fuel than it consumes. In some usages a reactor that produces the same kind of fissionable fuel that it consumes, regardless of the amount. The additional fissionable material is created when neutrons are absorbed in fertile materials. The process in both usages is known as breeding.

BURNABLE POISON   A neutron absorber (or poison) such as boron, which, when incorporated in the fuel or fuel cladding of a nuclear reactor, gradually "burns up" (is changed into nonabsorbing material) under neutron irradiation. This process compensates for the loss of reactivity in a reactor which occurs as fuel is consumed and poisonous fission products accumulate.

BURNUP   A measure of nuclear reactor fuel consumption. It can be expressed as either the percentage of fuel atoms that have undergone fission or the amount of energy produced per unit weight of fuel fissioned. In the latter case, it is usually expressed as megawatt-days per metric ton of fuel exposed.

BY-PRODUCTS MATERIAL   In atomic energy law, any radioactive material (except source or fissionable material) obtained in the process of producing or using source of fissionable material. Includes fission products and many other radioisotopes produced in nuclear reactors.

CLADDING   The outer jacket of nuclear fuel elements. It prevents corrosion of the fuel and the release of fission products into the coolant. Aluminum, stainless steel, and zirconium are typical cladding materials.

CONTAINMENT   The provision of a gastight shell or other enclosure around a reactor to confine fission products that otherwise would be released to the atmosphere in the event of a major accident.

CONTAMINATION   The presence of unwanted radioactive matter, or the "soiling" of objects or materials with "radioactive dirt."

CONTROL ROD   Rod, plate, or tube containing a strong neutron absorbing material (hafnium, boron, etc.) used to control the power of a nuclear reactor. A control rod absorbs neutrons, preventing them from causing further fissions.

COOLANT   Any fluid that is circulated through a nuclear reactor to remove heat. Common coolants are water, air, carbon dioxide, and liquid sodium.

CORE   The center of a nuclear reactor containing the fuel elements and usually the moderator, but not the reflector.

COSMIC RAYS   Radiation of many sorts that stem directly or indirectly from sources outside the earth's atmosphere. Cosmic radiation is part of the natural background radiation, and some of its constituents have extremely high energies.

CRITICALITY   The state of a nuclear reactor when it is sustaining a chain reaction.

CURIE   The basic unit used to describe the intensity of radioactivity in a sample of material. One curie equals 37 billion disintegrations per second, or approximately the radioactivity of 1 gram of radium.

DECAY   The spontaneous radioactive transformation of one nuclide into a different nuclide or into a different energy state of the same nuclide. Every decay process has a definite half-life.

DECAY HEAT   The heat produced by the decay of radioactive nuclides.

DECONTAMINATION   The removal of radioactive contaminants from surfaces, as by cleaning and washing with chemicals.

DESIGN BASIS ACCIDENT   The design basis accident is the most serious reactor accident that can be hypothesized from an adverse combination of equipment malfunction, operating errors, and other foreseeable causes.

DOPPLER EFFECT   A shift in the measured frequency of a wave pattern caused by movement of the receiving device or wave source. The moving receiver will intercept more or fewer waves per unit time depending on whether it is moving toward or away from the source of the waves. By analogy, in a reactor, since fission cross sections depend on relative velocity of the neutrons and the uranium atoms (neutron movement can be considered wave motion), vibration of the uranium atoms in a fuel element due to the increased operating temperature leads to the Doppler effect. This Doppler effect can vary the reactivity of the reactor.

DOSE   The amount of ionizing radiation energy absorbed per unit mass of irradiated material at a specific location, such as a part of the human body. Measured in reps, rems, and rads.

DOSE RATE   The radiation dose delivered per unit time and measured, for instance, in rems per hour.

DOSIMETER   A device that measures radiation dose, such as a film badge or ionization chamber.

ENRICHED URANIUM   Uranium in which the percentage of the fissionable isotope, uranium-235, has been increased above the 0.7% contained in natural uranium.

EXCESS REACTIVITY   Reactivity over and above that needed to achieve criticality. Excess reactivity is built into a reactor (by using extra fuel) in order to compensate for fuel burnup and the accumulation of fission-product poisons during operation.

EXCLUSION AREA   The area immediately surrounding a nuclear reactor where human habitation is prohibited in order to assure safety in the event of accident.

FALLOUT   Debris (radioactive material) that resettles to earth after a nuclear explosion. Fallout takes two forms. The first, called "local fallout," consists of the denser particles injected into the atmosphere by the explosion. They descend to earth within 24 hours near the site of the detonation and in an area extending downwind for some distance (often hundreds of miles), depending on meteorological conditions and the yield of the detonation. The other form, called "worldwide fallout," consists of lighter particles which ascend into the upper troposphere and stratosphere and are distributed over a wide area of the earth by atmospheric circulation. They then are brought to earth, mainly by rain or snow, over periods ranging from months to years.

FAST BREEDER REACTOR   A nuclear reactor that operates with fast neutrons and produces more fissionable material than it consumes.

FISSION PRODUCTS   Nuclei formed by the fission of heavy elements. They are of medium atomic weight, and many are radioactive. Examples: strontium-90, cesium-137.

FUEL CYCLE   The series of steps involved in supplying fuel for nuclear power reactors. It includes original fabrication of fuel elements, their use in a reactor, chemical processing to recover the fissionable material remaining in the spent fuel, reenrichment of the fissionable material, and refabrication into new fuel elements.

FUEL ELEMENT   A rod, tube, plate, or other geometrical form into which nuclear fuel is fabricated for use in a reactor.

FUEL REPROCESSING   The processing of reactor fuel to recover the unused fissionable material.

GAMMA RAYS   High-energy short-wavelength electromagnetic radiation emitted by nuclei. Energies of gamma rays are usually between 0.010 and 10 MeV. X rays also occur in this energy range, but are of nonnuclear origin. Gamma radiation usually accompanies alpha and beta emissions and always accompanies fission. Gamma rays are very penetrating and are best attenuated by dense materials like lead and depleted uranium.

GAS-COOLED REACTOR   A nuclear reactor in which gas is the coolant.

GENETIC EFFECTS OF RADIATION   Effects that produce changes in those cells of organisms which give rise to egg or sperm cells and therefore affect offspring of the exposed individuals.

HALF-LIFE   The time in which half the atoms in a radioactive substance disintegrate. Half-lives vary from millionths of a second to billions of years.

HALF-LIFE, BIOLOGICAL   The time required for a biological system, such as a man or an animal, to eliminate, by natural processes, half the amount of a substance which has entered it.

HEAT EXCHANGER   Any device that transfers heat from one fluid to another or to the environment.

HEAT SINK   Anything that absorbs heat; usually part of the environment, such as the air, a river, or outer space.

HETEROGENEOUS REACTOR   A nuclear reactor in which the fuel is separate from the moderator and is arranged in discrete bodies, such as fuel elements. Most reactors are heterogeneous.

HOMOGENEOUS REACTOR   A reactor in which the fuel is mixed with the moderator or coolant. Example: fused-salt reactor.

INDIRECT-CYCLE REACTOR SYSTEM   A nuclear reactor system in which a heat exchanger transfers heat from the reactor coolant to a second fluid which then drives a turbine.

ION EXCHANGE   A chemical process in which ions from two different molecules are exchanged.

IONIZATION   The process of adding electrons to, or knocking electrons from, atoms or molecules, thereby creating ions. High temperatures, electrical discharges, and nuclear radiation can cause ionization.

IONIZING RADIATION   Any radiation that directly or indirectly displaces electrons from the outer domains of atoms. Examples: alpha, beta, gamma radiation.

ISOTOPES   Atoms with the same atomic number (same chemical element) but different atomic weights. An equivalent statement is that the nuclei have the same number of protons but different numbers of neutrons. Thus, $^{12}_{6}C$, $^{13}_{6}C$, and $^{14}_{6}C$ are isotopes of the element carbon, the subscripts denoting their common atomic numbers, the superscripts denoting the varying atomic weights.

LINEAR ENERGY TRANSFER (LET)   The energy loss per unit path length, as given up by an ionizing particle to the medium it traverses.

LOW POPULATION ZONE An area of low population density sometimes required around a nuclear installation. The total number and density of residents is of concern in providing, with reasonable probability, that effective protective measures can be taken in the event of a serious accident.

MAXIMUM PERMISSIBLE CONCENTRATION That concentration of radioactive material in air, water, and foodstuffs which competent authorities have established as the maximum that would not create undue risk to human health.

MEGAWATT-DAY A unit used for expressing the burnup of fuel in a reactor; specifically, the number of megawatt-days of heat output per metric ton of fuel.

MEV One million electron volts.

MODERATOR A material, such as water or graphite, used in a reactor to slow down high-velocity neutrons, thus increasing the likelihood of further fissions.

NUCLEAR ENERGY The energy liberated by a nuclear reaction (fission or fusion) or by radioactive decay.

NUCLEAR POWER PLANT Any device, machine, or assembly thereof that converts nuclear energy into some form of useful power, such as mechanical or electric power. In a nuclear electric power plant, heat produced by a reactor is used to make steam, and the steam drives a turbine generator in the conventional way.

NUCLEAR REACTION A reaction involving an atom's nucleus, such as fission, neutron capture, radioactive decay, or fusion, as distinct from a chemical reaction, which is limited to changes in the electron structure surrounding the nucleus.

NUCLEAR REACTOR A device by means of which a fission chain reaction can be initiated, maintained, and controlled. Its essential component is a core with fissionable fuel. It usually has a moderator, a reflector, shielding, and control mechanisms.

NUCLEAR ROCKET A rocket powered by an engine that obtains from a nuclear reactor, rather than by chemical combustion, the heat for expanding a propulsive fluid, such as hydrogen.

NUCLEAR SUPERHEATING Superheating the steam produced in a reactor using heat from a reactor. Two methods are currently emphasized: recirculating the steam through the same core in which it is first produced (integral superheating) or passing the steam through a second, separate reactor.

NUCLIDE Any species of atom that exists for a measurable length of time. A nuclide can be distinguished by its atomic weight, atomic number, and energy state. A radionuclide is a radioactive nuclide.

ORGANIC-COOLED A nuclear reactor that uses waxlike organic chemicals, such as mixtures of polyphenyls and terphenyls, as coolant and usually also as moderator.

OVERPRESSURE The transient pressure over and above atmospheric pressure resulting from a blast wave from a nuclear explosion.

PHOTON A discrete quantity of electromagnetic energy. Photons have momentum but no mass or electrical charge.

PLUTONIUM A heavy, radioactive, metallic element with atomic number 94. Its most important isotope is fissionable plutonium-239, produced by neutron irradiation of uranium-238.

POWER DENSITY Rate of heat generated per unit volume of a nuclear reactor core.

POWER REACTOR A nuclear reactor designed for use in a nuclear power plant, as distinguished from reactors used primarily for research or for producing radiation or fissionable materials.

PRESSURE VESSEL A strong-walled container housing the core of most types of power reactors; usually also containing moderator, reflector, thermal shield, and control rods.

PRESSURIZED WATER REACTOR A power reactor in which heat is transferred from the core to a heat exchanger by water kept under high pressure to achieve high temperature and prevent boiling in the core. Steam is generated in a secondary circuit.

PRODUCTION A reactor designed primarily for large-scale production of plutonium by neutron irradiation of uranium-238. Also refers to a reactor used primarily for the production of isotopes.

PROMPT CRITICALITY The state of a reactor when the fission chain reaction is sustained solely by prompt neutrons, that is, without the help of delayed neutrons.

RAD The basic unit of absorbed dose of ionizing radiation. One rad is equal to the absorption of 100 ergs of radiation energy per gram of matter.

RADIATION The propagation of energy through matter or space in the form of waves. In atomic physics the term has been extended to include fast-moving particles (alpha and beta rays, free neutrons, etc.). Gamma rays and x rays, of particular interest in atomic physics, are electromagnetic radiation in which energy is propagated in packets called photons.

RADIATION PROTECTION GUIDE The total amounts of ionizing radiation dose over certain periods of time which may safely be permitted to exposed industrial groups and the public. These standards, established by the Federal Radiation Council, are equivalent to what was formerly called the "maximum permissible exposure."

RADIOACTIVITY The spontaneous decay or disintegration of an unstable atomic nucleus, accompanied by the emission of radiation.

RADIOACTIVITY CONCENTRATION GUIDE The concentration of radioactivity in air and water. Continuous consumption of air or water for 50 years, or until equilibrium is reached, yields "radiation protection guide" exposures. This Federal Radiation Council term replaces the former "maximum permissible concentration."

RADIOBIOLOGY The study of the scientific principles, mechanisms, and effects of the interaction of ionizing radiation with living matter.

RADIOISOTOPE An unstable isotope of an element that decays or disintegrates spontaneously, emitting radiation. More than 1,300 natural and artificial radioisotopes have been identified.

REACTIVITY A measure of the departure of a nuclear reactor from criticality. It is equal to the multiplication factor minus one and is thus zero precisely at criticality. If there is excess reactivity (positive reactivity), the reactor power will rise. Negative reactivity will result in a decreasing power level.

RECYCLING The reuse of fissionable material in irradiated reactor fuel which is recovered by chemical processing, reenriched, and then refabricated into new fuel elements.

REFLECTOR A layer of material immediately surrounding a nuclear reactor core which scatters back (reflects) into the core neutrons that would otherwise escape. The returned neutrons can then cause more fission and improve the neutron economy of the reactor.

REGULATING ROD A nuclear reactor control rod used for making frequent fine adjustments in reactivity.

RELATIVE BIOLOGICAL EFFECTIVENESS (RBE) The relative effectiveness of a given kind of ionizing radiation in producing a biological response as compared with 250,000 electron volt gamma rays.

REM Roentgen equivalent man. A unit of absorbed radiation dose in biological matter. It is equal to the absorbed dose in rads multiplied by the relative biological effectiveness of the radiation.

REP Roentgen equivalent physical. An obsolete unit of radiation dosage, now superseded by the rad.

ROENTGEN A unit of exposure of ionizing radiation. It is that amount of gamma or x rays required to produce ions carrying 1 electrostatic unit of electrical charge in 1 cubic centimeter of dry air under standard conditions.

SAFETY ROD A standby control rod used to bring about the rapid shutdown of a nuclear reactor in emergencies.

SOMATIC EFFECTS OF RADIATION Effects limited to the exposed individual, as distinguished from genetic effects. Large radiation doses can be fatal. Smaller doses may make the individual noticeably ill or may merely produce temporary changes in blood-cell levels detectable only in the laboratory.

TEMPERATURE COEFFICIENT (NEGATIVE) The change in reactor reactivity occurring when the operating temperature changes. The coefficient is said to be positive when an increase in temperature increases the reactivity, negative when an increase in temperature decreases reactivity. Negative temperature coefficients are desirable because they help to prevent power excursions.

TRITIUM A radioactive isotope of hydrogen with two neutrons and one proton in the nucleus. It is heavier than deuterium (heavy hydrogen). Tritium is used in industrial thickness gauges, as a label in tracer experiments, and in controlled fusion experiments.

URANIUM A naturally radioactive element with the atomic number 92 and an atomic weight of approximately 238.

The two principal naturally occurring isotopes are the fissionable uranium-235 (0.7% of natural uranium) and the fertile uranium-238 (99.3% of natural uranium), which by capture of a neutron and a decay sequence yields fissionable plutonium-239. Uranium-234 constitutes a minute fraction.

URANIUM HEXAFLUORIDE (UF_6) A compound of uranium and fluorine. UF_6 gas is the process fluid in the gaseous diffusion process.

URANIUM TETRAFLUORIDE (UF_4) A solid green compound; an intermediate product in the production of uranium hexafluoride gas, which in turn is used in the separation of uranium-235 in gaseous diffusion plants. Called green salt.

URANIUM TRIOXIDE (UO_3) An intermediate product in the refining of uranium. Called orange oxide.

VAPOR SUPPRESSION A safety system that can be incorporated in the design of structures housing water reactors. In a vapor suppression system, the space surrounding the reactor is vented into pools of water open to the outside air. If surges of hot vapors are released from the reactor in an accident, their energy is dissipated in the pools of water. Gases not condensed are scrubbed clean of radioactive particles by the bubbling.

WASTE, RADIOACTIVE Equipment and materials (from nuclear operations) which are radioactive and for which there is no further use. Liquid wastes are generally referred to as high-level (having radioactivity concentrations of hundreds to thousands of curies per gallon or cubic foot), low-level (in the range of 1 microcurie per gallon or cubic foot), and intermediate (between these extremes).

WATER BOILER A research reactor whose core consists of a small spherical container filled with uranium fuel in an aqueous solution. Heat is removed by a cooling coil in the core.

REFERENCES AND BIBLIOGRAPHY

REFERENCES

Atomic Energy Commission. Emergency core coolings report of the Advisory Task Force on Power Reactor Emergency Cooling. Clearinghouse for Federal Scientific and Technical Information, Springfield, Va. 22151 (1967).

Atomic Energy Commission. Safety standards for the packaging of radioactive and fissile materials, Chapter 0529 in AEC Manual. Superintendent of Documents, U.S. Government Printing Office, Washington, D.C. 20402 (August 22, 1966).

Beck, C. K. Requirements for transportation of nuclear fuels and wastes. Presented at ASCE National Meeting, Washington, D.C., July 25, 1969.

Belter, W. G. The management of radioactive wastes from fuel reprocessing. Presented at the Atomic Industrial Forum Workshop, Buck Hill Falls, Pa. (April 21, 1970).

Blanchard, R. L., H. L. Krieger, H. E. Kolde, and B. Kahn. Radiological surveillance studies at a BWR nuclear power station—estimated dose rates. Presented at the Health Physics Symposium on Aspects of Nuclear Facility Siting, Idaho Falls, Idaho, November 3–6, 1970.

Blomeke, J. O., and F. E. Harrington. Management of radioactive wastes at nuclear power stations, ORNL-4070. Clearinghouse for Federal Scientific and Technical Information, Springfield, Va. 22151 (1968).

Bond, V. P. The public and radiation from nuclear power plants. Presented at the Atomic Industrial Forum, Buck Hill Falls, Pa., April 22, 1970.

Code of Federal Regulations, Title 10 (Atomic Energy Commission) Part 20, Standards for protection against radiation, 10CFR20; Part 50, Licensing of production and utilization facilities. Appendix F, policy relating to the siting of fuel reprocessing plants and related waste management facilities, 10CFR50; Part 71, Packaging of radioactive material for transport, 10CFR71; Part 100, Reactor site criteria, 10CFR100 (December, 1968).

Coleman, J. R., and R. Liberace. Nuclear power production and estimated krypton-85 levels. Rad. Health Data Rep., 7:615–621 (November 1966).

Congressional Joint Committee on Atomic Energy. Hearings on environmental effects of producing electric power. Superintendent of Documents, U.S. Government Printing Office, Washington, D.C. 20402 (1969).

European Nuclear Energy Agency. Water cooled reactor safety, an assessment prepared for the Committee on Reactor Safety Technology, Paris, France (May 1970).

Federal Radiation Council. Background material for the development of radiation protection standards, Report No. 2. Superintendent of Documents, U.S. Government Printing Office, Washington, D.C. 20402 (September 1961).

Federal Radiation Council. Background material for the development of radiation protection standards, Report No. 5. Superintendent of Documents, U.S. Government Printing Office, Washington, D.C. 20402 (July 1964).

Federal Radiation Council. Background material for the development of radiation protection standards—protective action guides for strontium-89, strontium-90, and cesium-137, Report No. 7. Superintendent of Documents, U.S. Government Printing Office, Washington, D.C. 20402 (May 1965).

Freke, A. M. A model for the approximate calculation of safe rates of discharge of radioactive wastes into marine environments. Health Phys., 13:743–758 (July 1967).

Gamertsfelder, C. C. Regulatory experience and projection for future design criteria. Presented at the Southern Conference on Environmental Radiation Protection at Nuclear Power Plants, St. Petersburg, Fla., April 21–22, 1971.

Goldman, M. I., H. W. Calley, Jr., L. F. Garcia, J. A. Tiernan, and C. H. Poindexter. Effects of estimated radioactive effluents from the Calvert Cliffs Nuclear Power Station, NUS-TM-S-87. NUS Corporation, Rockville, Maryland 20850 (1969).

Goldstein, N. P., K. H. Sun, and J. L. Gonzalez. Radioactivity in fly ash from a coal burning power plant. Transactions of the American Nuclear Society 1971 Annual Meeting, June 13–17, 1971. American Nuclear Society, Inc., Hinsdale, Illinois (1971), pp. 66–67.

Heinisch, R. Transportation of nuclear fuel material in the United States, a report by Nuclear Assurance Corporation, Atlanta, Ga.

Illinois Pollution Control Board. Testimony of James M. Smith and Clifford Kent, General Electric Company, Springfield, Ill. December 9, 1970, and February 14, 1971.

International Atomic Energy Agency. Operation and control of ion-exchange processes for treatment of radioactive wastes, Technical Report Series #78. Vienna, Austria (1967).

International Atomic Energy Agency. Management of radioactive wastes at nuclear power plants, Safety Series No. 28. Vienna, Austria (1968).

International Atomic Energy Agency. Design and operation of evaporators for radioactive wastes, Technical Report Series #87. Vienna, Austria (1968b).

International Atomic Energy Agency. Treatment of airborne radioactive wastes. Parrish, E. C. and R. W. Schneider, Review of Inspection and Testing of Installed High-Efficiency Particulate Air-Filters at ORNL, STI/PUB/195. Vienna, Austria (1968c), p. 243.

International Commission on Radiological Protection. The evaluation of risks from radiation, ICRP Publication 8. Pergamon Press, Oxford, England (1965).

International Commission on Radiological Protection. Radiosensitivity and spatial distribution of dose, ICRP Publication 14. Pergamon Press, Oxford, England (1969).

Jacobs, D. G. Sources of tritium and its behavior upon release to the environment. AEC Division of Technical Information. Superintendent of Documents, U.S. Government Printing Office, Washington, D.C. 20402 (1968).

Joint Committee on Atomic Energy. Selected materials on environmental effects of producing electric power. Superintendent of Documents, U.S. Government Printing Office, Washington, D.C. 20402 (August 1969) pp. 223–320, 249–252, 507–521.

B. Kahn, R. L. Blanchard, H. L. Krieger, H. E. Kolde, D. B. Smith, A. Martin, S. Gold, W. J. Averett, W. L. Brinck and G. J. Karches. Radiological surveillance studies at a boiling water nuclear power reactor. PHS Publication, BRH/DER 70-1. Superintendent of Documents, U.S. Government Printing Office, Washington, D.C. 20402 (1970).

Kahn, B., et al. Radiological surveillance studies at a pressurized water nuclear power reactor. Environmental Protection Agency, Washington, D.C. (July 1971).

Keilholtz, G. W. Removal of radioactive noble gases from off-gas streams. Nucl. Safety, 8:155–160 (Winter 1966–1967).

Kelly, O. A., and W. C. T. Stoddart. Highway vehicle impact studies: tests and mathematical analyses of vehicle, package and tiedown systems capable of carrying radioactive material, ORNL-NSIC #61 (February, 1970). Superintendent of Documents, U.S. Government Printing Office, Washington, D.C. 20402. p. 139.

Leimkuhler, F. F., M. J. Karson, and J. T. Thompson. Statistical analyses of the frequency and severity of accidents to potential highway carriers of highly radioactive materials, NYO #9771, New York, N.Y. (July 1961).

Leimkuhler, F. F. Trucking of radioactive materials: safety vs. economy in highway transport. The Johns Hopkins Press, Baltimore, Md. (1963).

Lowman, F. G. The effects of the marine biosphere and hydrosphere upon the specific activity of contaminant radionuclides. Proceedings for the Symposium on Public Health Aspects of Peaceful Uses of Nuclear Explosives, Las Vegas, Nevada, April 7–11, 1969, pp. 436–459.

Martin, J. E., E. D. Harward, D. T. Oakley, J. M. Smith, and P. H. Bedrosian. Radioactivity from fossil-fuel and nuclear power plants. Proceedings of IAEA Symposium on Environ-mental Aspects of Nuclear Power Stations, Vienna, Austria (1969).

Merriman, J. R., et al. Removal of radioactive krypton and xenon from contaminated off-gas streams. Proceedings of the 11th Air Cleaning Conference, CONF-xxx-700816 (December 1970), pp. 175–203.

National Academy of Sciences. Radioactivity in the marine environment. National Academy of Sciences, Washington, D.C. (1971).

National Academy of Sciences–National Research Council. 1960. Biological effects of atomic radiation. National Academy of Sciences–National Research Council, Washington, D.C. (1956, 1960).

National Council on Radiation Protection and Measurements. Basic radiation protection criteria, Report No. 39. Washington, D.C. (January 15, 1971).

Oak Ridge National Laboratory. Annual progress report 1963, ORNL #3417. Oak Ridge, Tenn., p. 260.

Oak Ridge National Laboratory. Completion Report—Evaluation of the use of permselective membranes in the nuclear industry for removing xenon and krypton from various off-gas streams, Report #4522, Oak Ridge, Tenn. (April 1971).

Oak Ridge National Laboratory. Separation of noble gases from air using permselective membranes, ORNL #4145. Chemical Technology Division Annual Progress Report. Oak Ridge, Tenn. (October 1967).

Oak Ridge National Laboratory Staff. Siting of fuel reprocessing plants and waste management facilities, ORNL #4451. Clearinghouse for Federal Scientific and Technical Information, Springfield, Va. 22151 (July 1970).

Parsont, M. A., and M. I. Goldman. Effects of estimated radioactive effluents from the Davis-Besse Nuclear Power Station for the Toledo Edison Company, NUS-729, NUS Corporation, Rockville, Md. 20850 (November, 1970).

Offutt, G. F., and C. L. Bendixsen. Rare gas recovery facility at the Idaho Chemical Processing Plant, Publication No. IN-1221, Clearinghouse for Federal Scientific and Technical Information, Springfield, Va. 22151 (April 1969).

Pottier, P. E. Chemical treatment of radioactive wastes. Technical Report Series #89, IAEA, International Atomic Energy Agency, Vienna, Austria (1968).

Public Health Service. Public Health Service drinking water standards, revised 1962. PHS Publication No. 956, Superintendent of Documents, U.S. Government Printing Office, Washington, D.C. 20402 (August 1963).

Report to the Medical Research Council by its Committee on Protection Against Ionizing Radiations. Maximum permissible dietary contamination after the accidental release of radioactive material from a nuclear reactor, Brit. Med. J., 1:967–969 (1959).

Report to the Medical Research Council by its Committee on Protection Against Ionizing Radiations. Maximum permissible contamination of respirable air after an accidental release of radioiodine, radiostrontium, and cesium-137, Brit. Med. J., 2:576–579 (1961).

Rice, T. R., and J. P. Baptist. Ecological aspects of radioactivity in the marine environment. Presented at Environmental Radioactivity Symposium at the Johns Hopkins University, January 19–20, 1970.

Rogers, L., and C. C. Gamertsfelder. USA regulations for the control of releases of radioactivity into the environment in effluents from nuclear facilities. Presented at IAEA Symposium on Environmental Aspects of Nuclear Power Station, New York, N.Y. (August 1970).

Shappert, L. B., *et al.* A guide for the design, fabrication, and operation of shipping casks for nuclear applications, ORNL–NSIC #68. Clearinghouse for Federal Scientific and Technical Information, Springfield, Va. 22151 (February 1970).

Slansky, C. M., H. K. Peterson, and V. G. Johnson. Nuclear power growth spurs interest in fuel plant wastes. *Environ. Sci. and Tech.,* 3:446–451 (May 1969).

Southern Interstate Nuclear Board. Proceedings of Conference on Transportation of Nuclear Spent Fuel, Atlanta, Ga., February 5–6, 1970.

Stannard, J. N. Evaluation of health hazards to the public associated with nuclear power plant operations. Nuclear Power and the Public, Harry Foreman, M.D., Editor. University of Minnesota Press, Minneapolis, Minn. (1970), pp. 90–123.

Stephenson, M. J., V. R. Merriman, and D. I. Dunthorn. Experimental investigation of the removal of krypton and xenon from contaminated gas streams by selective absorption in fluorocarbon solvents. Publication No. K-1780, Oak Ridge Gaseous Diffusion Plant, Oak Ridge, Tenn. (1970).

Stewart, K. B. Rail accident statistics pertinent to the shipment of radioactive materials, HW #76299. Atomic Energy Commission, Washington, D.C. (January 21, 1963).

Subcommittee on Air and Water Pollution of the Senate Committee on Public Works. Hearings on underground uses of nuclear energy. Part I. Superintendent of Documents, U.S. Government Printing Office, Washington, D.C. 20402 (November 1969).

United Nations. Report of the United Nations Scientific Committee on the Effects of Atomic Radiation, Thirteenth Session, Supplement No. 17 (A/3838) (1958); Twenty-first Session, Supplement No. 14 (A/6314) (1966); Twenty-fourth Session, Supplement No. 13 (A/7613) (1969). United Nations, New York, N.Y.

Walchli, H. E. Nuclear fuel cycle and its transportation requirements. Presented at Conference on Transportation of Nuclear Spent Fuel, Southern Interstate Nuclear Board, Atlanta, Ga., February 5–6, 1970.

Wischow, R. P., *et al.* U.S. safeguards and experience in regulation and inspection of the private nuclear industry. Presented at the IAEA Symposium in Karlsruhe, Germany (July 6–10, 1971).

BIBLIOGRAPHY

Adcock, F. E., *et al.* Plutonium nitrate shipping packages, RFP-437. Atomic Energy Commission, Washington, D.C. (October 16, 1964).

American Nuclear Society. Procedures for considering abnormal reactor behavior in site evaluation. *Nucl. Engr. Bull.* (December 1963).

Atomic Energy Commission. AEC tentatively selects Kansas site for storage of radioactive wastes in salt mine, Press Release N-102. Atomic Energy Commission, Washington, D.C. (June 17, 1970).

Atomic Energy Commission. The Nuclear Industry—1970, Atomic Energy Commission, Washington, D.C.

Atomic Energy Commission. Nuclear terms. A brief glossary. USAEC Division of Technical Information Extension, Oak Ridge, Tenn. (April 1964).

Atomic Energy Commission. Policy statement regarding solid waste burial. Chapter 0511 of U.S. Atomic Energy Commission Manual (Immediate Action Directive 0511-21), Atomic Energy Commission, Washington, D.C. (March 20, 1970).

Belter, W. G. Ground disposal of radioactive waste in an ex-
panding nuclear power industry. Presented at the Annual Meeting of the American Institute of Mining, Metallurgical and Petroleum Engineers, Washington, D.C., February 18, 1969.

Belter, W. G., *et al.* New developments in radioactive waste management. Presented at the Fifteenth Annual Meeting of the American Nuclear Society, Hinsdale, Ill., June 16, 1969.

Belter, W. G., *et al.* The AEC's position on radioactive waste management. *Nucl. News,* Vol. 12, No. 11 (November 1969).

Blasewitz, A. G., *et al.* Interim status report on the waste solidification demonstration program, BNWL-1083. Battelle-Northwest, Richland, Washington (June 1969).

Blasewitz, A. G., *et al.* Status of the waste solidification demonstration program. Presented at American Nuclear Society Annual Meeting, Washington, D.C., November 18, 1970.

Blomeke, J. O. and F. E. Harrington. Management of radioactive wastes at nuclear power stations, ORNL-4070. Clearinghouse for Federal Scientific and Technical Information, Springfield, Va. 22151 (1968).

Bradshaw, R. L., and W. C. McClain. Project Salt Vault: a demonstration of the disposal of high activity solidified waste in underground salt mines, ORNL-4555. Superintendent of Documents, U.S. Government Printing Office, Washington, D.C. 20402 (to be published).

Campbell, B. F., *et al.* Current practice management of high-level radioactive waste in the United States of America. Proceedings of an International Symposium on Treatment and Storage of High-Level Radioactive Waste held by the International Atomic Energy Agency, Vienna, Austria, October 8–12, 1962.

Cochran, J. A. An investigation of airborne radioactive effluent from an operating nuclear fuel reprocessing plant. Northeastern Radiological Health Laboratory, Winchester, Mass. (July 1970).

Coleman, J. R., and R. Liberace. Nuclear power production and estimated krypton-85 levels. *Rad. Health Data Rep.,* 7:615–621 (November 1966).

Cowser, K. E., *et al.* Krypton-85 and ^{3}H in an expanding world nuclear power economy, ORNL #4168. Annual Report, Oak Ridge National Laboratory, Health Physics Division, Oak Ridge, Tenn. (July 1967).

Davis, W. K. Containment and engineered safety of nuclear power plants. Third United Nations International Conference on the Peaceful Uses of Atomic Energy, United Nations, New York (May 1964).

Dieckhoner, J. E. and G. E. Stigall. Radioactive waste management practices at nuclear power stations. PHS, Bureau of Radiological Health, Division of Environmental Radiation, Rockville, Md. 20852 (October 1970).

Federal Radiation Council. Background materials for the development of radiation protection standards, Report No. 1. Superintendent of Documents, U.S. Government Printing Office, Washington, D.C. 20402 (May 13, 1960); Report No. 2. (September 1961).

Federal Register. Latest revision of Code of Federal Regulations. Title 10 (Atomic Energy Commission) Part 20, standards for protection against radiation; Part 50, licensing of production and utilization facilities; Title 10 Part 100, reactor site criteria. Volume 35, No. 234 (December 3, 1970).

Federal Register. Siting of fuel reprocessing plants and related waste management facilities. Vol. 35, No. 222 (November 14, 1970).

Fowler, T. W., and D. E. Voit. A review of the radiological

and environmental aspects of krypton-85. Bureau of Radiological Health, Rockville, Md. 20852 (Revised June 1970).

General Electric Company. Design and analysis report IF 300 shipping cask, Docket No. 70-1220. General Electric Co., Schenectady, N.Y. (1969).

Goldman, M. I. United States practices in management of radioactive wastes at nuclear power plants, IAEA Safety Series No. 28, Vienna, Austria (1968).

Hearings before the Joint Committee on Atomic Energy. Environmental effects of producing electric power, Part 1. Superintendent of Documents, U.S. Government Printing Office, Washington, D.C. 20402 (October 28–31, November 4–7, 1969); Part 2 (Volumes 1 and 2), January 28–30 and February 24–26, 1970.

Hendrickson, M. M. The dose from krypton-85 released to the earth's atmosphere, BNWL-SA-3233A. Battelle-Northwest, Richland, Washington (July 1970).

International Commission on Radiological Protection. Recommendations of the ICRP, ICRP Publication 1. Pergamon Press, Oxford, England (September 1958).

International Commission on Radiological Protection. Report of Committee 11 on Permissible Dose for Internal Radiation (1959).

International Commission on Radiological Protection. Report No. 6 (1964).

International Commission on Radiological Protection. Principles of environmental monitoring related to the handling of radioactive materials, Report No. 7 (September 1965).

International Commission on Radiological Protection. The evaluation of risks from radiation, Report No. 8 (April 1965).

International Commission on Radiological Protection. Report No. 9. (September 1965).

International Commission on Radiological Protection. Report of Committee 4 on Evaluation of Radiation Doses to Body Tissues from Internal Contamination Due to Occupational Exposure (1968).

International Commission on Radiological Protection. Radiosensitivity and spatial distribution of dose, Report No. 14 (1969).

Irvine, A. R. Shipping-cask design considerations for fast breeder reactor fuel. Proceedings of the 16th Conference on Remote Systems Technology, American Nuclear Society, Hinsdale, Ill. (1969), pp. 197–204.

Joint Committee on Atomic Energy. Selected materials on environmental effects of producing electric power. Superintendent of Documents, U.S. Government Printing Office, Washington, D.C. 20402 (August 1969).

Kent, C. E., et al. Effluent control for boiling water reactors. Presented at the Symposium on Environmental Aspects of Nuclear Power Stations, International Atomic Energy, New York (August 1970).

King, L. J., A. Shimozato, and J. M. Holmes. Pilot plant studies of the decontamination of low-level process waste by a scavenging-precipitation foam separation process, ORNL #3803. Oak Ridge, Tenn. (1968).

Knapp, H. A. Cost and safety considerations in the transport of radioactive materials, TID-7569. Clearinghouse for Federal Scientific and Technical Information, Springfield, Va. 22151 (May 15, 1959), pp. 54–63.

Leonard, J. H., et al. Techniques for reducing routine release of radionuclides from nuclear power plants. University of Cincinnati, Ohio. (January 1971).

Lieberman, J. A. and W. G. Belter. Waste management and environmental aspects of nuclear power. Environ. Sci. and Tech. (June 1967).

Logsdon, J. E., and R. I. Chissler. Radioactive waste discharges to the environment from nuclear power facilities, BRH/DER 70-2. Clearinghouse for Federal Scientific and Technical Information, Springfield, Va. 22151 (March 1970).

Magno, P., et al. Liquid waste effluents from a nuclear fuel reprocessing plant. Northeastern Radiological Health Laboratory, Winchester, Mass. (November 1970).

Martin, J. E., et al. Comparison of radioactivity from fossil fuel and nuclear power plants. PHS, Bureau of Radiological Health, Division of Environmental Radiation, Rockville, Md. 20852 (November 1969).

Miner, S. Air pollution of radioactive substances, PB 188092. Litton Environmental Systems, Camarillo, Cal. (September 1969).

Morgan, K. Z. Acceptable risk concept. Lecture presented at the University of Florida, Gainesville, Fla. (November 4, 1969). Reprinted in hearings of the Joint Committee on Atomic Energy before the U.S. Congress on January 27–30, February 24–26, 1970, Part 2, Volume 1. Superintendent of Documents, U.S. Government Printing Office, Washington, D.C. 20402, pp. 1277–1301.

Morgan, K. Z. Adequacy of present standards of radiation exposure, testimony before the Subcommittee on Air and Water Pollution of the Senate Committee on Public Works. Superintendent of Documents, U.S. Government Printing Office, Washington, D.C. 20402 (August 1970).

Morgan, K. Z. Ionizing radiation: benefits versus risks. Health Phys. 17:539–549 (October 1969).

National Academy of Sciences–National Research Council. The disposal of radioactive waste on land, the report of the Committee on Waste Disposal of the Division of Earth Sciences, National Research Council, Publication No. 519. National Academy of Sciences–National Research Council, Washington, D.C. (April 1957).

National Council on Radiation Protection and Measurements. Recommendations of the NCRP, NCRP Report No. 39. National Council on Radiation Protection, Washington, D.C. (January 15, 1971).

Nuclear Power in the South, a Report on the Southern Governors' Task Force for Nuclear Power Policy, Atlanta, Ga., September 22, 1970.

Oak Ridge National Laboratory. Considerations related to the siting of fuel reprocessing plants and their associated waste management facilities, ORNL-CF-68-5-33, Rev. 1. Oak Ridge, Tenn. (August 14, 1968).

Oak Ridge National Laboratory. Siting of fuel reprocessing plants and waste management facilities, ORNL #4451. Oak Ridge, Tenn. (July 1970).

Office of Science and Technology. Considerations affecting steam power plant site selection, a report sponsored by the Energy Policy Staff, Office of Science and Technology. Superintendent of Documents, U.S. Government Printing Office, Washington, D.C. 20402 (December 1968).

Office of Science and Technology. Electric power and the environment, a report sponsored by the Energy Policy Staff, Office of Science and Technology. Superintendent of Documents, U.S. Government Printing Office, Washington, D.C. 20402 (August 1970).

Perona, J. J., et al. Comparative costs for final disposal of radioactive solids in concrete vaults, granite, and salt formations, ORNL-TM-664. Clearinghouse for Federal Scientific and

Technical Information, Springfield, Va. 22151 (October 23, 1963).

Proceedings of the Second International Symposium on Packaging and Transportation of Radioactive Materials, CONF-681001, Atomic Energy Commission, Washington, D.C. (October 14, 1968).

Regan, W. H., ed. Proceedings of the Symposium on the Solidification and Long-Term Storage of Highly Radioactive Wastes, AEC Report CONF-660208, Richland, Washington, February 14–18, 1966.

Rogers, L., and C. C. Gamertsfelder. U.S. regulations for the control of releases of radioactivity to the environment in effluents from nuclear facilities. Presented at IAEA Symposium on Environmental Aspects of Nuclear Power Stations, New York City, August 10–14, 1970.

Schneider, K. J. Status of technology in the United States for solidification of highly radioactive liquid wastes, BNWL-820. Clearinghouse for Federal Scientific and Technical Information, Springfield, Va. 22151 (October 1968).

Shappert, L. B. Cask designers' guide—a guide for the design, fabrication and operation of shipping casks for nuclear applications, ORNL-NSIC-68. Clearinghouse for Federal Scientific and Technical Information, Springfield, Va. 22151 (February 1970).

Sowly, F. D. Radiation and other risks. *Health Phys.* 11:879–887, 1965.

Starr, C. Social benefit versus technological risk. *Science* 165: 1232–1237 (September 1969).

Stewart, K. B. Rail accident statistics pertinent to the shipment of radioactive materials, HW-76299. Atomic Energy Commission, Washington, D.C.

Stigall, G. E., *et al.* Iodine-131 discharges from an operating boiling water reactor nuclear power station. Bureau of Radiological Health, Rockville, Md. 20852 (June 1970).

Strauss, S. D. Nuclear power plant safety. *Power* (January 1968) pp. 159–166.

Terpilak, M. S., C. L. Weaver, and S. Wieder. Dose assessment of ionizing radiation exposure to the population. *Rad. Health Data Rep.* (April 1971).

Terrill, J. G., *et al.* Public health factors in reactor site selection. *J. San. Engr. Div.* (June 1969).

Unruh, C. Present and projected sources of environmental radiation. Battelle-Northwest Memorial Institute, Pacific Northwest Laboratory, Richland, Washington (June 1970).

U.S. Department of Commerce. Maximum permissible body burden and maximum permissible concentrations of radionuclides in air and water for occupational exposure. Recommendations of the National Committee on Radiation Protection, NBS Handbook 69. Superintendent of Documents, U.S. Government Printing Office, Washington, D.C. 20402 (June 5, 1969).

U.S. Joint Congressional Committee on Atomic Energy. Environmental effects of producing electric power. Proceedings of hearings, Part 1. Superintendent of Documents, U.S. Government Printing Office, Washington, D.C. 20402 (1969).

U.S. Joint Congressional Committee on Atomic Energy. Environmental effects of producing electric power. Proceedings of hearings, Part 2. Superintendent of Documents, U.S. Government Printing Office, Washington, D.C. 20402 (1970).

Weaver, C. L. A proposed radioactivity concentration guide for shellfish. *Rad. Health Data Rep.* (September 1967).

Weaver, C. L. Radioactivity: sources and environmental distribution. Proceedings of a symposium held at the Johns Hopkins University (January 19–20, 1970) pp. 5–33.

Weaver, C. L. Tritium in the environment from nuclear power plants. *Rad. Health Data Rep.* 84:363–371 (April 1969).

Weems, S. J. The ice-condenser reactor containment system. *Nucl. Safety* (May–June 1970).

Working Group I(d)

Environmental Protection: Aesthetics and Land Use

INTRODUCTION

Without a regional planning process which includes land use guidelines, man has often failed to interact with his environment without abusing it. The incidence and degree of successful adaptation of his accelerating technology to his surroundings must be increased. We are in need of a comprehensive growth strategy that will tell us where to build and where not to build.

The prosperity of recent decades strengthened hopes for better conditions of life. But a new crisis has emerged.

This new crisis that was really an old one was a squib in the local newspapers at first—a footnote near the end of a television news broadcast. By the end of the 1960's, however, it captured headlines in newspapers and magazines in almost every nation. Man was destroying his environment.

For centuries men have been able to plan and build without the adverse impacts of their developments affecting more than a small fraction of their total environment. Nature's landscape could be treated as if it were only a tool for men's convenience, without the need for recognizing it as the key to their survival. However, with the accelerating population growth, coupled with accelerating use of nature's not inexhaustible resources, we must now take stock of the latent effect of such efforts on our life-support system.

We have advanced, but we have ignored the effects of these advances upon our life-support system. For example, urbanization and industrialization created countless opportunities, but they also created adverse side effects.

One of those opportunities for betterment of man's standard of living has been the development of electric power. With electric power, man's creative and productive energies have been multiplied many fold, while at the same time creature comforts have been increased. Such an opportunity has not been utilized without adverse effects.

Today, the American public is the willing, even demanding consumer of ever-increasing amounts of electric energy, to power necessities and luxuries, in a constant striving to upgrade the standard of living. During the past eight decades electric power demand has doubled every 10 years; it is estimated to increase sixfold by the year 2000. The time is past when we can make little plans, big though they be when compared to yesteryear, for supplying this constantly increasing demand. We must re-examine our goals and redefine our priorities, so that in haste we do not achieve a technically advanced and perfected life only to have ruined that which in our environment makes it really worth living.

1   MAJOR HUMAN VALUE PATTERNS IN SITING

PATTERNS OF HUMAN VALUE AND THEIR IMPLICATIONS TO POWER PLANT SITING AND ELECTRIC ENERGY TRANSMISSION

Utility transportation corridors in one form or another are necessary functional components of life in the mid-twentieth century. Each of the utilities (water, gas, telephone, power, and sewer) must be located in a manner appropriate to the specific system and within the fiscal and operational tolerances of the utility.

The siting of utilities to be developed in the future must be guided by reasonable and comprehensive siting criteria in order to foster harmony and suitability of each system in accordance with the single or multiple landscape characteristics of the area in which the utility is to be constructed.

The multiple impact on surface and subsurface landscape features plus the increased cost of service resulting from the implementation of a single corridor for each utility suggests the need to investigate the practicability of exploring the alternative of locating appropriate utilities within a common corridor. It must be recognized that criteria of location may differ for each system; the degree of variance has not been established, however. Research efforts should be undertaken to

1. Understand the optimum or acceptable site characteristics of utilities and transportation systems.

2. Determine the similarities and disparities of criteria established for the range of utilities investigated.

By using highway and utility corridors properly, the form and location of human impact can be shaped in harmony with nature's own patterns. Additionally, in determining where impact should be made, it should be remembered that there is more to life than just living, breathing, working, and eating. Man has a spirit that transcends the living of life for the sake of existing. It is important that patterns replenishing the spirit of man and offering hope for human well-being and happiness be preserved. There are many of these patterns, which are enumerated below. Their components require protection from impulsive action; they must be identified and protected from elimination for space. The individuals entrusted with the responsibility to site power plants or transmission rights-of-way must consider these human values and participate in their preservation by not eradicating them.

Historic Patterns

The United States is a nation of immigrants—a blending of differing ethnic groups, cultural values, and beliefs. Preserving these symbols of the American past will allow this generation and its posterity to understand the heritage from which they evolved. It is important to identify and protect these historic values as identified by regional and local historical societies or planning agencies.

Studies have indicated the variety of ethnic patterns and other such activities which are a part of our national heritage; an extensive variety of local architecture, cooking, handicrafts, customs, museums, and holidays exists within these cultural patterns. This

FIGURE 1   Historic patterns.

variety is important to environmental quality and needs continued recognition if it is not to be submerged in the current tendency toward conformity.

The historic identity serves not only as an important aspect of environmental quality and as a tie with the past, but serves also as an important recreational and tourist attraction. It is a heritage not to be exploited, but to be protected and valued. It can continue to help make life interesting and pleasant to both residents and visitors.

The physical–visual quality of historic resources (notably structures) should not be affected by positioning power plants or transmission lines in the visual context of historic features.

Landscape Personalities

Pattern combinations of water, topography, wetlands, and forests give a region its distinctive landscape char-

acters. Construction reflecting these personality patterns can help to preserve the visual integrity of the indigenous landscape.

The varied forms and combinations of man-modified natural resources can be identified in different parts of the landscape that give each area its distinguishing characteristics. The visual sum or result of these combined patterns of water, topography, wetlands, or forests is a unique series of regional personalities. The various three-dimensional visual patterns of agricultural production, urbanization (townscape), and transportation, also have their own unique personality patterns and add to the perceptual patchwork that is our environment, and must be preserved.

Environmental Corridors

Environmental corridors are areas relatively free from human impact; they are simply a pattern of "remaining

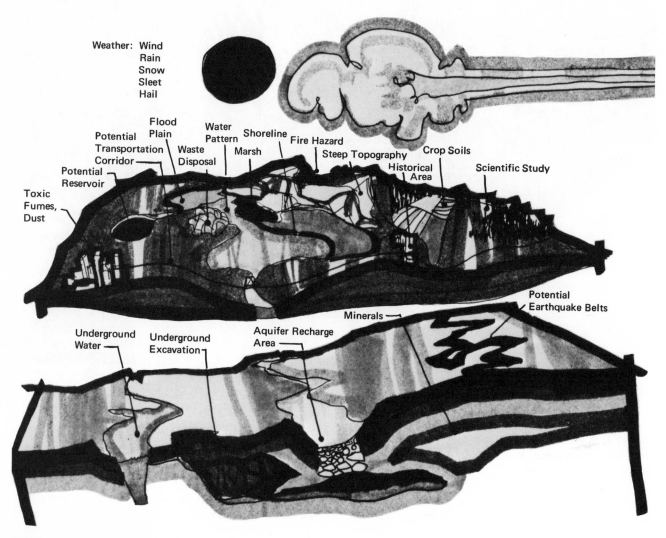

FIGURE 2 **Landscape personalities.**

FIGURE 3   Environmental corridors.

nature" as yet relatively untouched by the plow, saw, and blade or other form of human impact. Protected and developed wisely, these patterns offer future generations the same natural and cultural amenities that present generations enjoy today.

Within these patterns are found the unifying elements of nature (e.g., color, texture, pattern, and scale), a living organic and changing model that offers variety and serves as an all-pervading environmental network that makes the often haphazard misdesigns of man tolerable.

The natural elements within the corridors are the same as those brought back only at great cost in major redevelopment and renewal programs. On the pages that follow, the major environmental corridor patterns are illustrated and briefly described.

Land Form Corridors

These are comprised of a combination of "slopes" and "rims" that produce a ridge or land form corridor.

Slope Corridor   The basic feature that separates ridge tops from the valley floor is called *slope*. There has been scientific evidence that farming of slopes greater than 12½ percent often creates serious soil erosion problems.

The land may be too steep to plow, but for a man it is a place to scale either in body or mind. The fact that slope patterns have been difficult to scale has protected them from heavy human impact. These patterns remain unblemished in many parts of the country offering natural patterns of resource diversity for human enjoyment. The value derived from these resources should not be altered (by views of plants or the physical placement of structures on these features) unless no other viable alternative exists.

Rim Corridor   At the uppermost edge of a slope can be seen what might be called a *rim*. The rims often provide a pathway by foot or vehicle through an ever-changing variety of views and landscape forms—trees,

FIGURE 4   Slope corridor.

FIGURE 5   Rim corridor.

slopes, and rocks. Preservation of this natural pathway along its course will allow for a sequence of visual pleasure. These various characteristics can be classified. In one place along the rim one may find bare rock, in another place trees, all constituting a series of situations that provide a variety of experience.

Along the rim trails, natural observation platforms protrude from which one can view the surrounding landscape—down valleys and up nooks, spreading out into a quiltlike pattern.

Water–Wetland Corridors

Water–wetland corridors are the second major form of environmental corridor.

1. *Water*—Kept clean, water offers vast acreages of resource quality and open space.

2. *Wetlands*—Headwater marshes, wildlife habitat, sources of natural springs and aquifer recharge patterns produce wetlands.

3. *Floodplains*—Exceptional recreational and open space systems are offered by floodplains which offer little opportunity for safe man-made development.

4. *Sandy Soils*—When adjacent to water, sandy soils offer outstanding swimming areas.

These "surface patterns" of water, wetlands, floodplains, and sandy soils adjacent to water are in many cases enclosed by varying degrees of "slope."

Since slopes are subject to various degrees of erosion, they should be stabilized to prevent silting and pollution of the "surface" pattern below. Thus, any structures located on steep slopes, lakes, rivers, or streams should be stabilized immediately.

The rims of slope offer the best advantage spot from which to observe and contemplate the surface resources below. The rims also offer suitable traversing strips for bridle, hiking, and bicycle trails or parkways.

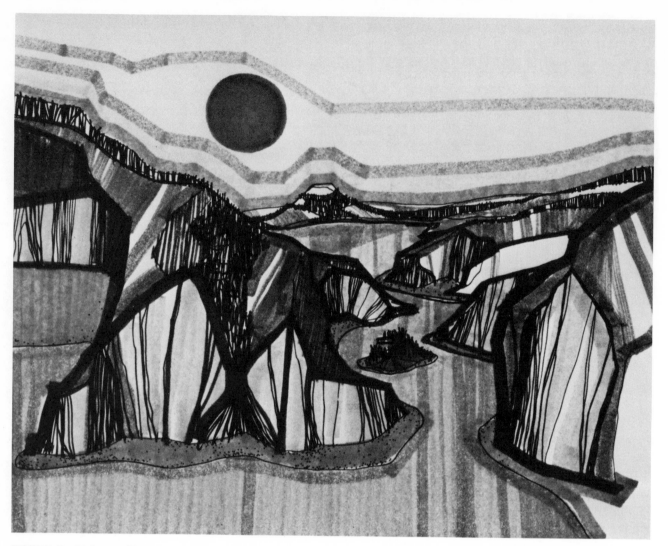

FIGURE 6 Water corridor.

FIGURE 7 Wetland corridor.

Water Corridor Springs, brooks, streams, and rivers are located here. In addition to serving as sources of recreation and transportation, these corridors provide a medium in which the world around them can be mirrored. Spoilage of lakes, rivers, or streams through careless action may destroy the regional habitat for fish and fisherman alike.

Wetland Corridor Adjacent to the springs, streams, rivers, and lakes, we may observe the wetlands. The wetlands are of value as a native environment for fish, animal and bird life. They provide a nesting, raising, and feeding ground for many types of wildlife. Some wetlands and certain soil types in and near our water systems provide *aquifer recharge* areas. These porous soils allow the surface water of an area to refill the underground water storage system for future use by man.

NATURAL AND CULTURAL VALUE PATTERNS

In addition to the mapping of corridor resources, there are many natural and cultural features in the landscape that are held in high esteem by the public as major environmental value patterns. Unlike corridor patterns which thread through the landscape, *most* of these supplementary resources have been found to occupy specific sites.

A study conducted by the State of Wisconsin * provided an opportunity to identify such features. A study of this kind can foster and expand awareness of the quality resources that must constitute an aspect of a comprehensive landscape planning program. These supplementary resources are outlined in Table 1.

---

* Philip H. Lewis, Jr., *Landscape Resource Inventory*, State of Wisconsin Department of Resource Development (May, 1964).

TABLE 1 Recreation Resource Values Identified in Wisconsin

I. WATER RESOURCES

Natural Resources
Intrinsic Resources

1. Waterfalls
2. Rapids, Whitewater
3. Bathing Beaches
4. Agate Beaches
5. Natural Springs, Artesian Flows
6. Canoe Routes
7. Wild Rice Areas
8. Exceptional Islands
9. Fish Habitat
10. Chasms
11. Trout
12. Muskellunge
13. Walleye
14. Bass
15. Northern Pike
16. Sturgeon
17. Catfish
18. Panfish

Man-made Facilities
Extrinsic Resources

19. Swimming Facilities
20. Boating Facilities, Ramps
21. Fuel, Repair, and Supplies
22. Marinas
23. Boating Areas
24. Outfitting Posts
25. Harbors of Refuge
26. Campsites
27. Canals
28. Dams, Fishways, Drainage Ways
29. Locks
30. Lighthouses

31. Fish Hatcheries
32. Mill Ponds
33. Reservoirs
34. Shelters for Ice Skating Areas

II. WETLAND RESOURCES

Natural Resources
Intrinsic Resources

35. Exceptional Wetlands
36. Wildlife Observation
37. Wildlife Hunting

Man-made Resources
Extrinsic Resources

38. Observation Platforms
39. Wetland Projects, Levees, Ditching and Diking
40. Wildlife Preserves
41. Hunting Preserves

III. TOPOGRAPHIC RESOURCES

Natural Values

(Unique Geological Formations)
42. Caves
43. Balanced Rocks
44. Castle Rocks
45. Exceptional Glacial Remains
46. Natural Bridges
47. Stones and Fossil Collection Areas
48. Mineral Ore Outcroppings
49. Outstanding Soil Conservation Projects—also farm conservation

Man-made Values

50. Ski Lifts
51. Ski Rope Tows

52. Ski Slope Structures
53. Snow Play Areas, Sledding, etc.
54. Ski Trails
55. Ski—(Cross Country)
56. Riding
57. Hiking
58. Nature Trails
59. Trail Shelters
60. Picnic Areas
61. Golf Courses
62. Youth Camps
63. Nature Camps
64. Day Camps

IV. VEGETATION RESOURCES

Natural Resources

65. Virgin Stands (Timber)
66. Rare Remnants
67. Outstanding Reforestation Projects
68. Wildflowers
69. Prairies
70. Specimens (Trees, etc.)
71. Unusual Crops
72. Orchards
73. Fire Towers
74. Fire Trails and Breaks

Man-made Resources

75. State Forests (Existing Potential)
76. County Forests (Existing Potential)
77. County Parks (Existing Potential)
78. State Parks (Existing Potential)
79. State Recreation Areas (Existing Potential)

TABLE 1—Continued

V. HISTORICAL AND CULTURAL
RESOURCES

Man-made Resources

80. Blacksmith Shops
81. Bridges (Covered, etc.)
82. Trading Posts
83. Old Mills
84. Taverns, Saloons
85. Old Mines
85A. Pioneer Church
86. Opera Houses
87. Historical Homes
88. Old Forts
89. Barracks
90. Lumber Camps
91. Battlefields
92. Historical Markers
93. Museums
94. Restaurants (Unusual
    Native Dishes)
95. Native Handicrafts
    (Draftsmen's Shop)
96. Local Festivals, Celebrations
97. Outstanding Farmers Markets
98. Modern Mines
99. Power Plants
100. Modern Mills
101. Interesting Industries
102. Commercial Fishing
103. Berry Picking
104. Ghost Towns
105. Rifle Shooting Ranges
106. Archery Ranges
107. Sugar Bush
108. Songbirds
109. Aesthetic Areas
110. Art Museums

VI. ARCHEOLOGICAL RESOURCES

111. Outstanding Buildings
112. Theaters

113. Ex. Public Lands
114. Ex. Private Lands
115. Prop. Public Lands
116. Prop. Private Lands
117. Effigy Mound
118. Sugar Bush
119. Petroglyph
120. Quartzite
121. Pipestone
122. Steatite
123. Quarry Flint
124. Copper
125. Lead
126. Quartz
127. Chlorite
128. Campsite
129. Village Site
130. Circular Enclosure
131. Square Enclosure
132. Rectangular Enclosure
133. Wild Rice
134. Cornfield
135. Garden Bed
136. Trail
137. Ford
138. Fort
139. Battlefield
140. Cache Pits
141. Workshop
142. Historic Village Sites
143. Provision Cache
144. Shell Heap
145. Ceramic Artifacts
146. Conical Mound
147. Mound Group
148. Mound—Round–Oval
149. Historic Cemetery
150. Prehistoric Cemetery
151. Stone Grave
152. Burial Ground
153. Grave

VII. WILDLIFE

Natural Resources

BIG GAME

154. Bear
155. Bobcat
156. Wolf
157. Deer
158. Red & Grey Fox

SMALL GAME

159. Pheasant
160. Quail
161. Woodcock
162. Hungarian Partridge
163. Ruffed Grouse
164. Sharp-tailed Grouse
165. Prairie Chicken
166. Muskrat
167. Beaver
168. Mink
169. Otter
170. Badger

WATERFOWL

171. Ducks
172. Geese
173. Swans

BIRDS

174. Eagles
175. Red-tailed Hawks
176. Herons
177. Great Horned Owls
178. Egrets
179. Osprey
180. Falcons
181. Goshawks
182. Cranes
183. Loons
184. Ibis
185. Hawks

THE COMMUNITY STAKE IN POWER PLANT LOCATION DECISIONS

Much of this report concerns technical issues relative to the aesthetics and land use considerations in the siting of power plants and transmission facilities. This section of the report seeks to bring out the importance of the citizen role in planning for power facilities. In this respect, there are three areas of citizen concern: a concern with resource use and the environment; a concern for public safety, primarily from nuclear plants; and a concern about the aesthetics of plant and transmission facility locations.

While each of these issues involves different technical considerations, they reflect a common need and responsibility on the part of power agencies to develop and maintain close communications with the community at large. Below is a summary of the issues and needs in the present situation, followed by a discussion of research approaches.

Present Situation: Issues and Needs

The past ten years have witnessed a marked change in public sensitivity to development activities in the private sector and, among them, the siting of power facilities. This sensitivity is an outgrowth of two fundamental and pervasive forces which have been in chain reaction since the turn of the century: (1) an exponential rate of acceleration in scientific endeavor, the generation of knowledge, and the creation of new technologies, and (2) a steady application of this know-how to the indus-

trial economy. Given the limitations of resources and a fixed land supply, it was only a matter of time before the effects of such powerful forces would become visible. Given also the rise of industrial complexes, governmental bureaucracies, and the remoteness and diffusion of policy-making and answerability in both sectors, it was inevitable that the public would be deeply concerned about the implications of these development pressures on the natural environment.

Viewed in this perspective, it is highly probable that public interest in the environment will increase rather than diminish. It therefore seems reasonable to expect that organized citizen groups will continue to seek more direct involvement in industrial and governmental decisions affecting the natural environment, public safety, and visual satisfaction. The new thrust appears to be carrying the public sector toward a more positive custodial role in matters of resource use and development, at the same time pressing for more responsibility and responsiveness to these issues from the private sector. As citizen concern becomes more organized, it is quite clear that the credibility of Madison Avenue approaches to such problems as the impact of oil or coal extraction on ecosystems or the effect of energy resource development and water use on the natural environment will be increasingly called into question, no matter how scientific the private sector's research or how careful the regulation of new facilities by the public sector may be. Clearly this is not a temporary flurry of citizen concern.

Customarily, the guardianship of the public interest in land use and other land development matters is vested in a planning commission which is established by the governing body of a given political jurisdiction on the basis of powers conferred by state enabling legislation. Planning commissions are authorized to prepare comprehensive plans for land development in their jurisdiction and recommend public improvement programs, zoning regulations, and other means of implementing these plans in the public interest. Depending on the provisions of state enabling legislation, which vary from one state to another, public hearings are held on public improvements (for example, rights of way for transmission lines, highways, and so on), zoning proposals and amendments (which, among other things, regulate the location of power plants, substations, and, in some ordinances, transmission towers), and other similar actions proposed by the governing body. These are usually one-time hearings, and because of the tremendous investment of staff effort which has gone into developing the plans, there is a tendency for public agencies to protect this "investment" and to seek to contain objections made at such hearings. Hearings tend to attract special interest groups and individuals whose interests are directly affected. Class actions by consumer groups are beginning to change the scenario. For these advocate groups and the public at large, the perfunctory aspect of the public hearing process is found wanting, especially in view of the vital stake the public has in the environment and the increasingly active role sought in decisions affecting the environment.

The concept of "the public" and who the public is needs to be re-examined. The social changes growing out of the sixties make it quite clear that hereafter, there can be no single public. There are multiple publics. Center-city people perceive the natural environment differently than do suburbanites, and metropolitan area residents see it differently than do residents of smaller places. There are regional variations—San Francisco Bay area residents see natural surroundings differently than Chicagoans or Philadelphians. Economic circumstances make a difference, and families in different stages of the life cycle see the environment differently. This diversity of value orientations and life styles means that the environment is perceived from a multiplicity of viewpoints. This variability is particularly marked in the aesthetic aspects of the environment.

Taken together, these considerations suggest that a new kind of approach is needed to decisions on power plant sites, especially where nuclear plants are proposed, and on the location of transmission rights-of-way, especially where overhead lines are proposed. In place of an emphasis on "educating the public," a new philosophy is needed for "involvement of the public"—involvement of all organized and unorganized segments of the community served by these facilities. In turn, "involvement" means a set of procedures for continuing opportunities in the communication and interchange between developers of energy and the various publics concerned. In lieu of "education," the accent is on a sequence of "consultations" as diagrammed in Table 2.

Traditionally the public sector has consisted of the planning agency and legislative body; in the future the concept of the public sector would be an expansion of governmental involvement, embracing a broader scope in communications. This entails a new concept of consultation—what is called here "community consultation."

The introduction of community consultations of this kind might be a joint effort of the planning agency and the power company in the affected jurisdiction. By whatever approach, joint or separate, it should be developed on a regional scale. In common with water resources, transportation, public open space, and other regional functions, the costs and benefits of particular structures and facilities may affect local jurisdictions differentially. This underscores the importance of regional solutions to resource use and development, closely coordinated with local plans for growth and development. In turn, it means also a coordinated approach to community consultations.

Community Consultation: Research Approaches

The notion of community consultation includes some of the traditional means of communication such as regular information reports through the mass media, public hearings, and formal governmental reviews and approvals. However, two additional modes of communication are proposed—one especially suited to unorganized groups and the other to organized groups. Both serve a purpose in gathering information on perceived qualities of the environment, the extent and basis of local anxieties about safety, the preference structure about development alternatives, and the visual aspects of strong concern. At the same time, in the process of consultation, factual information about needs, plan alternatives, and cost–benefits can be communicated.

For the unorganized community-at-large, a survey research approach of the kind being used increasingly in assessing housing preferences and environmental satisfaction can be used effectively. A professionally designed sequence of surveys, administered to a probability sample of household heads and spouses, can establish perceived concerns among persons of different life styles, identifying sources of anxiety and getting at preference patterns. Using a panel-type survey in which respondents are revisited and consulted in the successive cycles indicated in Table 2, panel members, in effect, become consultants in the process of developing plans and evaluating them. Results can be statistically ana-

TABLE 2   Schematic of Consultations on Siting Considerations

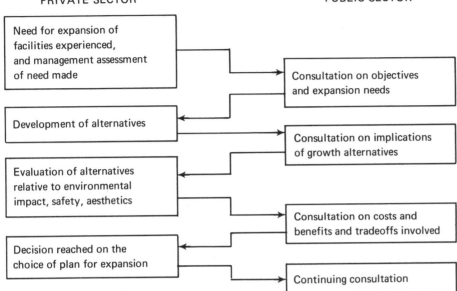

PRIVATE SECTOR

PUBLIC SECTOR

Need for expansion of facilities experienced, and management assessment of need made	Consultation on objectives and expansion needs
Development of alternatives	Consultation on implications of growth alternatives
Evaluation of alternatives relative to environmental impact, safety, aesthetics	Consultation on costs and benefits and tradeoffs involved
Decision reached on the choice of plan for expansion	Continuing consultation

lyzed and interpreted, and through creative use of mass media television can be used to extend the effectiveness of survey contacts in the community. In some cities, TV public service presentations tied in with neighborhood forums have been effective in communicating with citizens.

For organized community groups, a more formalized recurring series of consultations can be particularly effective. Techniques for recording the visual environment in its predevelopment state, for simulating changed states, and for taking a subject through the modified environment as it would be seen on foot or as seen by a passenger in an automobile are now being used in experimental form. Applicable as an interactive device in the above-mentioned TV–neighborhood forum mode of communication, environmental simulation of this kind has particularly promising possibilities for use in gaming a location decision in the course of working with small groups from citizen organizations. By exchanging roles, the representative of the power company, the city planning professional, and the citizen can learn much about the other's perceptions. This promises to be a means of anticipating controversies that would arise over power facility sitings and of developing modifications to planning proposals for resimulation and retest in gaming situations.

HUMAN ATTITUDES AND POWER COMPONENTS: A SUGGESTED SURVEY RESEARCH METHOD

Besides the system or method which manipulates variables and their representative sets, critical establishment and input of data into the system is essential. Human attitudes and desires are an essential part of the selection and implementation process. Methods for infusing human attitude data into the planning process are most difficult. The following survey research method, developed for measuring attitudes about the urban-suburban visual characteristics, appears worthy of use in determining human attitudes about transmission facilities in relation to their environmental context. The general method has been adapted to the particular problem of transmission line location.

The Applied Survey Research Method

A fundamental assumption underlying an improved and environmentally sensitive alignment methodology is that the individual is important in the total milieu of contemporary society. As an individual views his world, he may feel that objects are very much the same as he journeys from place to place: a tree here; a building there; a power structure; a hillside; and so forth. Closer observation, however, reveals that the volume and relationship of these objects can be perceived in vastly different ways. In his book, *Townscapes,* Gordon Cullen (1964) states:

In fact, there is an art of relationship just as there is an art of architecture. Its purpose is to take all the elements that go to create the environment; buildings, trees, nature, water, traffic, advertisements, and so on, and to weave them together in such a way that drama is released.

At first thought, the question of what physical elements are required to satisfy human needs may seem to be easily resolved, particularly with reference to such a generalized concept as "human needs." The diverse composition of a national population exceeding two hundred million, however, suggests that daring to state a definition that includes all people requires a monumental generalization and no little nerve.

It must be recognized that the characteristics of environmental quality or lack of quality can be as different as (1) professional dogma, (2) lay attitudes, or (3) an interpretation founded upon assimilation of lay and professional quality interpretation, which will surely detract from the completeness of interpretation.

The Process and Value of Survey Research Method

In the absence of documentary evidence of lay attitudes towards environmental conditions, it is necessary to survey or invite every possibility for public participation to individuals from that portion of society. This will provide a broad representation of attitudes.

The potential value to be derived from survey research (or social measurement) will be only as good as the quality of research questions, questionnaire administration, and data analysis.

In any survey research project, the researcher is interested in gaining information about a population for which information on a specific subject is nonexistent or unreliable. In selected instances, the survey is designed to engender information about a population, great or small in number. What is sought in the present instance is the nature and scope of negative visual reaction to power transmission facilities among various groups of society.

The type and quality of information acquired through survey research stems from the structuring and organization of the research design. The *process,* therefore, is a most important aspect of an attitude survey. The recognized order of research is outlined below:

1. Determine the information desired.
2. Determine the population from which that information is desired.
3. Prepare a questionnaire designed to gain the desired information.
4. Conduct a questionnaire pretest.
5. Administer the questionnaire.
6. Evaluate the returns.
7. Develop recommendations to correct the problem or maximize an opportunity.

The determination of what comprises visually and environmentally acceptable or unacceptable power transmission facilities may differ by geographical region as a result of diverse factors including (1) the conditioning of a population to the presence or absence of structures and conductors, (2) the degree of visual access to structures and conductors, (3) the density of structures and conductors, and (4) their design and configuration.

In order to accurately articulate public responses to transmission forms, a variety of population groups (age, sex, income, etc.) by geographical region must be surveyed to exact individual response to the question of environmental quality and visual compatibility. Following a random sampling of a region's population, a collective interpretation of environmental quality can be accomplished.

A survey research effort should consider the following:

1. On what basis do various age, sex, income, and education groups judge the appearance of transmission form?
2. Do people judge transmission form unsightliness as a product of quantity?
3. In what instances does isolated form (structure leg, body, insulator, conductor) surpass the importance of quantity and vice versa, and why?
4. Do certain combinations of landscape and transmission forms evoke variant value responses, and, if so, why?
5. Do regional and geographic differences result in different responses?
6. Are urban and rural values apparent in responses? (disparities and similarities)

To better understand some of the inherent characteristics of the environment, the recognition that the professional must know more about lay attitudes concerning power transmission forms suggests the possible adaptation of a graphic measuring technique to the analysis of the various visual elements comprising the observed scene. This is known as the environmental analyzer technique.

The distinct advantage of the environmental analyzer lies in the possibility it presents for the calculation of the percentages of the total visual field that the various significant components of the environment (including structures and conductors) comprise. By using percentages, the designer is able to determine what differences exist in one view as contrasted with a different view. This technique enables the user to analyze the environment systematically, and therefore to develop a better understanding of some of the factors which contribute to pleasing or displeasing visual stimuli.

The environmental analyzer enables the form giver to say, for example, that view A consists of 3% power structure and view B consists of 15% power structure. This results in more precise communication than merely stating that structures are more dominant in view B than view A.

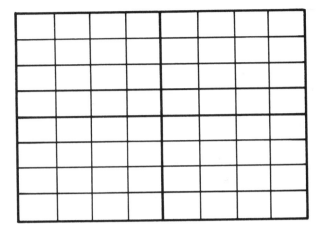

The measurement grid is an exact duplicate of the image area of a 35-mm slide. It comprises the most important aspect of the environmental analyzer technique, and consists of 64 equal-sized rectangles. The total area of the grid represents 100% of the human field of vision. Each rectangle therefore represents approximately 1.56% of the visual field.

In order to use the environmental analyzer, one must first determine what views are to be analyzed and from what vantage point. Then one decides upon the visual components that are likely to be present in each of the views.

Significant visual components include controlled open spaces (around buildings); natural open spaces (such as fields and prairies); buildings; nonsurface components such as utility poles, signs, and street lights; pedestrian paved areas; and vehicular paved areas. In the analysis, each of several 35-mm slides is projected onto the measurement grid, and the percent-age of each significant component in each view is estimated by noting the amount of coverage of the component within each of the rectangles where that particular component appears.

Lay Participation in the Environmental Analyzer Technique

The desire to understand lay values as they relate to environmental quality involved the exposure of a sample population from a specified area to 35-mm slides or pertinent visual features. Responses can be recorded on response forms like the one below:

VISUAL RESPONSE FORM

Directions:

Rating Scale	Comments
1 = very undesirable view	In this column you are
2 = undesirable view	asked to write a brief de-
3 = neutral (no opinion)	scription of why you rated
4 = desirable view	the view as you did.
5 = very desirable view	

The analysis of participant responses is comprised of three stages: determining the average response to each view as recorded on the attitude response form; reviewing the verbal responses and determining why the test population responded as it did; and projecting each view onto the measurement grid. The ratio of significant visual components can then be related to the value and rationale response representing the average.

In an era of increasing population and large-scale design, design professionals must not lose sight of the importance of the individual and the need for diversity in planning and testing. This task can be more adequately accomplished by establishing communication between the physical designer and the public. Herein lies the importance and the value of the attitude survey.

3 ENVIRONMETRICS: THE COLLECTION OF NATURAL AND CULTURAL DATA

Data variables relevant to specific regional landscape characteristics—both natural and cultural—can be stored in several two-dimensional formats, including a computer data bank system and standard graphic overlays.

There is necessarily a complex array of aspects related to the problem of locating power plants and transmission lines, depending upon location, potential economic investment, and other varied constraints which are relevant in a particular situation. Since the transmission facility is by nature both linear and extensive, and therefore is in physical contact with more environment, the specific environmental diversity transected is varied and the management relationships are complex.

As negative and positive aspects accompany interstate highway construction, for example, similar plus and minus effects are associated with power plant siting or transmission rights-of-way. A report published by the professional office, Landscapes Limited, states:

Historically, the impact and resultant cultural and environmental change from a transmission facility has not normally been a measurable part of the locational process. Admittedly, the quantifiable impact of an energy transmission system and an interstate highway is not of the same magnitude. But even though the magnitude of the problem between highway and transmission linkages is not the same, some recent work with highway location methodologies indicates that many of the same principles and approaches may be applicable. Complexity, prediction, and resultant environmental–ecological ramifications of impacts are a critical aspect of a method of locating applicable linkage systems. As the result of an involvement with the location of an interstate linkage (Highway 57) between Green Bay and Milwaukee, Wisconsin, a methodology has evolved which is capable of coping with complex, interactive decisions.

The method is dependent upon storing relevant and objective engineering, cultural, and environmental data in a computer data bank format. Since the transmission location problem is spatial (effecting units of land), the data is compiled and stored in a cellular manner and when combined spatially depicts the context of features and linkages as they exist in the landscape.*

Although the optional format of data storage relates to the scope and complexity of a given problem or landscape, the primary component of comprehensive problem-solving lies in the development of a data bank as comprehensive as financial and intellectual resources permit.

Although the scope and quality of current information is far below limits of acceptability in many instances, every effort should be directed to improve the quantity and quality of data that must be used to make power plant location and transmission alignment decisions.

SUGGESTED INFORMATION VARIABLES: NATURAL AND CULTURAL

Relevant, current, and compatible data are a basic requirement for intelligent decision-making related to comprehensive planning. Diverse and current data are absolutely essential to aid in the decision(s) of siting power plants and transmission rights-of-way.

---

* Landscapes Limited, *A Suggested Transmission Line Location Methodology,* Madison, Wisconsin, 1970.

TABLE 3 Suggested Cultural Data Types

I. TRANSPORTATION
  A. Highways—All
    1. Over 8 Lanes
    2. 6–8 Lanes
    3. 5–6 Lanes
    4. 4 Lanes
    5. 2 Lanes (Heavy Use)
    6. 2 Lanes (Medium Use)
    7. 2 Lanes (Light Use)
    8. Dirt Road
    9. Highway Interchanges
  B. Railroads
    1. Single Track
    2. Multiple Track
  C. Air
    1. Airports
    2. Heliports
    3. Landing Fields
  D. Transportation Characteristics
    1. Volume
    2. Origin–Destination
    3. Scenic Highways

II. UTILITIES
  A. Gas Lines
  B. Telephone Cables
  C. Water
  D. Sewers
  E. Radio Towers

III. POWER LINES (SCE & OTHERS)
  A. Power Lines (Existing)
  B. Power Lines (Proposed)
  C. Substations
  D. Pumping Stations

IV. GOVERNMENTAL UNITS (Boundary Lines)
  A. Federal
  B. State
  C. County
  D. Municipal

V. SCHOOL DISTRICT BOUNDARIES

VI. LAND OWNERSHIP
  A. Public
    1. Federal
    2. State
    3. County
    4. Local
  B. Private
    1. Individual
    2. Corporate
    3. Noncorporate (large land holdings)

VII. PARKS AND RECREATIONAL AREAS
  A. National Forests
  B. State Parks
  C. County & City Parks
  D. Regional Parks (Existing)
  E. Regional Parks (Proposed)
  F. Military Bases
  G. Historical Sites
  H. Educational Areas
  I. Wilderness Areas
  J. Environmental Corridors
  K. Recreational Facilities
    1. Golf Courses
    2. Public Campgrounds
    3. Private Campgrounds
    4. Race Tracks
    5. Rifle Ranges
    6. Surfing Areas
    7. Boat Launching or Rental Areas

VIII. LAND USE
  A. Residential
    1. Single Family
    2. Medium Density
    3. High Density
    4. Estates
  B. Planned Unit Development
  C. Local & Neighborhood Business
  D. General Business (Restricted Commercial)
  E. General Commercial & Heavy Commercial
  F. Shopping Centers
  G. Business & Professional Offices
  H. Commercial & Manufacturing
  I. Manufacturing Research
  J. Industrial
    1. Light Industry
      a. Mineral Extraction
      b. Petroleum Tank Farm
      c. Oil Fields
    2. Medium Industry
    3. Heavy Industry
    4. Industrial Parks
  K. Public and Quasipublic
    1. Stadiums & Coliseums
    2. Fairgrounds
    3. Schools
    4. Churches
    5. Public Parking
    6. Hospitals
    7. Postal Stations
    8. Municipal Buildings

IX. LAND USE—PROPOSED

X. RESIDENTIAL LAND VALUES
  A. $0–$20,000
  B. $20,000–$25,000
  C. $25,000–$35,000
  D. $35,000–$50,000
  E. $50,000 & above
  F. Under $30,000
  G. Over $30,000

XI. POPULATION
  A. Population (Existing and Projected)
  B. Population Profile (Existing)
    1. Age
    2. Sex
  C. Population Profile (Projected)
    1. Age
    2. Sex

An important precursor to generation of a list of relevant information (data) is determination of the area actually and potentially affected by the decision—in this instance, power plant siting or transmission rights-of-way.

The "study area" scale and scope should reflect the landscape system potentially modified by the proposed electric power-related land use. Therefore, some natural boundary system should be employed to determine the boundaries of the study area. Subsequently, data can be developed appropriate to the characteristics of the area boundaries defined by the study team. These boundaries can be expanded or contracted as dictated by the natural and cultural characteristics of the area. An understanding of these systems will develop as the study team becomes more familiar with the region; thus study area expansion or contraction will be developed.

Implementation of a power plant (and associated uses) or transmission right-of-way must be founded on the following considerations, (1) characteristics (natural and cultural) suitable for the facility in question and (2) location of the facility in a manner that will result in least (or nil) negative impact on natural or cultural systems.

It must be recognized that types and location of characteristics, appropriate or inappropriate, vary to

TABLE 4   Suggested Natural Data Types

I. ABIOTIC
   A. Climatic Factors
      1. Macroclimate—Restrictive Ranges
      2. Microclimate—Restrictive Areas
      3. Microclimate—Unique (air pollution patterns)
   B. Ground Water & Geohydrological Factors
      1. Principal Aquifers
      2. Transmissibility of Principal Aquifers
      3. Water Quality
      4. Underground Water Table
         a. Depths
            1) shallow (0'–10')
            2) intermediate (10'–50')
            3) deep (50' or greater)
   C. Surficial Water Factors
      1. Watershed Order
      2. Ocean
      3. Lakes & Streams
         a. Lake—Width of Narrower Axis
            1) less than 600 feet
            2) 600 feet to 1,200 feet
            3) greater than 1,200 feet
         b. Stream—Width of Narrower Axis
            1) less than 600 feet
            2) 600 to 1,200 feet
            3) greater than 1,200 feet
         c. Wetland—Width of Narrower Axis
            1) less than 600 feet
            2) 600 to 1,200 feet
            3) greater than 1,200 feet
      4. Intermittent Streams
      5. Intermittent Rivers and Ditches
      6. Ditches, Aqueducts, Canals
      7. Springs
      8. Floodplains
      9. Estuaries
     10. Wetlands
     11. Aquifer Recharge Areas
     12. Unique & Scarce Features
   D. Geologic Factors
      1. Bedrock Geology
         a. Depths
            1) shallow (0'–10')
            2) intermediate (10'–50')
            3) deep (50' or greater)
      2. Earthquake Zones
      3. Extractive Deposits
      4. Unique & Scarce Deposits
      5. Topography
         a. Slope
            1) 0–12.5%
            2) 12.5–50%
            3) 50% or greater
      6. Landform
         a. Main Ridge
         b. Secondary Ridge
         c. Main Ravine
         d. Secondary Ravine
         e. Pits and Depressions
      7. Physiographic Factors
         a. Landslide Areas

         b. Unique Feature
         c. Surface Drainage Patterns
         d. Slope
         e. Orientation
   E. Soil Factors
      1. Soil Groups
      2. Permeability Patterns
      3. Erodability Patterns
      4. Stability Patterns
      5. Productivity Patterns
      6. Zone of Eluviation (less than 2')
         a. fine
         b. sandy
         c. coarse
      7. Zone of Illuviation (4"–8")
         a. fine
         b. sandy
         c. coarse
      8. Parent Material
         a. fine
         b. sandy
         c. coarse

II. BIOTIC
   A. Vegetation
      1. Age
         a. less than ten years
         b. ten to one hundred years
         c. more than one hundred years
      2. Size (dbh)
         a. less than 6"
         b. 6" to 18"
         c. greater than 18"
      3. Dominant Seasonal Color
         a. green
         b. yellow–brown
         c. orange–red
      4. Future Succession
         a. high
         b. low
      5. Unique Species
      6. Ecological Communities
         a. Unique communities
         b. Unique species
      7. Percent of Canopy Cover
      8. Percent of Shrub Layer Cover
      9. Percent of Ground Flora Cover
     10. Scientific Areas
     11. Forests
     12. Scrub Brush
     13. Orchards
   B. Wildlife Factors
      1. Wildlife Habitats
      2. Unique Behavioral Units
      3. Unique Species
      4. Wildlife Refuges
      5. National Wilderness Areas
      6. National Forests
      7. Areas of Interest to the Fish & Game Department

greater or lesser degree from one area of the country to another. Thus, regional idiosyncrasies of data generation will occur; it is positively unacceptable to generalize data requirements for decision-making from one area of the country to another, just as it is erroneous to generalize electric energy requirements from one region to another.

Although the breadth and detail of a data bank (information base) is determined by financial resources, it is essential to generate as broad an information base as financial resources permit; thus the more information the better without sacrificing quality.

Consideration of functional, ecological, social, and economic aspects of power plant siting or transmission alignment is necessary in order to generate appropriate data variables implicit in (1) characteristics suitable for the system and (2) least negative impact on natural or cultural systems. These strategies (considerations) are the alignment or power plant location of

1. Least impact on the natural ecosystem
2. Least visual impact
3. Least cost of acquisition
4. Least cost of construction
5. Least impact on human settlement

Tables 3 and 4 list the natural and cultural information types (variables) and suggest the breadth and detail of information required to accommodate the functional, engineering, economic, and social dimensions of energy facilities and to accomplish least disruption of environmentally sensitive systems, natural and cultural. Implementation of these strategies can be accomplished only if pertinent variables are interpreted (on the basis of capability of various natural and cultural systems to withstand the type and intensity of impact) and plotted in a spatially relatable fashion, i.e., graphic overlays or computer printouts.

VISUAL COMPATIBILITY MATRIX

The physical component of power transmission is most significantly manifested visually in lattice structures and conductors. In recognition of the visual impact of these components, the staff of Landscapes Limited utilized the matrix method to analyze the visual characteristics of towers and conductors on a staged basis. The results of this evaluation are documented on the Visual Compatibility Matrix (Figure 8).

Stage I

The first stage of the analysis was the determination of components of both the structure and the right-of-way. Observation produced a four-part subdivision of lattice structure towers: (1) the leg, (2) the body, (3) the cross-arms, and (4) the conductors. These four elements were isolated and evaluated independent of the other components, thereby facilitating a more specific evaluation than is possible in a general observation of the total unit. The right-of-way was similarly subdivided

VISUAL COMPATIBILITY MATRIX — 6 □ Whole No. — 9 □ Fractional No.

CULTURAL

Visual Compatibility	Residential	1. Low Density	2. Medium Density	3. High Density	Commercial	1. Clustered	2. Isolated	Public & Quasi-Public	Industrial	1. Heavy Industry	2. Medium Industry	3. Light Industry	4. Power (sub-stations)	Recreational	1. Parks	2. Playgrounds	3. Open Space	Agricultural	1. Nursery	2. Croplands	a. orchards	b. grains & grasses	Transportation	Structural Elements (man-made)	1. Walls	2. Fences	3. Drainage Canals	Means
Structure																												
Leg	30				15			40	31					33						31			28	30	30			29
Body	24				10			15	18					18						29			19	13				19
Cross Arms																												
Conductors	23				30			32	13					17					19	20			15	13				22
R.O.W.																												
Central Area	26							40	14					32						31			30	38				31
Edge	22							33	14					34					22	22			11	37	30			25
Means	24							32	18					26					20	27			26	26	30			25

FIGURE 8 Visual compatibility matrix.

to include (1) the central area, and (2) the edge of the right-of-way.

Stage II

Stage II involved a review of 35-mm slides of existing rights-of-way in order to determine the types of land uses normally adjacent to the power corridor. The most frequent adjacent uses include residential, commercial, public and quasipublic, industrial, parks and open spaces, agricultural, and transportation.

Stage III

Stage III involved the actual evaluation of components itemized in Stage I in the context of each land use iterated in Stage II. This evaluation of visual compatibility was conducted by three professionals in Landscapes Limited, including a landscape architect, an environmental designer, and a physical planner. The slides of existing conditions were organized by land use and projected on a viewing surface. The staff then evaluated the *leg* in the context of each land use. Next the *body* was evaluated in the context of each land use, and so on until all components had been analyzed on the basis of visual compatibility.

The staff established a five-level rank of visual compatibility or incompatibility, as follows:

1 = very incompatible
2 = incompatible
3 = neutral
4 = compatible
5 = very compatible

Each of the three respondents assigned a numerical value to each component (leg, body, etc.) as viewed in the context of each land use. Subsequently, a mean value was calculated and recorded in the appropriate box in the matrix.

The visual compatibility matrix reveals a wide range of values spanning a continuum from a low of 1.0 to a high of 4.0. The matrix also includes a calculated mean of each component and each land use. This rigorous process resulted in highly objective and concentrated observation of extant visual characteristics. The consultants have also supplemented their documentation of quantitative value assignments with the following list of other observed conditions:

1. Hazy atmosphere and low lighting levels are the most favorable conditions for viewing lattice tower structures.

2. Keyed to visual matrix, the context in which lattice tower is viewed is a critical factor in acceptability of appearance.

3. Distance is also a critical factor; distant structures are more acceptable.

4. The density of units seen together has great bearing on their acceptability; many towers seen closely together are most unacceptable.

5. Areas of the right-of-way that lack distinct boundaries (e.g., walls or fences) have a higher degree of acceptability.

6. The use of vegetation at the edges of the right-of-way is visually more successful than the use of man-made elements such as walls or fences.

7. The leg of a structure is more compatible as distance increases.

8. As the light levels vary the appearance of towers changes. A tower silhouetted against the sky is incompatible, but the *base* of the tower disappears and becomes moderately compatible.

9. Atmospheric haze and fog are favorable conditions; lower contrast of structure against background improves appearance.

10. The more visually pleasing the foreground, the more compatible the base and leg elements of the tower.

11. The context or setting in which a tower is seen or placed is the essential problem.

12. A mix of tower forms and distribution lines is highly undesirable.

Following this evaluation, the staff identified the most significant visual incompatibilities, based upon the mean value factor. These illustrative sketches follow in the chapter on Communications (p. 257).

USE OF COLOR IN THE SITING OF GENERATING PLANTS AND TRANSMISSION LINES

The modern thermal electric generating station is so large that even the proper use of color cannot hope to effectively reduce its size or blend it into the landscape. Color can, however, make more pleasing the spatial relationships of its various forms. At the very least, a number of color schemes should be made and studied in conjunction with the plant surroundings.

Transmission lines and substation structures, on the other hand, can be rendered much less conspicuous by proper color selection. Just the dulling of galvanized lattice towers and of aluminum conductor will often tend to blend the line into its surroundings. Some have found that the use of rusting steel presents a pleasing, less obtrusive tower style. The color duration of the line's surroundings during the year must be considered before a blending color can be selected.

If a structure or a piece of equipment such as a trans-

	APPEARANCE	Very Pleasant	Pleasant	Undecided	Unpleasant	Very Unpleasant
POWER PLANT SITE						
Power Plant Structure						
Utilization of Utility Owned Space Surrounding the Power Plant Site						
Transmission Right-of-Way						
Right-of-Way						
Right-of-Way (Edge)						
Transmission Structures						
Lattice Structure						
Steel Pole Structure						
Wooden Pole Structure						
Steel "H" Structure						
Wood "H" Structure						
Special Transmission Structures						
Aesthetic Structure (Dreyfus)						
Aesthetic Structure (Other/Specify)						
Conductors/Insulators						
Access Road						

FIGURE 9   Power structure/right-of-way appearance matrix. This questionnaire is designed to elicit lay response to the appearance of component aspects of (1) the power plant per se and associated land uses, (2) aspects of transmission rights-of-way, and (3) components of high-voltage transmission. The respondent is asked to record his (or her) judgment of the appearance of the components listed (as actually observed by the respondent) on a continuum from very pleasant to very unpleasant.

former cannot be blended with like colors, perhaps a firmly contrasting color that does not try to hide an item will prove best. Such equipment is not lacking in interest, and unashamedly highlighting them through color will stimulate that interest.

POWER STRUCTURE/RIGHT-OF-WAY QUESTIONNAIRE: INFORMATION GRIDS

Presented herein are five grids (Figures 9–13) that can be utilized as individual response forms or to tabulate the collective responses of the sample population participating in the attitude exercise. Physical aspects of power plants and transmission alignment included in the information grids are

1. Power structure and right-of-way appearance.
2. Power plant site olfactory condition matrix.
3. Power plant site/corridor alignment adjacency.
4. Power structure/right-of-way land value effect.
5. Power structure/right-of-way, human attitude response matrix (a composite of the above aspects).

Before suggestions for improved power plant siting and transmission alignment can be offered, a broader base of information must be developed for utilization as a base of understanding leading to the development of sound criteria that consider (1) requirements and constraints associated with all aspects of energy generation and transmission, (2) optimization of plant siting and generation in harmony with the natural and cultural characteristics of a region or regions, and (3) attitudes of the general public on a variety of aspects that can or do affect people's lives in various ways.

The following questionnaire grids have been developed in an attempt to afford the general public an opportunity to respond to several aspects of power plant siting and transmission alignment. The survey research design for administration of the questionnaires has not been resolved. These questionnaires must be administered throughout the various regions of the country, ensuring response in areas of the country where the effects included in the questionnaire do, in fact, exist, and individuals have established opinions based upon actual experience.

	OLFACTORY	Odor from Emissions	Very Pleasant	Pleasant	Undecided	Unpleasant	Very Unpleasant	Odor from Fuel Residue	Very Pleasant	Pleasant	Undecided	Unpleasant	Very Unpleasant
POWER PLANT SITE													
Power Plant Structure Emissions													
Utilization of Utility Owned Space Surrounding the Power Plant Site													

FIGURE 10 Power plant site olfactory condition matrix. An aspect of power plant siting that may or may not concern individuals is the matter of odors associated with the power plant site. These odors may emanate from the waste materials associated with energy production in the form of (1) stack emissions or (2) fuel residue on the site. The respondent is asked to respond to odors from emissions and odor from fuel residue on a continuum from very pleasant to very unpleasant.

	ADJACENCY	Would locate immediately adjacent	with visual access	without visual access	Would locate at inter-mediate distance	with visual access	without visual access	Would locate at great distance	with visual access	without visual access	Would not locate at any of the above
POWER PLANT SITE											
Power Plant Structure											
Utilization of Utility Owned Space Surrounding the Power Plant Site											
Transmission Right-of-Way											
Right-of-Way											
Right-of-Way (Edge)											
Transmission Structures											
Lattice Structure											
Steel Pole Structure											
Wooden Pole Structure											
Steel "H" Structure											
Wood "H" Structure											
Special Transmission Structures											
Aesthetic Structure (Dreyfus)											
Aesthetic Structure (Other/Specify)											
Conductors/Insulators											
Access Road											

FIGURE 11 Power plant site/corridor alignment adjacency matrix. The question of human settlement in the context of power plants and transmission rights-of-way stimulated the design of this questionnaire in order to determine the physical proximity individuals will settle in relation to these physical components of electrical energy supply.

The value continuum ranges from "would locate immediately adjacent (with visual access)," to "would locate at a great distance with no visual access."

	LAND VALUE	Significant + Effect	Positive Effect	No Effect	Negative Effect	Significant − Effect
POWER PLANT SITE						
Power Plant Structure						
Utilization of Utility Owned Space Surrounding the Power Plant Site						
Transmission Right-of-Way						
Right-of-Way						
Right-of-Way (Edge)						
Transmission Structures						
Lattice Structure						
Steel Pole Structure						
Wooden Pole Structure						
Steel "H" Structure						
Wood "H" Structure						
Special Transmission Structures						
Aesthetic Structure (Dreyfus)						
Aesthetic Structure (Other/Specify)						
Conductors/Insulators						
Access Road						

+ = Positive Effect   − = Negative Effect

FIGURE 12 Power structure/right-of-way land value effect matrix. Personal attitudes (or actual conditions) regarding the effect of power structures and rights-of-way on land values led to the formulation of this questionnaire.

A detailed study of actual property sales records of residences (and other land uses) at various distances from power plants and rights-of-way would document actual characteristics of land value conditions.

This study, in concert with public attitudes, would serve to document (1) public concerns and (2) whether or not those concerns are justified.

FIGURE 13 Power structure/right-of-way human attitude response matrix. The range of actual numerical responses to each of the questionnaire aspects would be recorded on this composite information grid (according to region) as a basis for establishing land-use and aesthetics criteria (by region) representing the attitudes of a sample population in the various regions participating in the survey.

UNDERGROUND TRANSMISSION

The American public today is recipient of increasing levels of electrical energy to power both the required and convenience elements often referred to as contributory to the high standard of living enjoyed in the United States.

The reliance upon electrical energy in contemporary society is manifested in many forms, including television sets, electric can openers, food mixers, electric hair dryers, and other trappings of an increasingly affluent society.

In addition to these individual user requirements for electric power, the shift in population from rural to urban areas, the shifts in population from one urban area to another geographical area, the creation of new towns and the establishment of second homes in the country, and increasing industrial activity require the conveyance of energy to developing areas.

New corridors continually are being carved from our landscape resource to create new transmission lines. The presence of poles, towers, and wires obtrudes on the landscape, moving many of us to wonder about the feasibility of underground transmission that would eliminate these visual discords.

The ever-increasing amounts of electric power that must be transmitted from points of generation to points of consumption has necessitated the utilization of progressively higher transmission voltages. Since 1930, line-to-line voltages have increased from 110 kV to 220 kV, 345 kV, 500 kV, and in 1970 to 765 kV. During the next decade or two, electric power transmis-

sion is predicted to reach 1,000-kV or 1,250-kV levels. The higher transmission line voltages in free air require progressively higher supporting towers and wider rights-of-way. In 1968 the United States had 300,000 miles of overhead power transmission line. The estimated land area in the right-of-way totaled about 7 million acres—an area roughly the size of Belgium. By 1980 this will be multiplied 3-fold.

The technological, economic, and aesthetic pressures for undergrounding electric transmission lines have been augmented both by the continued exponential growth of energy consumption and by the rapid extension of populated areas. As a result, there is now a resurgent interest in the development of transmission lines employing compressed gas as the basic insulating medium with periodically spaced solid insulators for the conductor support.

The past three decades have seen the steady development of underground high-voltage cable for voltages of 138 kV, 230 kV, and 345 kV. These conventional cables utilize spiral-wrapped oil-paper insulation. This solid insulation introduces dielectric capacitance and losses along with thermal resistance which result in limitations of power capacity and permissible length of line. The installed cost of conventional cable is typically 20 times that of an overhead line in open country at the same voltage. Nevertheless, for cities and critical short lengths, these oil-paper cables are indispensable; about 3,000 miles of such underground cable are now installed. Such cable has been developed to 500 kV but not yet put into actual use. The oil-paper cable approach, because of its voltage and power limitations, is not technically or economically encouraging for future underground power transmission at 500 kV and above.

A variety of new technologies has been proposed, ranging from the almost immediately available to the long-range. Most available and practical among these are cables which utilize compressed gases as the insulating medium.

APPROPRIATE LAND USES

Inclusion of land uses that are appropriate to power plant sites and their corresponding transmission line rights-of-way represents an aspect of power plant siting that should be regarded as an opportunity rather than another problem. Involvement with land uses in addition to problems associated with siting will undoubtedly add to the difficulties of planning, negotiating, designing, and scheduling relating to the overall project. But it may be one of the most satisfying features of the project to the local community and is therefore very important.

No particular effort will be made to designate land uses as applying to thermal, hydro, or other type of power facility except as might be restricted by the use. However, it should be recognized that most future power plants will be thermal, utilizing either nuclear or fossil fuel.

Appropriate land uses properly should consider social needs, political or public acceptance, and aesthetics, as well as economic factors. These are mostly complex factors. Problems arise in the process of establishing the need, identifying the community, and determining who expresses public acceptance.

More progress has been made to date in joint- or multiple-use of transmission rights-of-way than in joint plant site use. Dual uses associated with the power plant, or site-associated, are more likely to involve much larger economic considerations, are more complex, and therefore have not occurred to a similar extent.

Rights-of-Way Uses

The current policy of most utilities today is similar in permitting the maximum use by others of the utility-owned rights-of-way, provided such use and the manner in which it is exercised will not interfere with the owner's rights or endanger its facilities. This policy would apply to land held in easement or in fee; the latter may provide some restriction in use, however.

An important consideration to the concept of shared usage is that it can be permitted after the fact. Adverse reaction to existing transmission lines, for example,

might be ameliorated by improving a particularly objectionable section of right-of-way.

Joint Utility Usage

Usage of common right-of-way utility corridors for power, water, gas, telephone, sewage, and heat has substantial public appeal. It has been used sparingly in this country; greater use of this concept is made in Europe. However, utilities have shared rights-of-way with railroads for years, especially in the metropolitan areas. Here, electric distribution lines, telephone and gas lines, in some combination, have been placed alongside the railroad tracks.

Availability of separate rights-of-way in conjunction with simplified planning and operation has slowed development of utility corridors. Additionally, such corridors will present much more of an aesthetic problem. Greater use of this concept can be anticipated with decrease in available land, especially in the urban areas. Use of utility corridors will require substantial long-range regional planning. This would involve close coordination among cities, counties, states, the federal government, and utilities. The effect such corridors will have on reliability of services will also have to be determined.

The inclusion of highways with utility corridors is somewhat controversial on the basis of aesthetic considerations. As more effort is made to improve the aesthetics of both highways and transmission lines, it becomes more difficult to incorporate them in the same right-of-way. However, there can be many areas where joint use is the most practical, considering the net benefit to a community.

Energy Corridor

A variation of common utility use, the energy corridor, has been given some consideration in the San Francisco Bay area. There, in addition to basic utilities, transport of industrial products, including petroleum products, fiber, coal and ore slurries is considered.

Recreation

Various forms of recreation have been incorporated within transmission rights-of-way. Riding trails, hiking trails, and areas opened up for use as bike trails (both power-driven and self-propelled) are probably the most common. Where the properties front on waterways, wildlife sanctuaries, game preserves, or shore frontage developed for fishermen are possible.

Perhaps the most rewarding, however, are urban recreation areas. These facilities can include such features as Little League ball parks, tennis courts, hard-surfaced basketball courts, picnic areas, playgrounds for youngsters, plus others limited only by the imagination and capabilities of the participating parties.

For the social benefits accruing from installing such recreational facilities, the cost is often quite modest. Depending on the type of agreement, the utility obligations may be limited to preliminary grading, landscaping, drainage, and subsequent leasing to a community agency. The agency is usually responsible for installation of the recreation equipment and future maintenance of the area. It is important that some acknowledgment regarding supervision of the facility be spelled out to help maintain the installation. Neglect and vandalism could destroy the aesthetic value of the facility and actually produce a worse effect than the open right-of-way.

Commercial

Present commercial uses of rights-of-way include limited-use storage, including auto parking and agriculture. The latter represents by far the greatest joint use with transmission lines and is common throughout the nation. Often, the right-of-way was obtained as an easement through existing farmland with little or no change in the operation of the farm. Christmas tree farms have recently been added as a "crop" to the long list of farm products using rights-of-way, in addition to conventional garden nurseries. Livestock grazing is common with some farm operations, but in some cases stock has been brought in after the right-of-way was cleared.

Aesthetic Improvement

In many urban areas, as well as certain scenic areas, programs of beautification and screen planting have been under way for several years. While this in itself is not technically a "use," where such a program is initiated in urban areas, land can provide natural community play areas, with natural trails and walkways.

Plant Site Related Uses

Joint uses associated with power plants or projects have tended to be limited, particularly in the past, to some form of recreation. This has been particularly true of hydroelectric facilities where recreational features have been an important benefit of project development.

Functional, dual uses have not been common but are being discussed with some development at an increasing rate today.

Recreation

All types of power plants are inherently interesting to many people and are natural educational or sight-seeing attractions. This usually starts during the construction period and continues after the plant is placed in operation. Most utilities are aware of this and provide a wide range of information centers during the various plant phases.

These information centers can have a wide range of interests. Commonly, they provide informational source material and rest facilities. Others can have special features: In the gold country of California, an old ore stamp mill was restored and placed in the visitors' area located near the power plant. Others, such as are being constructed at nuclear power plants, can have exhibits (for example, artifacts from the plant excavation), displays, films, meeting rooms, and overlooks.

Hydroelectric sites usually afford the opportunity for outdoor recreational pursuits, picnic areas, campgrounds, boating (including launching ramps), and fishing. The latter sport is often assisted by construction of fish ladders and hatcheries to maintain or improve the original fish population.

Where a nuclear plant has cooling water ponds associated with it, water-oriented recreation may be developed. Sport fishing can be especially attractive in these cases since there fish are usually fastgrowing and larger.

Where a thermal plant is located on an ocean, it is often possible to develop any beach potential. In regions of especially cold water, it is conceivable that the cooling water discharge could be directed toward the beach so as to warm the water for swimming.

Commercial

Commercial uses are mostly associated with warm water effluent from thermal plants.

Studies are under way to check the beneficial uses of heated water in agriculture. It is considered that the warm water will stimulate and enhance plant growth and will protect some crops from frost. This can be by direct irrigation or soil heating. The most probable development will be where the water can be used to provide environmental control for greenhouses or poultry houses. Location of such structures in the exclusion area of a nuclear plant has the advantage of shortening the water-transport distance and providing revenue. Present studies indicate that such facilities would be economic for several kinds of vegetable crops.

Warm water is also used in "aquiculture" or marine-culture. Many experiments have already been conducted to show that warm water discharge can be used to produce higher yields of various fish, shellfish, and crustaceans. Some of these experimental programs, such as those on oysters, shrimp, and lobster are becoming commercial operations. Fish such as catfish and trout have shown large increases in yield per acre.

Perhaps one of the most ambitious research projects has recently been entered into by the Ralston Purina Company. A five-year program will study the feasibility of utilizing heated water from power plants in farming areas for seafood.

Where thermal plants are located within or near industrial areas, there may be opportunities to sell steam. This can be for such uses as meat processing or heating.

Dual Uses

Dual-use facilities, especially those concerned with nuclear plants, have several inherent problems. Negotiating, licensing, design, and construction are more complex. The combination of a large desalination and power plant (usually nuclear) has been given serious consideration, although none has been constructed to date. The desalination unit uses steam from the power plant and therefore must be located nearby. Since both are large facilities, this makes it a little more difficult to make aesthetically pleasing.

Location of such an installation today is limited to the Pacific Southwest, where the Bolsa Island plant was carried to the preliminary design stage. More recently, the State of California has authorized a feasibility study to be made for a 50,000-acre-ft/yr desalination plant that would be tied to a nuclear plant now under construction.

Where waste waters, industrial or agricultural, represent a pollution problem, it may be possible to use them in the thermal plant cooling system and decrease the quantity to be disposed of.

Use of garbage incinerators to generate steam for delivery to a nearby steam plant has been used on a very limited scale in this country but is quite common in Europe.

The combination of large industrial plants with a thermal plant is similar to the desalination–nuclear combination. For example, the Midland (Michigan) nuclear power plant, in this case, will supply process steam to the Dow Chemical Company.

IMPLICATIONS OF POWER PLANT SITING AND CORRIDOR ALIGNMENT TO POTENTIAL FUTURE DEVELOPMENT: REGIONAL–LOCAL

The history of the electric utility industry has been that of keeping abreast of the growth of electrical load.

Small generating plants located in and surrounded by their load areas have given way to larger, more efficient and economical but more remote plants often serving several load areas, interconnected by high-voltage transmission lines.

The electric industry sites its hydro plants on rivers in accordance with the availability of economical sites. Thermal plants are most often located near rivers, often closer to load and fuel supply. Nuclear plants also need cooling water, but can be remote from their fuel sources. High-voltage transmission then links these generating plants with the load and with each other, often with some redundant lines to assure system reliability.

The time is long past when we could site these plants and lines without regard for their impacts on our life-support systems and our aesthetic sensibilities.

What impacts do these generating plants, lines, and substations exert on their surroundings? Do they act to accelerate, or to retard, "growth"? Since they almost always are built to satisfy a need seen to be developing, they cannot be said to accelerate the growth of electric load. Only failure to construct a facility, in violation of a utility's responsibility, could actually retard the growth that the facility was to serve.

It is seldom that a generating plant will inhibit the growth of industrial development in its immediate area, and often residential growth will occur. A thermal power plant, with adequate control of particulate and gaseous emissions and noise, appropriately separated by a greenbelt and landscaping, is not a bad neighbor either to industry or housing.

Since we must assume that a power plant is a less desirable neighbor than a park, or another house, once having selected the general area for a plant site, we should look to other compatible uses which will make for optimum area development. For instance, could a modern incinerator occupy a nearby area and make its waste heat useful for power generation? Can the plant's heated effluent be used for irrigation or as a first stage in desalination? We should always look beyond the immediate siting problem to achieve a better utilization of our land resources.

The sharing of rights-of-way is an important concept that must not be overlooked and toward which more effort and, particularly, cooperation must be devoted. Natural gas and oil pipelines, telephone cables, water and sewage pipelines, railroads, and even highway routes all may offer potential for development of common corridor.

Transmission line corridors through undeveloped areas often can provide greenbelt separation for future urban and suburban development. These corridors could be left in their natural state for trails and wildlife habitat or perhaps be made into park and recreation areas. In urban areas they often can be used for parking and access corridors. The rights-of-way offer little impediment to agriculture or pasture land except under the towers themselves, but they are inimical to the growth of forest products except, in some areas, for Christmas tree production.

Land use potential may be affected by aesthetic considerations. With power plants and transmission lines there is no general agreement even among experts as to what constitutes a pleasing structure. Many of the so-called outdoor power plants have been regarded favorably by some modern architects. Lattice-type transmission towers are functionally correct and in many applications are often much less obtrusive on a scenic vista than some of the more expensive, heavy, pipe-type structures that will find acceptance in other locations.

The "low profile" substation or the use of box girder takeoff structures will find application in urban environments but may be unnecessary where the substation can be surrounded by sufficient screening.

Optimum land use is always of paramount importance. Since growing electrical loads require more plant sites and transmission corridors, these must fit the best possible utilization of our existing land resource to achieve an optimum balance among diverse existing and potential land uses. Some, like Lewis (1967) and McHarg (1969), have developed methods of identifying such uses and compatibilities. Such studies, however, must include only those variables truly associated with siting conditions, which then must be given a scale of importance commensurate with their optimum public value. For example, one of the variables that must be considered in routing a transmission line is the necessity for suitable soil conditions to provide reliable tower foundations. This could dictate use of the more stable ridge lines, even though such use may violate aesthetic criteria.

The public must always be brought into the planning process at an appropriate point, a point that is often hard to establish. Several alternative plant sites or line routes should be presented for the public's consideration, yet their preliminary nature must be emphasized to overcome immediate opposition. It probably never will be possible to get full concurrence with any siting decision by the people who are immediately affected, even though the growth that eventually comes about demonstrates that the then-existing structures are, indeed, compatible.

5   COMMUNICATIONS

DESIGN CONSIDERATIONS AND VISUAL COMMUNICATIONS

In the preceding chapters various rationale and ideas have been expressed concerning potential methods of reducing or eliminating the impact of power plant sites and transmission systems. These ideas have ranged from burying the conductor systems to managing the rights-of-way in special ways to preserve environmental diversity. There are unlimited numbers of potential solutions when they are considered abstractly. The following diagrams and illustrations reflect examples of abstract solutions which are not necessarily applicable to the transmission-line location problems in all regions of the country, but which represent ideas which will evoke response and additional thinking and provide a basis for considering alternative solutions.

These alternatives must be communicated to those individuals responsible for the ultimate implemented action of related alternatives, to the general public potentially affected by potential implementation of these and other alternatives, and to professionals competent to assess the desirability of implementing (or not implementing) concepts that would seem to have potential benefit but require professional evaluation.

The engineering, environmental, economic, and social implications of implementing concepts must be understood prior to the extensive application of land use and aesthetics alternatives.

Maintenance of Residential Quality

As the demand for power increases, physical power components will become visible to more people.

It is becoming increasingly apparent that the benefits of power are offset by environmental problems of various description, including the visual–physical component of power transmission and distribution.

In order to protect the investment of individual property owners, the service industry must assume significant responsibility for creating or preserving desirable environmental conditions. In instances when the right-of-way can be seen from nearby residential streets, vegetation should be introduced to screen the structures from view as effectively as possible.

Rights-of-Way for Human Use

Existing and future rights-of-way can serve as fine pedestrian open spaces and recreation areas in or near cities. If these uses are encouraged, however, some measure must be taken to minimize or eliminate the visual dominance of the tower structures. Earth mounds and plant materials have a dual value in screening structures and defining space interestingly.

Neighborhood playgrounds have been developed in some areas of the country. The visual quality of such areas is diminished, however, by the dominant lattice tower structures and conductors.

This problem can be remedied by placing play areas closer to a structure and introducing an admixture of plant materials to screen the structure from view.

Structure with Shelter for Recreational or Play Activities

In those instances of structure location within rights-of-way utilized publicly, design consideration could be directed to the visual and functional improvement of the structure base.

Structure on Open Easement

Pole or lattice structures exist in a variety of contextural relationships ranging from rural rights-of-way, removed from direct human presence, to urban areas wherein rights-of-way constitute potential open-space areas as a place for human activity.

Easement Edge Defined by Vegetation

Edges of rights-of-way in urbanized portions of the United States are often defined by vegetation in the backyards of adjacent residences. This buffer typically terminates abruptly at the edge of the right-of-way.

Suggested Treatment

An integration of residential open-space and right-of-way area can be achieved by installing vegetative species of variant height characteristics at maturity. Taller species can be introduced at the edge, progressively lessening in height toward the center of the right-of-way. This accomplishes visual continuity while respecting conductor clearance requirements and sway.

Structure on a Tangent with the Ridge

Location of the right-of-way and structures on a parallel with the ridge, on the slope and down from the rim, can partially conceal (1) the easement cleared of vegetation, and (2) a considerable portion of the structure body.

Structure Perpendicular to a Road

Cleared rights-of-way and structures perpendicular to the highway are frequently visible as a motorist travels throughout some areas of the country.

Visual Access from the Highway

There has been increasing visibility of power structures of one form or another along the nation's highways. This situation will occur more frequently as the motoring public and the demand for more power grow.

Visual Screening

Figure 15 depicts a method of screening a structure from view by introducing plant materials adjacent to the highway. Cooperation between the responsible highway department and the corporation installing the structures could be beneficial in creating a more pleasing view from the highway while simultaneously screening power structures from view. In instances where this practice is employed, utilities (such as telephone) might be ganged together to maximize the value of the screen.

Visual Context

The massing of structures and conductors in close proximity to each other results in visual discontinuity. This situation should be avoided whenever possible, but particularly in areas adjacent to or within visual access from residential areas, parks, and other land uses that involve large numbers of people.

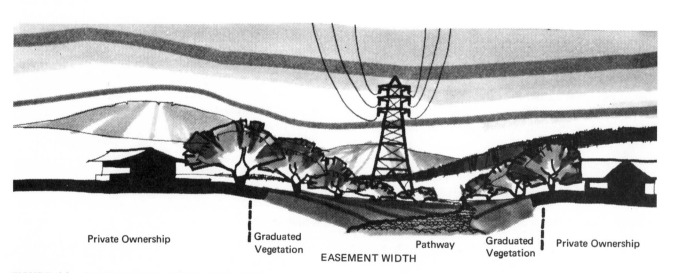

Private Ownership | Graduated Vegetation | EASEMENT WIDTH — Pathway | Graduated Vegetation | Private Ownership

FIGURE 14 **Suggested treatment for rights-of-way.**

Gang Up on Utilities Behind Screen

VISUAL SCREEN

FIGURE 15   Visual screening of rights-of-way.

Right-of-Way Management

It is desirable not to define the edge of the right-of-way by means of natural or man-made materials. Greater visual and landscape continuity is maintained in rural areas by not creating artificial boundaries.

Structure Placement

Figure 16 illustrates the placement of a structure in an open area in clear view to the motorist.

The lower sketch depicts an alternative to the situation depicted in the top sketch in which the power

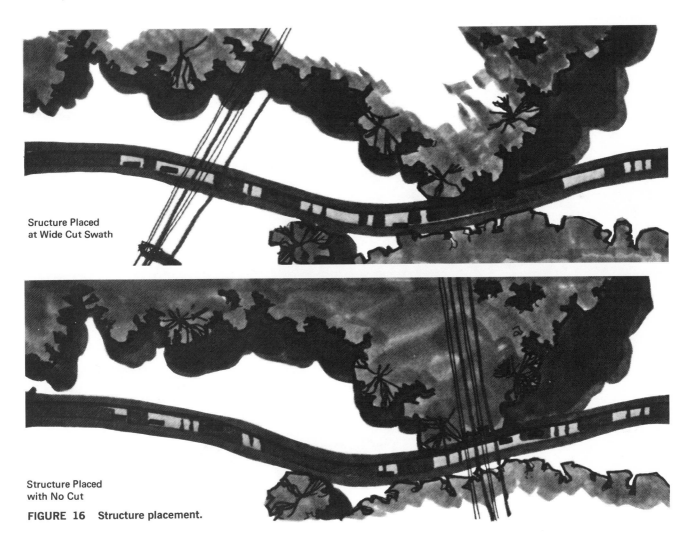

Sructure Placed
at Wide Cut Swath

Structure Placed
with No Cut

FIGURE 16   Structure placement.

components are clearly in view. Providing the vegetation is within the prescribed level of tolerance for vertical clearance of the conductors, the tower could be positioned in an existing clearing at the point that will direct the conductors over the highway where partially screened by vegetation.

Scientific Area

A regional landscape usually contains a substantial variety of physical–biological characteristics, including areas of actual or potential scientific value. As power requirements increase, the probability of locating power rights-of-way adjacent to those areas will increase. In the event that rights-of-way and structures are located in the visual context of scientific biological areas, the structures should be screened from view with varieties of plant materials appropriate to the plant communities present.

Visual Screen in the Context of a Quality Resource

Figure 17 suggests a means of screening a structure from view. The use of earth mounds and vegetation between the valued resource and the power structure will do a great deal to preserve the visual integrity of the historic node. The inclusion of vegetation, in addition to its inherent beauty, constitutes a visual obstruction which directs attention away from the screen and toward an object of greater interest—in this instance, a historic building.

Easement with Access Road

Public access should be provided on a tangent through transmission easements in those instances wherein

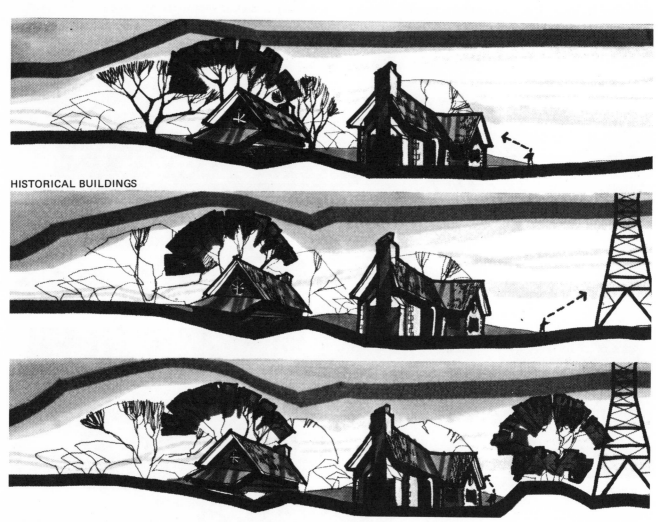

HISTORICAL BUILDINGS

FIGURE 17   Visual screening of transmission towers.

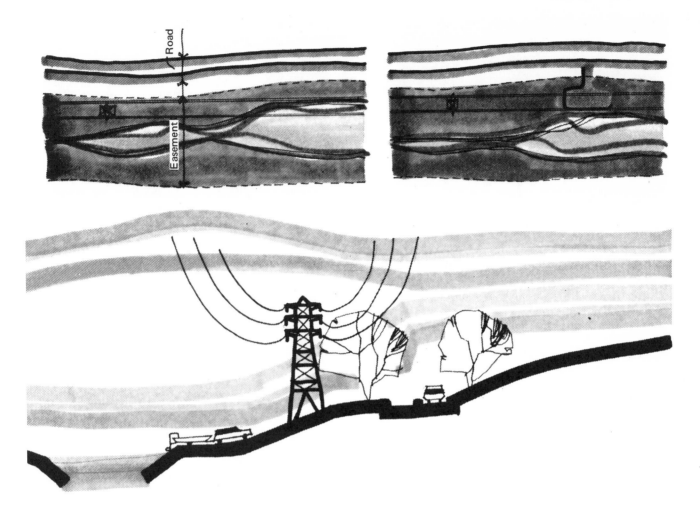

FIGURE 18   Easement with access road to accommodate boat launch.

public use of resources is sanctioned (i.e., public use of water resources, see Figure 18) by an appropriate governmental unit and whenever the health, safety, and welfare of an individual or group is not jeopardized. Utility rights-of-way should not constitute barriers between a public domain, such as a public highway, and a public resource, unless extenuating circumstances dictate otherwise.

Earth Mound as a Buffer

Substantial development of several types has occurred in many areas of the country in recent years. Medium-density residential areas have been built adjacent to highway rights-of-way with visual access to power transmission components. Alternative earth mounding and screening practices aid as a buffer to highway noise and views while simultaneously screening tower components from the highway and residential areas.

Earth Mound and Vegetation as Buffer Elements

Figure 19 suggests a means of effectively concealing structures and right-of-way from view. The creation of an earth mound topped with vegetation outside the highway right-of-way can provide an effective visual concealment of structures placed on a slope with scarce vegetation.

Figure 20 suggests various ways in which screening may be articulated in order to enhance visual experiences along a pedestrian trail.

Full View

Forced View from a Distance
Through Earth Mounding and
Planting

FIGURE 19 Buffer elements.

FIGURE 20 Implementing buffer elements.

Towers Located on Top of Rims

FIGURE 21 Towers on tops of rims.

Visual Context of Electric Power Structures

The context in which a lattice tower is viewed is a critical factor in the acceptability of its appearance.

Towers Located on Top of Rims

The rims of topographic relief often constitute the locus of transmission rights-of-way and structures (Figure 21). In some regions of the country the vegetation associated with vertical relief is sparse and provides minimal (if any) screen of transmission structures. Furthermore, structures placed on rims are viewed in silhouette with the sky. The starkness of this visual context situation depends upon the intensity of

light and color which varies according to season and atmospheric determinants.

Structures Placed on the Slope

Placing structures on the slope prevents the visual impact illustrated in Figure 21.

A semipermeable helicopter pad (to allow drainage) could be provided in areas difficult to service from an access road. In regions characterized by steep topography, the construction of access roads can be difficult and damaging to ground flora. The alternative recommended in Figure 22 suggests a means of enabling structure maintenance in areas of difficult access.

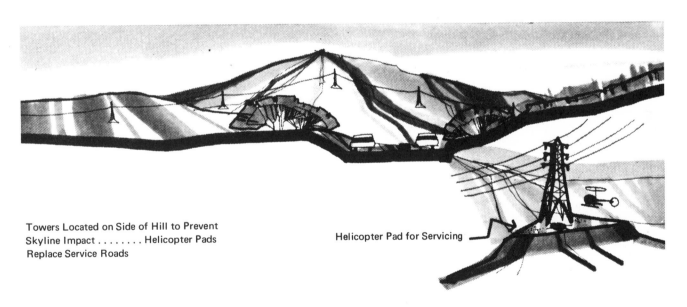

Towers Located on Side of Hill to Prevent
Skyline Impact . . . . Helicopter Pads
Replace Service Roads

Helicopter Pad for Servicing

FIGURE 22 Construction on slopes.

6

CONCLUSIONS AND RECOMMENDATIONS

Preceding sections have outlined general approaches and concepts applicable to protection of the landscape. But there is so great a degree of uniqueness in each situation that it is not feasible to set forth specific recommendations on a nationwide basis. Rather, the siting of each facility should take into account appropriate landscape criteria, and site planning should be based on the relevant information. To aid site planners in this, Tables 5 and 6 exemplify such criteria and information needs.

TABLE 5   Criteria for Power Plant Siting: Aesthetics

Suggested Criteria	Information That Must Be Obtained before These Criteria Can Be Established as Realistic Criteria
1. The siting of a power plant should attempt to avoid those physical conditions which would make a power plant site stand out *singularly* or noticeably as a stark development.	1. An analysis of the natural and cultural elements of the study region, such as vegetation, topography, land forms, architecture, building materials, colors, should be made. When *singularity* cannot be avoided in the resource analysis process, then landscaping, architectural design variation should be emphasized in reducing the visual impact of the site.
2. The siting of a power plant should avoid a multiplicity of seemingly randomly placed facilities or adjacent forms or developments. *Form simplicity* should be an objective in power plant siting and design.	2. An inventory and analysis of the natural and cultural elements or design components of the study region should be made at both the macro and micro scales. Structure designs, site planning, and landscaping should be directed toward the creation of visual simplicity.
3. Power plant siting should avoid or minimize the disruption of exceptional natural edges or natural *continuity* (e.g. ridgelines, streams, river or lake shorelines, tree lines or cultural edges such as city skylines, exceptional panoramic views, etc.).	3. Visual landscape studies should identify significant skylines, shorelines, ridgelines, panorama, etc., which should be preserved.
4. The overall *domination* of visually sensitive areas by a power plant should be avoided, particularly areas of recreation, residential or exceptional scenic quality.	4. A visual landscape analysis identifying land forms, physiography, and vegetative patterns of areas should be made to help determine siting which will have the least dominating visual influence on regions.
5. The *viewpoints* or visibility and public exposure of a site should be considered in the siting of a power plant. Areas characterized by intense public usage and many scenic views should be avoided.	5. A visual landscape analysis is again required to identify significant viewpoints, visual sensitivity, and types of exposure.
6. *Seasonal variations* in vegetative foliage, color, and textures, as well as public exposure (recreation demand cycles) should be considered in the siting of a power plant.	6. Studies of seasonal variations in scenic character and public exposure should be made.

TABLE 6   Criteria for Power Plant Siting: Land Use

Suggested Criteria	Information That Must Be Obtained before These Criteria Can Be Established as Realistic Criteria
1. Power plant sites should avoid those land uses which cannot reasonably absorb waste heat releases.	1. Studies should be made of the region's physical capabilities to absorb waste heat and the possible utilization of waste heat in various land uses found within the area.
2. Power plant sites should avoid those land uses which would be adversely affected by atmospheric releases.	2. Studies should be made of the effects of atmospheric releases on the various land uses found within the area. Such land uses should be inventoried and mapped. In addition, meteorological patterns which would affect the dispersion of such releases should be identified.
3. Power plant sites should avoid areas of geologic instability, or areas of recurrent catastrophic phenomena such as tsunamis, storms, floods, dam failure, etc.	3. Studies should identify those land patterns which are characterized by recurrent or potential catastrophic phenomena as well as the impacts that a well-designed facility can withstand.
4. Power plant sites should be located in proximity to transportation (road, rail, air, water).	4. Transportation analysis should be made to determine the most efficient, reliable, and suitable routes to meet all requirements of power plants.
5. Power plant sites should be located in proximity to load centers to reduce the impact of transmission facilities on the land resource.	5. Systems analysis of load requirements and flows should be made.
6. Power plant sites should be located in proximity to main transmission grids or transmission corridors to reduce the impact of transmission facilities on the land resource.	6. Existing and potential transmission corridors should be located.
7. Power plant sites should be compatible with existing and projected land use plans and zoning ordinances.	7. Where available, land use planning data should be utilized in land use suitability analyses. The proposed location, layout, and design parameters should be coordinated with appropriate local planning agencies to assure maximum compatibility between facilities and existing and future land use.
8. The availability of water for power plant makeup, domestic services, and for use in cooling processes should be analyzed for impact on the water resource and effect on associated land uses.	8. Regional water supply/demand studies should be made to determine impact of a proposed plant on this resource.
9. The thermal effect on receiving water and its biota should be considered in siting.	9. Detailed biological/ecological studies should be made of heated water discharges.
10. Marine sites must be located to minimize the adverse effects on estuaries and other major shellfish and finfish resources.	10. Water uses within a three-to-five mile radius from power plant intake or outfall should receive detailed ecologic analysis. Estuaries and ecologically sensitive areas should be identified in site analysis studies and should be avoided.
11. Inland sites should be located to minimize the adverse effects on fish migration routes, seasonal and migratory fish, and biota.	11. Water uses within a three-to-five mile radius from power plant intake or outfall should receive detailed ecologic analysis. Significant fish migration routes and biota should be identified, their habitat located and considered in the site analysis procedures.
12. Existing recreation resources activities and facilities should be avoided by power plants.	12. Comprehensive resource analysis studies should be made of the existing recreation and associated cultural resources of the region and their compatibility with potential influence of a power plant.
13. Areas of exceptional visual and cultural diversity with high potential for recreation development should be protected from the potential influence of power plants as should wetland areas, scientific areas, and exceptional croplands.	13. Comprehensive resource analysis studies should be made of the region's intrinsic and extrinsic resources and their compatibility with the potential influence of power plants.

REFERENCES

Alexander, Christopher, and Serge Chermayeff, *Community and Privacy; Toward a New Architecture of Humanism,* Anchor-Doubleday, New York, 1963.

Committee on Resources and Man, *Resources and Man,* National Academy of Sciences, Washington, D.C., 1969.

Crowe, Sylvia, *The Landscape of Power,* The Architectural Press, London, 1958.

Cullen, Gordon, *Townscape,* Reinhold Publishing Corporation, New York, 1964.

Department of the Interior and Department of Agriculture, *Environmental Criteria for Electric Transmission Systems,* U.S. Government Printing Office, Washington, D.C., 1970.

Edison Electric Institute, *Electric Power Facilities and the Environment* (prepared by the Plant Siting Task Force, Committee on Environment, February 1, 1970).

Geigen, R., *Climate near the Ground,* Harvard University Press, 1965.

Hutchinson, G. E., "The Biosphere," Scientific American, September 1970.

Landscapes Limited, *Indiana Dunes National Lakeshore,* Lake-Porter County Regional Transportation and Planning Commission, Crown Point, Indiana, 1970. Pp. 44–53.

Landscapes Limited, *Public Policy and the Environment,* Public Land Law Review Commission, Washington, D.C., 1970.

Lewis, Philip H., Jr., State of Wisconsin Department of Resource Development, *Landscape Resource Inventory,* May 1964.

Lewis, Philip H., Jr., *Regional Design for Human Impact,* Thomas Publishing Company, Kaukauna, Wisconsin, 1967.

McHarg, Ian, *Design with Nature,* Natural History Press, Garden City, New York, 1969.

Pugh, Paul F., *Underground Distribution Cable Systems,* Institute of Electrical and Electronics Engineers, 1970.

Whyte, William H., *The Last Landscape,* Doubleday & Company, Inc., Garden City, New York, 1968.

Working Group II

Systems Approach to Site Selection

267

INTRODUCTION

The requirement for electric power generation is becoming more critical because of the increased dependence of our economy upon electrical energy and the rising affluence of our society coupled with population growth. An increasing concern over the degradation of our environment as a result of our increasingly industrial society has raised questions concerning the need for continued rapid growth in power demands and the potential impact of power generation facilities upon the environment. Therefore, it is necessary to give equal consideration to satisfying environmental constraints as well as satisfying the "traditional" requirements of providing future power generating facilities in an economical and timely manner. Today's power plant must be planned, engineered, and constructed under the severe constraints imposed by technological change, economics, environmental protection, population growth, and urbanization, and with consideration for social consequences.

Many companies may have to change the methods they use to plan, design, construct, and operate their facilities to fulfill these dual requirements. Minimizing adverse effects on the environment and providing for energy needs of society must now be regarded as dual objectives right from the beginning. Public acceptability and related political considerations are no longer "external forces." They are often powerful and even decisive, and need to be included in any generalized analysis that deals with the total problem.

The utility is confronted by a complex set of technical and social forces pushing in different directions and has the challenging and difficult task of working out a balanced solution to any requirement for new generating capacity. The use of a systems approach to power plant siting may substantially ease this problem. Through such an approach, all factors, their ranges of constraints and feasibility, and their interrelation, which influence decisions to construct generating facilities, can be weighed and the preferred solution to meet overall objectives determined. It is the purpose of this report to discuss the systems approach as it relates to siting power plants.

The systems approach to power plant siting is presented in this report as a four-step process.

1. Forecasting electrical energy demand
2. Long-range planning activity directed at identifying and committing power plant sites for future use
3. Identifying and selecting a specific site/generation alternative to meet the next increment of energy demand
4. Obtaining acceptance by the public and approval by regulatory authorities of the site/generation alternative

The first phase in a systems approach to power plant siting is to develop a long-range forecast of the demand for electrical energy in a utility system area. Such a forecast is discussed in Chapter 1 of this report. Along with inputs regarding unit outage rates, transmission system reliability, pooling arrangements, and desired system load-carrying capability, the energy requirements forecast enables the utility to assess the need for new generating sites, as well as the need for expansion of generation capacity at existing sites.

The next phase requires a long-range planning effort in order to identify and commit a sufficient number of sites with adequate potential generating capacity to meet this demand. The utility company, government planning and regulating agencies, and the interested public should participate in this planning effort. The factors to be considered and a conceptual site ranking model are discussed in Chapter 2. Application of such a model will indicate the sound technical, economic, and environmental considerations required to identify and commit sites most suitable for power plant facilities.

Once the requisite inventory of sites to meet the anticipated future demand has been acquired, it becomes necessary to initiate specific expansion plans to meet this demand. Selection of a specific site/generation alternative is the next step, which is discussed in Chapter 3. This section identifies the current and future generation alternatives, their potential environmental impacts, and the pollution abatement technology available and anticipated to mitigate these impacts. An economic and environmental impact assessment model for ranking the specific alternatives is outlined. The main object of the site/generation alternative selection process should be to insure that the overall environmental impact of the facility will meet acceptable standards and that the desired generating capacity will be achieved while optimizing total plant cost.

The site acceptance and approval phase is discussed in Chapter 4. The history, current practice, and future trends of regulatory and social concern with power plant siting are presented. It is evident that there is an increasingly large segment of the public interested in the siting process. An "open planning" policy concept, which gives the public an opportunity to express its attitudes, is presented in this section.

Chapters 5 and 6 discuss potential research and development activities in the areas of power plant siting and generation, and the conclusions and recommendations of the Working Group, respectively.

1   FUTURE ELECTRICAL ENERGY REQUIREMENTS

GROWTH OF ELECTRICAL POWER DEMAND

The U.S. utility industry is facing a continuous growth in electrical power demand. During the past 30 years, loads have grown at an average rate of approximately 7 percent per year, doubling every 10 years. However, many experts believe the use of electrical energy cannot continue to increase indefinitely, and that growth in per capita consumption of electrical energy may decrease in succeeding decades. An electrical energy forecast prepared by the Regional Advisory Committees to the Federal Power Commission National Power Survey predicts an annual electric power consumption of approximately 6 trillion kilowatt-hours by 1990. To meet this demand for electrical energy, it is estimated that installed capacity must increase from 340,000 MW in 1970 to 1,261,000 MW by the year 1990, indicating that over 900,000 megawatts of electrical generating facilities must be installed. This estimate of future growth, as well as other estimates, are discussed in Chapter 2 of the Working Group III report entitled "Energy and Economic Growth." While each estimate varies in the rate of growth projected, it is obvious that additional generating facilities will be necessary to meet future power needs.

The forecaster, when developing future power estimates, must of necessity make assumptions, many of which significantly affect the results of his study. These include

1. Population
2. Gross national product
3. Standard of living

4. Technology
5. Price of electricity
6. Shift in energy use
7. Utilization efficiency

These same problems must be faced by each utility planner when he is trying to determine future growth of his own system. Because of the uncertainty of many important variables in the forecasting process, growth curves must be constantly updated. While the estimate of increased capacity requirement may be in error, it is obvious that in any case there is a need to construct a great many generating stations to meet increases in total demand and to replace existing units.

It is the purpose of each utility planning department to determine what that demand will be and the preferred way to expand the utility system to meet it.

LOAD FORECASTING

The first step in any power system planning activity is preparing the load forecast. In a proper system study, forecast of the load variation throughout the year determines the total energy to be supplied to customers, as well as the loading on individual units at any given time. Seasonal and daily load fluctuations must also be considered when evaluating total system capacity.

Short-range forecasts are used for operation planning, and intermediate or long-range forecasts are best suited for capacity planning. These forecasts must be compatible.

Preparation of a short-range forecast should take into consideration (a) the outlook of economic conditions nationwide with emphasis on the service area, and (b) the effect weather conditions and seasonal patterns may have on demand variations.

Intermediate and long-range forecasts are developed by using economic, demographic, resource, and probabilistic models.

The economic model considers such variables as national, state, and regional gross product, personal income, monetary policy, unemployment rate, etc. This model provides a measure of the consumers' probable demand for electric power.

In developing a population forecast, the rate of natural increase (births over deaths) and the net migration are forecast. These forecasts in turn depend on expected social and economic conditions.

The amount of resources and the rate of depletion must be considered when forecasting the rate of electrical power growth. Available land, water, fuel, etc., are essential for the continued growth of an area.

The various models are combined to develop a deterministic forecast on which planning is based. However, a probabilistic model is needed to forecast uncertainty. Factors considered should include those being used to develop the forecast, such as economic levels, weather and seasonal conditions, population growth, etc. A further uncertainty to which the model must address itself is the relationship between the forecast factors and electrical demand. These probabilities indicate the degree of confidence that can be placed in the forecast.

PLANNING CRITERIA STANDARDS

Before an expansion plan can be prepared, the desired level of reliability or quality of service must be established. Every electrical system must have enough facilities to provide power generating capability in excess of normal power demands. This reserve generating capacity is necessary to replace the capacity of facilities which are taken out of service on a scheduled basis to perform maintenance, refueling, etc., or which are shut down on an emergency basis due to equipment failures. The more reserve generating capacity a system has, the lower the probability that a scheduled or unscheduled outage of units will cause a power shortage. This reliability level is determined by balancing economy and reliability. The goal is to achieve the maximum economies of scale with a commensurate level of reliability, where reliability is evaluated in the total context of the system in terms of types of units,

their probability of outage, interconnections, and other emergency resources. Similarly, more redundancy built into the transmission system and interconnections with other utilities decreases the likelihood that any outage will cause interruption to a customer or cause unacceptable voltage reductions.

The reliability requirement is based on past experience and management's assessment of what the escalating system need for this requirement will be in the future. This requirement is also subject to sensitivity analysis, but it is generally used as a guideline for planning to assure the same minimum reliability standard for each plan. The planner then can assume that whenever reliability deteriorates below the acceptable level, improvement in the system is required.

For the generating system, the reliability requirement is a complex probability evaluation. A unit by unit representation of existing and future units and simulation of load characteristics are necessary in a probabilistic calculation. This method takes account of not only the peak loads but the off-peak loads and the scheduled and expected unscheduled outages of each unit. The index of reliability may be expressed in the form of loss of load probability and might, for example, have a value of one day in 10 years, which means that the utility accepts the criterion that through a 10-year period on the average there will be only one time when the available capacity is not adequate to meet the expected load.

For the transmission system, the reliability requirements normally set the acceptable rating of transmission lines, transformers, and switchgear under different conditions, as well as the type of contingency the system must withstand without interruption of service, overloading of facilities, or unacceptable voltage. Formulation of a general long-range transmission plan starts with the model of the existing system, complete with all the lines, transformers, switching stations, and generating stations, and includes those additions to the transmission system which will be required to meet future energy demands. Several load levels are simulated on the system model by a computer program for alternative transmission networks, and each network must transport the necessary electrical energy into the existing utility system while maintaining overall system reliability.

This information on the timing and size of system capacity and transmission expansion is vital to the utility in assessing the need for new sites and transmission rights-of-way. It is at this point that the utility may find that it must begin to identify and commit those new sites and rights-of-way required to meet the forecasted load demand.

Upon completion of evaluating future system growth and expansion requirements, it is necessary for the utility, in close cooperation with the applicable regulatory agencies and the public, to identify and commit acceptable sites to meet these requirements well in advance of their anticipated dedication dates.

This long-range planning phase should be undertaken with an emphasis upon selecting sites that provide the utility the greatest amount of flexibility in developing an expansion plan. Based upon the considerations delineated in this section, the utility is faced with selecting a sufficient number of sites to meet the anticipated demand. Since the considerations are fairly general in nature, it is necessary for the utility to select sites with the greatest adaptability to a number of alternative generation modes in order to ensure the ultimate high potential of the sites for future development. During the long-range planning phase of the siting process, all characteristics (either negative or positive) of sites under consideration should be identified in consultation with regulatory agencies and representatives of the concerned public.

There are two principal differences between the procedures historically followed by the utilities and those described above. First, the driving force for the commitment of power plant sites can no longer be principally an economic one. More important than the availability of an inexpensive site with the requisite technical characteristics, such as abundance of cooling water and adequate transportation access, is compatibility of such site development with current and probable future area land utilization, as well as with minimi-

zation of adverse environmental impacts. Economic considerations are not a sufficient index, since the expected environmental impacts are also highly relevant to ultimate utilization of the site. By evaluating each potential site on identical bases, it is possible to select the most favored locations.

The second difference in the siting process is that regulatory agencies and the public need to be considered and involved at an early stage in site identification and commitment. Whereas today these groups play their most prominent role in the licensing and acceptance phase, the need for cooperation on the part of both the utility and the public, from the early stages of system planning, is becoming increasingly apparent. In a free and open atmosphere of mutual concern for both the energy and the environmental needs of a region, the utility and the public, through the responsible government agencies and interested groups, can reach a common ground for agreement on expansion plans that will transcend the confrontations that in some cases characterize such plans today. It is incumbent upon the utility to recognize the concerns of the public. Thus, an informed public, through consultation with the regulatory agencies and the technical forces of the utility, will be able to evaluate potential sites on the bases of the considerations listed, weighing all relevant considerations, and to support the selection of those sites that have greatest potential for future power plant site development.

Having identified sites that most satisfy the general siting criteria, using the suggested evaluation process, these sites should be set aside by the utility or by a state or other governmental body, if such a procedure has been established, for further detailed evaluation and possible future use. In certain cases, it may be desirable for the utility to acquire those sites. This commitment of sites for future use is a most important output of the long-range planning process. Recent trends and possible future directions for implementing this process are identified in more detail in Chapter 4, Site Acceptance and Approval.

POWER PLANT SITING CONSIDERATIONS

Determination of acceptability of a site for future electric generating stations must consider both the technological and economic needs of the plant itself as well as the environmental, energy, and land use requirements of the region or community in which it is located and the area which it serves. Recently, assessment of these community needs has become increasingly complex.

To be acceptable, a plant site must have the following physical characteristics:

Area—A plot size sufficient to wholly contain the generating station and its associated facilities, which may include powerhouse facilities, emission abatement facilities, fuel receiving/storage facilities, switchyard and transmission line access, cooling water facilities, and exclusion area.

Soils—The soil must possess adequate load-bearing capacity and be of a workable consistency.

Natural Phenomena—The facility should be capable of being reasonably protected from natural hazards.

In addition to physical characteristics related to the land area itself, the site location must have:

Construction Access—During the construction phase, there must be access to the site for heavy materials and construction equipment. There must also be an available pool of manpower with requisite construction skills.

Operational Access—During plant operation, access for fuel must be provided, whether by barge, pipeline, railway, or highway, as well as a reliable source of cooling water (for thermal generation plants). In addition, transmission rights-of-way must also be secured.

It is not appropriate to site a power plant only on the basis of its physical needs. The needs of the communities or region to be served or affected by the plant must also be considered. Besides the need for

electrical energy, the economic and environmental needs of the area are important. These needs are coming under ever-increasing scrutiny, especially in the case of remotely sited plants that are not generating a significant quantity of energy for the community in which they are located. The principal interest that must be served is supplying energy needs of a region in a manner that creates the greatest overall benefit and minimum overall detrimental impact for that entire region.

One way in which an area can realize a net benefit from a power plant siting is through economic enhancement of the area. A site should be evaluated for the community benefits of increased manpower needs in the area, due both to direct employment by the utility and also to the employment spin-off created by an increased availability of inexpensive energy. In addition, the effect of increased tax revenue to the area should be included in the site evaluation. Siting of a power plant can appreciably improve the economic status of certain areas.

It is not apparent at this time if any electric generating station can be justified on the single consideration of its economic impact. Every electric generation process known to man has potential detrimental impacts upon the environment, since each disturbs the existing state of nature. Therefore, the process that minimizes this net detriment, either by minimizing the generated waste products or by transforming them into beneficial by-products, while maximizing regional benefits of the new energy and of the economic stimulation, is the most favored one. Hence, the utility must carefully evaluate the environmental characteristics of each site and select only those sites that have the greatest potential for minimizing impact, as well as the flexibility to meet a variety of ever-changing constraints.

The land must be zoned appropriately, and its use must be compatible with future plans of the area. If the plant is readily accessible to the public, special consideration must be given to the potential for constructing a physically attractive structure. In the case of nuclear power plants, the site must also be sufficiently removed from populated areas, considering the anticipated as well as the current demographic composition. The site should also be assessed for its impact upon natural resources, such as historical, archaeological, and recreational characteristics, and the utilization of nonrenewable or nonabundant fuel resources.

Effects of the plant upon the atmosphere must also be evaluated. Although these effects will be highly dependent upon the specific generating/emission abatement system selected, certain general characteristics can be assessed to determine relative acceptability for the purpose of eliminating unfavorable sites. The approximate quantities of airborne combustion products and the effect of local meteorology upon their dispersion should be considered. In addition, the effect of the plant upon local meteorology, for example, by the influence of cooling towers, must be indicated. Effect of the plant upon the water bodies being utilized, both in terms of thermal discharges and liquid effluents, must also be assessed, as well as the effect upon the aquatic community.

These factors must all be considered in identifying and committing sites for future power plant developments. Only those sites that can be developed with a minimum impact upon the environment should be retained for further investigation.

SITE EVALUATION METHOD

The systematic identification and commitment of power plant sites utilizes a ranking procedure in which the weighted impacts of different effects are combined to yield an overall measure of the site. Table 1 illustrates an example of this procedure, in which several sites are evaluated for their potential development as fossil-fueled, nuclear, hydro, and geothermal power plants. In this example, a list of sites is evaluated on the following bases:

1. *Economic Considerations*
Site Development Costs: The incremental or differential dollars per kilowatt cost for site development for both the initial and the ultimate site capacity.
2. *Environmental Considerations*
Air: Assessment based upon site meteorology and dispersion potential and site remoteness.
Water: Assessment based upon effects of thermal and waste discharges upon the water supply. Also, the effect of disturbances upon the aquatic community.
Land: Compatibility of site with general area development plan, including aesthetic factors and noise and population proximity for nuclear plants.
3. *Community Considerations*
Recreation: Enhancement/withdrawal of areas for recreation.
Economic Improvement: Benefits to the area from installation of the unit.

The example proceeds by listing all of the sites potentially available to the utility for each generation alternative. It is obvious that certain site/generation combinations are not feasible, and these alternatives are eliminated.

The remaining alternatives are described in more detail, both on an economic basis, in terms of the dif-

ferential cost in dollars per kilowatt for site development, and on an impact basis, in qualitative terms (relative ranking) primarily utilizing sound engineering judgment. Differential costs of development are converted to relative rankings that are based upon the specific dollar range involved, according to the following conversion basis:

$/kW	Rating
0–5	4
6–10	3
11–15	2
16–20	1
21–25	0
25	Prohibitive (P), not considered to be a viable alternative

The three factors are then combined, utilizing weighting factors to determine final site rankings.

The approach outlined above, combining both the tangible features of the site (the dollars to develop it as a feasible site) with the intangible features required to make it acceptable, is a fundamentally sound one. One of the principal aspects of the systems approach is its consideration of all relevant variables in order to reach a decision, and this evaluation model enables these variables to be logically combined into a single figure of merit for each site alternative.

However, development of the weighting factors used in the combination of these features into one measure of goodness is difficult. For example, the importance the local community would place upon environmental features may be different from those of the regulatory agency or the utility. However, when these features are weighted by a panel consisting of utility, regulatory agency, and public representatives, it is expected that a greater sensitivity to environmental impacts would be manifested, and a consensus on the weighting factors developed. In cases where the cost differentials for developing alternative sites are relatively minor in comparison to the total plant cost, environmental and public acceptance factor weights may be predominant. On the other hand, the number-one-ranked site/generation alternative in the example, an ocean nuclear site, becomes only the seventh-ranked choice if its site development costs are found to be $15/kW, indicating the need for well-developed input data in the model. The example, however, is readily amenable to adjustment by changing weighting factors as a result of consensus evaluation.

It is apparent that most of the utility industry's problems to date, outside of those directly related to changing regulatory constraints (the evaluation of which cannot be appreciably improved prior to the site/generation selection phase), have resulted from an inability to accurately assess the degree of importance of differing

TABLE 1 Site Evaluation Matrix

	Site Development		Environmental Impacts			Community Benefits		Total (rating × weight)	Final Rank
	$/kW	Rating	Air	Water	Land	Recreation	Economic		
Weighting Factor		30	15	15	20	10	10	100	
Site 1—Desert Location (New)									
Fossil	5	4	1	3	2	2	4	280	7
Nuclear	10	3	4	2	3	2	4	300	4
Hydro		P						—	
Geothermal		P						—	
Site 2—Urban Location (Existing)									
Fossil	10	3	0	3	2	3	2	225	11
Nuclear		P						—	
Hydro		P						—	
Geothermal		P						—	
Site 3—River									
Fossil	15	2	2	3	2	3	3	235	10
Nuclear	10	3	4	2	3	3	3	300	4
Hydro	5	4	4	2	3	2	3	320	3
Geothermal		P						—	
Site 4—Ocean Location (Existing)									
Fossil	10	3	2	3	3	3	3	285	6
Nuclear	5	4	4	2	3	4	3	340	1
Hydro		P						—	
Geothermal		P						—	
Site 5—Geothermal Field									
Fossil	15	2	2	2	3	2	4	240	9
Nuclear	15	2	4	1	4	2	4	275	8
Hydro		P						—	
Geothermal	5	4	3	4	2	2	4	335	2

environmental impacts upon a siting proposal. It is incumbent upon the utility to incorporate public attitudes into the assessment in order to provide a greater measure of certainty that the sites will be chosen in the expressed best interests of the public. Such attitudes may be secured from key representatives of the local community, including local chambers of commerce and service organizations as well as local conservation and environmental leaders. Although the utility can provide the technical specifications regarding generation alternatives, it cannot independently evaluate the impacts. The relationships comparing, for example, coal and oil emissions, may not be perceived by the public to be linear. If the public felt that all emissions into the existing atmosphere were bad, then low-sulfur fuel oil and high-sulfur coal may in fact be considered as having an approximately equal bad rating. That is, fossil-fuel emissions at any level would be regarded as violating the minimum constraint. To have different weights would be to avoid the real issue—that some group considers emission levels from use of either fuel as greater than some threshold value, and hence unacceptable. With the help of the public, realistic appraisals of such relationships can be developed that take into account all the interests involved in the site identification process.

An improvement in the example for determining specific environmental impacts is the use of probability measures. The probability index should indicate the likelihood that an impact may result in the delay or even cancellation of a site. Having developed these indices for each impact, they can be multiplied together to yield a probable value for the ultimate development of a site/generation alternative. Note that an urban nuclear site would receive a zero-probability value for the population proximity factor, since 10 CFR 100 essentially rules out such a location, resulting in an overall site probability value of zero, regardless of its other characteristics. This treatment insures that any site which cannot meet regulatory constraints will be eliminated from consideration at an early stage. These probabilities could be independent of an assessment of economic feasibility in which there are only technically and legislatively unfeasible concepts, but no economically unfeasible ones. Great expense (for example, underground nuclear plants) would be reflected in a low, but nonzero probability, the utility thus being provided with a measure of the ultimate development potential of the site, which is the information it requires during the site identification stage.

Since most sites that are considered more than cursorily are likely to have a fairly high potential for one form of generation or another, the most attractive sites will be those that have high potential for more than one generation method. The use of probability indices introduces a measure that is readily amenable to compilation. Combinatorial probabilities can be used to merge the different generation probabilities at a given site into a single goodness measure. These single measures then allow a ranking of the sites under consideration. Note that a site with only one generation possibility would score the value of that probability, while sites with several alternatives would achieve a rank higher than that of any single alternative. Hence, flexibility can be sacrificed for high individual generation alternative potential, but in most cases flexibility will be preferred.

The example clearly indicates the importance of ranking, and considering a wide spectrum of different alternatives. The use of public and regulatory attitudes to develop probability measures of these impacts provides the utility with an index that reflects the overall ranked desirability of the sites being evaluated, in a framework relatively free of economic constraints. This should enable the utility to make a decision to commit the required sites for future development, at which time a much more detailed site selection procedure must be initiated.

3   SELECTION OF SITE AND MODE OF GENERATION

The site identification and commitment process provides an inventory of potential sites for future utility system expansion. When the forecast of system demand indicates the need for additional or replacement generation capacity, the site/generation alternative selection phase is initiated. In this phase, cost estimates of each potential generation mode at each site are developed. In addition, and equally important, the environmental impacts of each alternative are assessed in detail, based upon both the current state-of-the-art and anticipated near-term developments for pollution abatement systems.

This section proposes a framework for selecting the preferred site/generation alternative. The traditional technical/economic factors and the public interest/environmental impact factors are considered the primary criteria. Appendix A, Technological Assessment of Generation Alternatives, provides background material for this section. The Appendix identifies current and future energy generation technology and the potential environmental impacts of these various modes of generation. Pollution abatement system technology for the various environmental impact areas is also discussed in the Appendix.

ECONOMIC AND ENVIRONMENTAL IMPACT ASSESSMENT OF ALTERNATIVES

Detailed cost estimates of the alternatives should include several different factors. First to be considered are the fixed and variable costs of design, construction, operation, and maintenance of the facility. These costs should also include the expenses incurred to comply with regulatory standards. The cost of transmission, including system losses, which is a highly site-dependent factor, is then added to develop a total energy cost.

Basically, this procedure is similar to that currently being followed by the utilities in planning system expansion. However, the need to develop these estimates for each site and potential generation method, including the cost of pollution abatement equipment, adds a considerable degree of complexity to such estimating.

In addition to developing these detailed cost estimates, an assessment of the environmental impacts of each scheme must be prepared. While the utility is responsible for developing information regarding total levels of impact, it should not, in general, independently assess the effects of these levels. Representatives of the public must also have the opportunity to evaluate the environmental impacts, particularly in view of the lack of basic scientific knowledge in many instances and the judgmental nature of such assessments. The utility should undertake full investigations and monitoring programs at potential sites to establish a baseline for assessing environmental impacts. This evaluation must consider the preoperational, startup, and operational environmental impact conditions.

To select a most-favored or least-impact alternative, it is also necessary to develop tradeoffs among various environmental impacts. The most obvious approach is to use weighting factors, based upon the opinions or consensus of professionals whose expertise is multidisciplinary. Such professionals are currently being employed by utilities in increasing numbers. Similar expertise is being retained by the social-interest groups. Consensus-based decision making is suggested by this approach.

On the basis of such weighting factors, the alternatives can be ranked for each environmental impact and then for each site. The result is a list of site/generation combinations ranked on the basis of environmental impacts.

At this point the utility must select one of these combinations for its next unit additions, considering in its decision both the economic ranking and the environmental impact ranking. Unless there exists a scheme that is most favored in both respects, that is, a dominant alternative, the utility must exercise considerable judgment and imagination in making this selection.

While it is clear that many members of the public are vitally concerned over adequate protection of the environment, it is questionable if the public is really pre-pared to pay the costs of total environmental impact abatement. The utility must, therefore, evaluate the costs it would incur to change the rankings of the alternative schemes, in terms of environmental impact. These may be discretionary costs, in the sense that they are above those costs absolutely necessary to comply with the applicable regulatory standards. But they are also basic costs in that they may be required to satisfy the social needs and concerns of the public. Such costs might, in terms of underground substations and transmission and distribution lines, recreational facilities, zero-release nuclear installations, desulfurized fuel, and a number of other possibilities, establish the range of flexibility that each site/generation alternative possesses. Utility executives and regulatory authorities must then exercise their judgment to responsibly carry out their charter and their responsibility to supply the public with energy, at a level of environmental impact that is acceptable to the public and economically feasible.

4   SITE ACCEPTANCE AND APPROVAL

Increasing concern for environmental quality has caused state legislators to give state regulatory commissions and other state agencies greater authority in controlling environmental effects of electric power plants.* The majority of states have commissions that must grant authority to a utility within its jurisdiction before the utility is permitted to construct a power plant or transmission line and, before an approval certificate is issued, the commission considers a wide range of environmental factors, including aesthetics, conservation, pollution, and recreation. Some states have also established new agencies with specific responsibility for setting and enforcing environmental controls.

CURRENT REGULATORY FRAMEWORK

The two points of view that have been in conflict—the need for ample supply of electric power at a reasonable cost and the strong desire to improve environmental quality—are both matters of major political importance to elected officials. Accordingly, elected officials closely observe and attempt to assert varying degrees of control and influence in these areas. This means that at an early stage in the planning for new generating facilities, regulatory officials at all levels should be apprised of the planning and can be expected, dependent upon local conditions, to exert strong influences directed toward assuring adequate electric power to the region with a minimum impingement on the environment.

Since this dual objective is not easily obtainable, it requires some adroitness and spirit of compromise on the part of those responsible for the planning, not only in the state government but also within the electric utility industry, and it is important that the cooperation of representatives of environmental and conservation organizations is solicited. It requires some discussions and accommodations by each group prior to a public forum on any new facility; otherwise, public discussion can easily force key individuals from each group into positions not easily changed. It also makes more necessary some final decision-making body acceptable to all parties to resolve any deep-seated conflicts that may arise.

Of particular sensitivity is the position of the legislator representing the immediate area in which the proposed facility will be built. The additional tax revenues that will accrue to the immediate township may not be a sufficient inducement to the local citizens to put up with the inconveniences and disadvantages resulting from the plant being located in their immediate area. He, as the representative of the local citizenry, is generally the individual most sensitive to their desires. Accordingly, his opposition should be given great weight, and any suggestions he may have for making the facility more acceptable to his constituents should be given careful consideration.

The growing concern for the environment has had its effects also at the grass roots level. Frequently, legislation is being introduced at the local town council and county levels to require new transmission lines to be placed underground or to prohibit the building of

---

* See Appendix B: Site Acceptance and Approval—Historical Background.

additional fossil-fueled or nuclear power plants. In municipal areas, numerous permits and certificates must be obtained from various agencies, such as the health department, the department of water, fire department, and similar organizations, during construction and prior to operation of a new plant. Failure to obtain a necessary permit could prevent construction or operation of the plant. Electric utilities thus are increasingly under threat that construction will be subject to costly delays or, when completed, failure to obtain a needed permit will prevent plant operation.

The increasing conflict between those advocating preservation of the environment and those responsible for meeting the increasing demand for electric power has had its effect also at the federal level. Although the federal government originally had taken little or no active part in regulating fossil-fueled electric generating plants, concern for environmental matters has recently changed the picture.

The National Environmental Policy Act of 1969, the Water Quality Improvement Act of 1970, and the Clean Air Act Amendments of 1970 (amendments to the Clean Air Act of 1967), while not pre-empting the environmental field, have placed the federal government in a dominant role with respect to regulating environmental matters which, in turn, greatly affect the siting and operation of power plants. The Air Quality Act of 1967, as amended, and the Water Quality Improvement Act of 1970 have given the federal government the authority to set national air and water quality standards that the individual states are required to implement within a specified time under threat of having their authority in these regulatory areas abrogated. The National Environmental Policy Act of 1969 declares the prevention or elimination of damage to the environment to be a national policy and requires "all agencies and all federal officials with a legislative mandate and responsibility to consider the consequences of their actions on the environment."

Under the National Environmental Policy Act of 1969, among other requirements, all federal agencies, "to the fullest extent possible" must, on all major federal actions significantly affecting the quality of the environment, include a detailed statement on

(i) the environmental impact of the proposed action,

(ii) any adverse environmental effects which cannot be avoided should the proposal be implemented,

(iii) alternatives to the proposed action,

(iv) the relationship between local short-term uses of man's environment and the maintenance and enhancement of long-term productivity, and

(v) any irreversible and irretrievable commitments of resources which would be involved in the proposed action should it be implemented.

In addition, the Act requires that

Prior to making any detailed statement, the responsible Federal official shall consult with and obtain the comments of any Federal agency which has jurisdiction by law or special expertise with respect to any environmental impact involved. Copies of such statement and the comments and views of the appropriate Federal, State and local agencies, which are authorized to develop and enforce environmental standards, shall be made available to the President, the Council on Environmental Quality and to the public. . . .

The federal government has licensing jurisdiction for nuclear power plants and hydroelectric plants through the Atomic Energy Commission and the Federal Power Commission, respectively. Section 10 of the Rivers and Harbors Act of 1899 also requires a construction permit from the Corps of Engineers for facilities to be constructed on navigable waters. Under rules published on April 7, 1971, the Corps of Engineers now requires a discharge permit under the Refuse Act of 1899 for electric steam plants using navigable waters for cooling purposes. The Federal Fish and Wildlife Coordination Act and the National Historic Preservation Act also have application to new generating facilities.

No matter what type of generating plant or what site is proposed, some recourse must be made to a federal agency prior to construction and operation. An electric utility planning new generating and transmission facilities must also be prepared to satisfy the provisions of the National Environmental Protection Act of 1969, the full application of which has not yet been fully determined. This includes applications to the Department of the Army Corps of Engineers involving river construction permits, and even the Federal Aviation Agency in connection with location and height of smokestacks.

A public utility planning a new generating facility and transmission line today has no one single regulatory agency to approach for approval of its plans. Construction of the facility by itself requires 3 to 8 years, and usually approval of the site must precede start of construction. Having successfully met local zoning restrictions, the electric utility company must still face a multiplicity of municipal, state, and federal reviews continuing through the construction period, any one of which conceivably could prevent plant operation. In major municipalities, such as Los Angeles and New York, this involves as many as 30 or more specific approvals from varying governmental agencies. Compounding the problem for the public utility is the fact that substantial sums of money are committed to purchase and preparation of the site and the construction of the plant prior to the final government permit being issued for plant operation.

Even assuming the electric utility company has successfully worked its way through the multiple municipal, state, and federal agencies that have participated in the review and final approval of its selected site and proposed facility, and it has received all necessary certificates and permits, it still has no assurance that it may proceed without worry. In addition to having had opportunities to make their reviews known during public hearings held by the various regulatory agencies reviewing the proposed plant, opponents to the plant still have recourse to the courts. Thus, actions of the various reviewing agencies are subject to additional judicial review. After extensive testimony and legal argument, the courts may determine that the government agency erred in its determination and send the matter back for rehearing. There are some dedicated environmentalists who strongly advocate using the judicial review of regulatory agency decisions as a means of delaying, if not preventing, new electric generating facilities, and who hope that the threat of such action may forestall the utility from its proposed action.

Regulatory and other restraints on siting and constructing new generating facilities and transmission lines are quite numerous and not yet clearly defined. The long time period required from the point of initiating action, through completing all necessary regulatory and judicial action and constructing the plant, when environmental laws and their interpretation are in a state of rapid change, makes it extremely difficult to know when or if a new generating plant may go into operation.

FUTURE TRENDS

The need for new electric generating facilities to meet the predicted increasing demands for electric energy, as well as to replace obsolete equipment, will require identifying and acquiring additional sites for power plants, together with the accompanying rights-of-way for transmission lines, whether overhead or underground. The conflict between those responsible for electrical energy supply and those wishing to restrict and limit further electric production for environmental or other reasons will become more acute in the foreseeable future.

At the local level, whether it be urban, suburban, or rural, it will become more and more difficult to obtain sites for construction and operation of power plants. The desire to have the benefits of electric energy, but wanting the disadvantages associated with its production and distribution "in somebody else's backyard," will become more prevalent unless some overriding public decision-making process is developed to pre-empt local opposition or to ration the use of electric energy. If future brownout and blackout occurrences are not to

be national in scope, decision-making bodies, preferably at the state level, undoubtedly will have to be developed with near-absolute powers in deciding if and where new power plants are to be built, or to what extent electric usage is to be curtailed.

The principal problem that must be solved is how to structure the decision-making process to cope effectively with the dilemma caused by the conflict between concern with the environment and the increasing demand for electricity. Accordingly, there must be an attempt to integrate the planning process with a regulatory process that assures proper attention is given to environmental effects of additional generating facilities, at the same time permitting construction of needed facilities in a timely manner.

It seems logical that efforts will be made to consolidate the decision-making process as nearly as possible in one agency—a one-stop process—in which both the need for electricity and the environmental effects will be considered and where the proponents and opponents will be given the opportunity to be heard. However, it is already obvious that local, county, village, and municipal interests will strongly resist any attempt by the state to set up a one-stop siting agency that does not permit some kind of veto at the local level. Also, environmentalists and others will insist that any agency or decision-making body that may be formed must have representation by those associated with their beliefs and objectives. They will strongly oppose such authority being granted to the Public Service Commission or other existing regulatory agency presently responsible for overseeing the electric utility companies, since they believe such agencies are oriented more toward meeting electric power demands than toward protecting the environment.

From the utility company point of view, a one-stop process would be favorable for a number of reasons but, in fact, represents a utopia. It does not appear feasible that in the foreseeable future one agency will have the authority to completely handle the power plant siting issue. Consequently, what is sought is a final arbiter, in other words, a proper authority where, after all evidence has been adduced and after conditions have been weighted, a final decision may be rendered. It is important that the utility industry and the public maintain access to such a decision-making body so that future electrical demands can be met while maintaining a quality environment.

California, Maryland, Vermont, and Washington have already adopted special procedures for certification of power plant sites. Several other states are currently considering legislative proposals and can be expected to follow suit.

The State of California has established a Power Plant

Siting Committee within its Resources Agency. The Committee had its origin in August of 1964, when the Administrator of Resources for the State appointed an Ad Hoc Committee to review certain proposals of the Pacific Gas and Electric Company to acquire sites for nuclear power plants. The Ad Hoc Committee, which was composed of representatives of the Departments of Fish and Game, Parks and Recreation, Water Resources, Conservation, and the Water Quality Control Board, functioned so well that shortly thereafter a standing committee was created and charged with responsibility of reviewing and making recommendations on all proposed nuclear sites in California. Since then, the Committee's jurisdiction has been broadened to include sites for all thermal power plants, and its importance has been further increased through action of the California State Assembly directing the Resources Agency, in concert with California electric utilities, to undertake the preparation of a long-range plan for establishing power plant sites to meet the projected growth within the State. As an outgrowth of that, the State Legislature in 1970 gave the State Resources Agency responsibility for developing a plan indicating the optimum location of all electric generating power plants expected to be constructed in the State over the next 20 years. The plan is to include new plants and additions to existing plants and to identify locations deemed suitable from an environmental standpoint. The State Resources Agency is also responsible for establishing a program of research on improved methods of power plant siting.

New York State likewise has made an effort to participate in long-range site planning and, like California, initially approached the problem through formation of an Ad Hoc Committee to help acquire future sites for nuclear power plants. The Governor, in 1970, requested the New York State Atomic and Space Development Authority to prepare an inventory of sites compatible with protection of the environment to insure that there would be sufficient sites in the future to meet New York State's electric power needs.

The Authority set up a Nuclear Power Siting Committee made up of State government officials, including representatives of the Public Service Commission, the Department of Commerce, and environmental and conservation agencies. In addition, there is a representation on the Siting Committee from the New York Power Pool and a nuclear research organization composed of the major investor-owned utilities. While recommendation of a site by the Siting Committee is no guarantee to the utility that the site will be acceptable to the multitudinous government agencies involved in licensing a facility, New York's efforts in this regard are a major step forward in involving state government in the site selection process.

In New York, the Atomic and Space Development Authority is authorized to acquire and hold sites for future use by public utility companies, and Maryland recently empowered a State agency to acquire sites for future needed electric generating facilities. To date, however, neither state has exercised this authority.

While California, Maryland, Vermont, and Washington have adopted special procedures for certification of power plant sites, and other states are considering similar legislative proposals, it is generally true that an electric utility company cannot obtain approval of a site prior to incurring heavy expenses in the purchase of land and engineering studies associated with the site. Some recent examples of utilities that invested large sums of money in site acquisition and site preparation and, after long delays involving extensive regulatory reviews, were not granted approval for use of the sites are Pacific Gas and Electric—Bodega Head; City of Los Angeles—Malibu; New York State Electric & Gas —Bell Station. These were proposed nuclear plants requiring a construction permit from the AEC. A proposed site for a pumped storage project by Consolidated Edison Company at Cornwall, New York, was submitted to the Federal Power Commission in January 1963 and as of 1971 is still being pursued through the regulatory process, including a second round of judicial review, with no assurance it will be approved.

Unfortunately, final governmental approval cannot be assured until purchase of site has been made, extensive and costly engineering studies have been undertaken by the utility, and long regulatory reviews have been completed. Long-range planning for site acquisition and approval may well require, in some instances, setting up land banks to which approved sites can be designated and held 20 or more years in advance of their use. The cost of holding such sites would have to be included in the sale or lease charges to the utility and ultimately be borne by the rate payer.

Today, utilities are increasingly seeking to obtain public understanding and general acceptance of their proposals to construct new generating plants or transmission lines. This includes prior meetings with key representatives of the local community, including local chambers of commerce and service organizations, at which time the utility representatives can describe in detail their proposed plans and means for meeting environmental and other problems. Efforts should also be made to bring local conservation and environmental leaders into early discussions prior to completing detailed plans in an effort to make a plant more acceptable aesthetically as well as minimizing adverse environmental effects.

In some areas of the country, utilities have been successful in obtaining assistance from representatives of local and national conservation and environmental organizations in helping to select sites for future plants.* Such groups can often furnish useful information on proposed sites and means for making the proposed facilities more environmentally (including aesthetically) acceptable. However, this cooperation appears to be the exception rather than the rule.

If effective methods do not evolve at the state or regional level to compromise the conflict between those opposing additional electric generating sites and those wishing to meet the increasing demand for electric power, then the federal government undoubtedly will move more actively into this area, particularly if the resulting shortage of electric power becomes a national problem.

Environmental restraints that have evolved at the federal level to date have been for the most part in the nature of guidelines and criteria for the states to follow in establishing environmental standards, and only if the states do not promulgate or maintain such standards does the federal government have jurisdiction. Even though electric utility companies are required to file long-range plans with the Federal Power Commission under a rule adopted on April 10, 1970, questions pertaining to land use planning are still matters primarily of state concern. However, various legislative proposals pertaining to land use planning and specifically to power plant siting have been introduced in the U.S. Congress, and it would seem clear that if proper mechanisms are not developed at the state and regional level, the federal government will seek to dominate, if not to pre-empt, this field.

As local villages, counties, and municipalities will oppose the state government having exclusive determination of land use for electric needs, so might we expect the state governments to oppose any effort of the national government to obtain pre-emptive authority in this field. Even the electric utilities, who are caught in the crossfire between agencies of state government having conflicting responsibilities, in many cases may be expected to favor state rather than federal regulation as being more understanding of the local problem.

Since more than one federal agency is usually involved in any plans for new electric generating facilities, some effort to centralize their actions may be expected in conjunction with the reviews required by the National Environmental Act of 1969. The danger exists, however, that rather than simplifying and shortening the decision-making process, future federal actions directed during siting of electric generating facilities will add additional levels of review to those already existing at the state level and thus compound the problem.

---

* See Appendix C: Northern States Power Approach to Open Planning.

5   RESEARCH AND DEVELOPMENT ACTIVITIES

Previous sections of this report identify a great many technical areas, ranging from air quality to advanced generation techniques, in which new or additional research can help significantly to make electrical generating and transmission facilities more compatible with the environment and to optimize use of natural resources. In this section and Appendix D, these areas are summarized and additional areas are indicated. Together, these represent an extensive research undertaking that requires proper coordination in order to achieve the success so vital to the nation's health and development.

RESEARCH AND DEVELOPMENT PRIORITIES

Since it is currently impossible, from the standpoint of availability of both trained manpower and funds, to carry out simultaneously all the research suggested, it is necessary to establish a set of priorities to reach the goal of compatibility in an orderly manner. For this reason, the following general priority categories have been established.

1. *Minimizing Effects on Human Health*—There can be little question that problem areas that potentially have a direct harmful effect on health require the highest priority for research.

2. *Conserving Nonrenewable Resources*—Vital to the future of the nation is the optimum use of its resources—the various waters and their associated aquatic and marine life, fuel supplies, and land. In some cases the problems are immediate, but, in general, develop-

ment of efficient resource utilization programs can be slightly longer-ranged than solutions to health-related problems.

3. *Increasing Productivity, Comfort, and Enjoyment* —Although highly desirable, it is incongruous to accord certain problems relating to enhancement of enjoyment or productivity the same priority as health problems or resource preservation, when limitations must be placed on research efforts.

Each of the above categories has been subdivided into individual research areas with an internal priority ranking. A discussion of these individual research areas is presented in Appendix D. It should be noted that work performed in connection with certain processes actually may produce benefits in several areas. However, description of the research appears under the area of greatest potential benefit. The breakdown of each category, with its research areas, is as follows:

1. Minimizing Effects on Human Health
    a. Air Quality
        SO_2 removal
        NO_x removal
        Particulate removal
        Identification of minor gas constituents
        Investigation of stack gas constituent effects
    b. Radiation Sources
        Reduction of radioactive waste releases during plant operation
        Reduction of radioactive waste releases during fuel reprocessing

285

2. Conserving Nonrenewable Resources
   a. Water and Marine Life
      Alternative methods of heat disposal
      Effects of thermal discharge
      More efficient generating processes
      Thermal discharge utilization
   b. Fuel
      Nuclear breeder reactor
      Fusion
      Geothermal energy
      Solar energy
      Tidal energy
   c. Land
      New siting techniques
      Underground transmission
      Dispersed generation
      Advanced substation designs
3. Increasing Productivity, Comfort, and Enjoyment
   a. Modification of Facilities
      Architectural improvements
      Noise control
   b. Joint Ventures with Other Industries
   c. More Efficient Use of Electricity

ORGANIZATION

Because the scope of the research suggested and the magnitude of the financial support required, there must be a coordinated effort between industry and government in all areas of research and development. Several different types of organizations have been proposed to achieve this coordination. A careful study of all alternative arrangements should be made before committing to a particular approach. Two possible approaches are:

(1) a new federal government agency similar to NASA and (2) an industry–government partnership similar to the Electric Research Council.

Creation of a new federal agency to coordinate electric utility research would be costly, if historical trends hold true. A number of political obstacles would have to be overcome in order to establish the agency, and delays of 3–5 years would be both likely and expensive. In addition, there would be continuous political pressure in areas where decisions should be made on technical merit (e.g., past decisions regarding the location of major NASA facilities, and the operation of AEC facilities), often resulting in much more expensive solutions. Also, the costs associated with allocating and administering research funds would be greater for this type of organization.

An organization with a directing board made up of individuals from all segments of the electric utility industry, such as the present Electric Research Council,* closely coordinated with interested government agencies, including FPC, AEC, etc., interested manufacturers, and knowledgeable individuals representing environmental concerns, would appear to offer a suitable forum for the resolution of research needs and priorities. Such an organization could plan and oversee research programs developed to meet a series of specific goals, including both technological and environmental areas of concern. Leadership of the organization could rotate periodically between public and private utility representatives, with a full-time director and staff performing administrative functions.

---

* The Electric Research Council consists of representatives from the investor-owned utilities, Tennessee Valley Authority, American Public Power Association, Rural Electric Cooperatives, and Department of the Interior.

6 CONCLUSIONS AND RECOMMENDATIONS

CONCLUSIONS

This report has reviewed the present situation with regard to electric energy needs and power plant siting. The situation is characterized by a rapid increase in the number and size of power plants and by an increasing conflict between such facilities and a new public concern for the quality of the environment and the use of land and natural resources.

Based upon a detailed review of this situation and the alternatives available to utilities, the following conclusions are drawn:

1. The systems engineering approach for selecting and locating generation and transmission facilities has been used by electric utilities for many years. This approach is even more applicable today and will receive increasing emphasis in the future. Importantly, as identified in this report, the input criteria must be expanded to include the new set of social values concerning environment.

2. While it is recognized that electric energy usage cannot continue to grow indefinitely, it does not appear that the rate of growth over the next 10-year period will differ significantly from the present rate. Growth during the following decade, to 1990, may perhaps be at a reduced rate. However, a significant number of new facilities will have to be constructed to replace existing facilities and to meet increased energy demands.

3. Utilities are currently faced with a chaotic situation with respect to the availability of suitable and adequate fossil fuel supplies for present and future generating stations. This situation greatly compounds the difficulties of planning and siting new facilities. Further, there appears to be an impending shortage of uranium enrichment facilities within a decade or less. This shortage imposes a corresponding limitation on the ability of a utility to choose nuclear (light water reactor) power as a satisfactory alternate.

4. Since nuclear power plants discharge a minimum amount of air pollutants, do not require nearby fuel sources or major fuel handling facilities, can be made aesthetically attractive, and remove a minimum amount of land from public use since exclusion areas may be used for public recreation or other purposes, this mode of generation will provide an ever-increasing percentage of future electrical capacity, providing there is an adequate source of nuclear fuel.

5. Although not free from environmental impacts, hydroelectric power provides a means of generating electricity that causes no contaminants to be discharged into either the air or water environs. However, because of the limited availability of hydroelectric sites in most parts of the United States and the large amount of land required for reservoirs, this method of generation will not provide a major portion of future electrical requirements.

6. Pumped storage hydroelectric plants provide an effective means for utilities to maximize use of existing generating facilities to meet peak load demands. Although not an energy resource, they provide a utility with increased generating capacity. Throughout the United States there are a significant number of sites on which such plants could be constructed. However, due to the large amount of land required for reservoirs and the inability of these systems to operate without re-

ceiving energy from other generation facilities, pumped storage plants will provide only a small portion of future electrical capacity.

7. There exists a developed technology for extra high voltage transmission for both alternating and direct current, which allows for great flexibility in generating plant location, including remote siting. The principal noneconomic constraint in the development of such transmission is public reaction to the presence and appearance of these facilities.

8. The probability of significant applications of new methods for producing electric power during the next 10 to 15 years is extremely remote. While it is expected that progress will be made in various areas of advanced generation technology, it is unlikely that their development will reach the level necessary to provide industry with viable options to the current practice of using large-scale, steam-driven turbine-generators. Accordingly, all planning in the next decade, and most probably for the second decade, should contemplate utilizing the present large-scale central station format for utility plant expansion.

9. There is a high probability of achieving solutions to the sulfur dioxide problem within the next decade. Research is now in progress on various methods of sulfur removal, including fuel desulfurization and flue gas scrubbing systems. Long-range planning can and should assume that the technology necessary to solve the sulfur problem will become available.

10. Development of two-stage combustion has demonstrated that the production of nitrogen oxides may be reduced to approximately 50 to 25 percent of previous levels in many large fossil boilers, depending upon the boiler design or fuel being used. However, it does not appear that new techniques allowing significant reductions of nitrogen oxide emissions will be developed in the near future. This factor, coupled with new ambient air quality standards, will preclude siting new fossil plants or expanding existing plants in a number of large cities in the United States.

11. All steam generating facilities reject heat into the atmosphere. Condenser cooling water discharged into another body of water transfers heat to the atmosphere by evaporation, radiation, convection, and conduction. When cooling towers are employed, heat is rejected directly to the atmosphere, primarily by evaporation in wet cooling towers or by convection in dry cooling towers. Each type of system has its environmental advantages and disadvantages, which must be thoroughly investigated before deciding which method of cooling should be employed at a particular generating site.

12. Electrical generating facilities and their associated transmission and distribution systems require significant amounts of land. Although this requirement can be reduced by undergrounding portions of these systems, large amounts of land will still be necessary to meet projected capacity and transmission needs. When constructing such facilities, efforts must be made to make them as compatible as possible with surrounding structures and land areas.

13. A great deal of research and development is being performed to provide methods of minimizing the environmental impact of electrical facilities. In many instances, there is duplication of research effort between various institutions and corporations within the United States. Often, the results of the R&D programs are not being adequately disseminated to allow maximum use of available information. Because of the limited amount of financial resources available and the necessity to solve current environmental problems in a timely manner, it is essential that a coordinated, well-planned research and development program be undertaken.

14. A method of effectively determining true public interest factors in power plant siting studies is to maintain an active, open, and continuing interaction with all appropriate, interested parties from the selection of a site until the plant is in operation.

15. As a result of the increased concern for the environment by local, state, and federal governments, a great amount of new legislation is being considered. It is important that any such legislation reduce to the maximum extent possible the difficult task of securing permits, licenses, and approval from the many agencies, boards, and districts at all governmental levels and that it approach as closely as feasible a one-stop review process to consider the public interest as a whole.

RECOMMENDATIONS

To meet the increased demand for electrical energy and maintain a quality environment, many of the current conflicts between utilities, governmental organizations, and environmentalists must be resolved. To assist in resolving some of these conflicts, the following recommendations are made:

1. Governmental policy decisions should encourage the development of such energy sources so as to permit the power industry to retain the option of installing various types of generation, dependent only on future technological and environmental control developments.

2. All steam generating stations discharge heat into the environment. Various methods have been developed to effectively dissipate that heat into the environment. Because of the uniqueness of each prospective generat-

ing site, all methods should be investigated prior to deciding which of the available alternatives is preferred. It is not advisable to rule out any particular method for a region on the basis of its not being the preferred system for a particular site.

3. The power industry should continue to plan on the basis of large central station generation during the next two decades or until such time as new advanced methods of generation are developed sufficiently to provide viable options to current generation techniques.

4. In planning for new generation for the next two decades, the planning process should anticipate that the problem of the sulfur compounds will be greatly reduced, thereby minimizing the importance of this consideration in the fuel selection and plant location aspects of planning.

5. The increased development of rational zoning laws and community land use plans, together with such other state and federal policies regarding their respective public land, would alleviate the power plant siting problem and to a great extent reduce controversy over plant location.

6. It is recommended that transmission rights-of-way be used for other compatible purposes whenever possible in such a manner as to make the presence and character of the right-of-way community assets.

7. To resolve power plant siting issues in the total public interest, a mechanism should be established wherein controversial cases would be brought before a final arbiter, who, after considering all factors, would render a judgment in the total public interest.

8. To solve many of the present technical environmental problems, an enlarged nationwide research program should be planned along with a central coordinating organization to develop the program and oversee its successful conduct. Such an organization could be composed of representatives from the electric utility industry and would be closely coordinated with interested governmental agencies, interested manufacturers, and knowledgeable individuals representing environmental concerns. It would be responsible for planning and overseeing research programs developed to achieve specific technological and environmental goals.

9. Because of limitations in both available manpower and funds, the following general priority categories and their order should be established for direction of a research program:

    a. Minimize effects on human health

    b. Conserve nonrenewable resources

    c. Increase productivity, comfort, and enjoyment

Within each category, a more specific breakdown of individual research areas is provided in Chapter 5.

STATEMENT OF EXCEPTION

Theodore J. Nagel

In general, I concur with most of the substantive contents of the report prepared by Working Group II on Systems Approach to Power Plant Siting, including essentially all of its stated conclusions and all of its stated recommendations. I strongly disagree, however, with certain portions of the report dealing with the planning process in power plant site selection and with the degree of involvement by the public at large and by governmental agencies in the early stages of that process. Considering the fundamental nature of the issues involved and the importance of presenting in this report as balanced a judgment on these issues as possible, I believe it essential to have my views clearly stated, on the record, in those areas where they vary substantially from the views expressed in the report. Specifically, my exceptions relate to the following:

1. The report seems to argue that regulatory agencies and the public need to be directly and intimately involved, *from the early stages of system planning,* in site identification and commitment. I do not believe that this—as a practical matter—is either possible or desirable.

Power system planning—as many other disciplines in engineering and science—involves the application of complex, specialized knowledge that cannot be obtained except through extensive training and experience. Because of its highly technical nature, system planning cannot be undertaken in an "open forum" without the risk of total confusion and interminable delay. While the interests of the public in specific siting proposals must be considered, the important question is at what

stage of the planning process the public's views can be focused and intelligently brought to bear on the choice of power plant sites. In my judgment, such public involvement can be most meaningful only *after* the utility has selected its sites and prepared in full its technical, economic, and environmental evidence for public review. Earlier involvement would result in useless rhetoric and an evasion of responsibility by the utility, with a resultant failure to meet its overall public obligation.

It is unrealistic to assume that responsible decisions in the public interest can be made, and opposition to power plant siting reduced significantly, by public and regulatory involvement in the earliest stages of utility planning before full study is made of alternative sites, of their suitability for power generation purposes, and of their possible environmental impact. Also, power plant sites cannot be studied in isolation but are just one part of an interrelated power supply system. Their selection has a significant effect on the entire pattern of transmission development.

In an orderly society, the various segments of that society must meet their public responsibilities within such constraints as may apply. It is the electric utility industry's responsibility, which must not be abdicated to premature public review, to provide an adequate, reliable, and economical supply of electric power to its customers within the framework of overall national goals, including environmental goals, that are established by the society at large.

2. In its Chapter 3, the report states that "only those sites that can be developed with a *minimum* im-

290

pact upon the environment should be retained for further investigation" (emphasis added). This theme is repeated on several occasions in the report. I do not believe that the public interest at large will be best served by siting power plants to *minimize* the impact on the environment and by ignoring, in effect, the economic costs and the power supply reliability considerations involved. Important as the environmental considerations are, they must be balanced against a number of other considerations that are equally important to society.

Siting of power plants is closely interwoven with the entire pattern of power system development in both generation and transmission and could involve, therefore, economic penalties and reliability handicaps well beyond those associated with a particular plant site. These economic penalties and power supply reliability handicaps may well be greater than what society at large is prepared to incur for *minimizing* the impact on the environment. It is a fact that no human activity can be undertaken without some impact on the environment. The basic question involves the relative benefits and costs of environmental improvement rather than the absolute minimization of environmental impact. As previously indicated, it is the responsibility of the electric utility industry to fulfill the electric energy requirements of its consumers in the most reliable and economical manner, *while at the same time complying in full with established environmental standards.* Such standards need to be arrived at by the expression of the public will through the established processes of government, taking into account the relative costs and benefits.

3. The report contains a detailed description of a "site evaluation method," utilizing "a ranking procedure in which the weighted impacts of different effects are combined to yield an overall measure of the site." In my judgment, such an approach is both unrealistic and simplistic. Power plant site selection and evaluation is a complex iterative process involving detailed study which needs to take into account all significant and relevant factors and which terminates in a searching appraisal of the most suitable choices. It is not amenable to a box-score assessment utilizing *ad hoc,* consensus-based weighting factors, such as suggested in the report in its conceptual site ranking model.

The solution to the energy-environment dilemma can only be based on the establishment of rational standards for air, water, and land use, with the power industry prepared to assume its full responsibility to plan its systems within the constraints of such standards.

APPENDIX A: TECHNOLOGICAL ASSESSMENT OF GENERATION ALTERNATIVES

To meet the increased demand for electrical energy, utilities throughout the United States are faced with the problem of siting large electric generating facilities. Currently, these facilities utilize either fossil or nuclear fuels or harness the energy of falling water. Other methods of producing electrical energy, such as using geothermal steam, are viable. However, during the next decade, they could provide only a small portion of the additional electrical energy needed to support our society.

Each type of generating facility has an impact upon the environment. These impacts are varied, and, consequently, design of a facility for a particular site requires detailed understanding of the site parameters and of methods available to reduce the impacts.

It is the purpose of this appendix to (1) review the methods of generation available to the utility, and (2) summarize the environmental effects of power generating facilities and the alternative methods and status of technology to reduce those effects.

THERMAL TURBOELECTRIC POWER PLANTS

Methods of Generation

Thermal turboelectric generation involves the use of a heat source to generate steam or hot gases that expand in a turbine to drive an electric generator. Until recent years, the only significant generation by this method was by combustion of carbonaceous materials of fossil origin to generate steam for use in a turbine. More re-

cently, nuclear power and, to a small extent, geothermal steam have taken their place in the generation methods selected by utilities.

Fossil Fuel Plants

The essential features of a typical fossil-fired generating plant are a steam-generating unit (boiler), with associated forced and induced draft fans; the flue gas system, consisting of a stack and dust collectors as necessary; the turbine-generator unit and its auxiliaries, of which the condenser is of principal interest; and the fuel system. In many cases, particularly in the South and Southwest, all equipment is outdoors. In more severe climates, all are indoors or semienclosed.

A modern fossil-fired plant might typically consist of a boiler house about 250 feet in height by 150 feet square in plan; a turbine hall 250 feet long, 100 feet wide, and 100 feet high; and a number of auxiliary structures of lesser dimensions. Land use, including cooling towers, fuel storage, etc., is about 1,000 acres.

The simplest fuel to fire is natural gas. Gas is piped at high pressure (up to 600 psig), reduced in pressure by throttling valves, and fired directly into the furnace of the boiler. Since gas is free of ash and sulfur, the flue gas system consists only of a relatively short stack extending about 50 feet above the boiler. Gas firing normally produces lower levels of nitrogen oxides than oil or coal because of lower *average* flame temperatures. However, gas at present and in the foreseeable future is in such short supply that few, if any, plants will be constructed in the United States for gas only. Liquefied natural gas (LNG) firing is of some interest in several seaport areas on the East Coast. Such plants would be identical to natural-gas-fired plants except for the presence of LNG storage tanks, which often are constructed below ground.

An oil-fired plant represents the second level of boiler plant complexity. Until recently, the only oil burned in power stations was a heavy residual oil commonly known as ASTM No. 6 Fuel Oil, or Bunker C. This material is a low-priced refinery by-product that was used only where a supply was available from nearby petroleum refineries. Bunker C oil contains all the ash and much of the sulfur contained in the crude oil fed to the refinery, as well as a small amount of fixed nitrogen. Sulfur content of residuums used in this country ranges from 2 to 6 percent. Recently, utilities have been buying low-sulfur crude oil, desulfurized fuel oil, or a blend of residuum and distillate oil with a reduced sulfur content so as to reduce the amount of sulfur dioxide emitted in the flue gas. These oils are all at premium prices over residuum. In a well-designed boiler utilizing NO_x reduction techniques, the nitrogen oxide level with oil firing has been on the order of 200–300 ppm in the flue gas. Considerable work has been done to reduce NO_x levels by modifying firing conditions and by varying the primary and secondary air in the furnace. Until some means of removing NO_x from flue gas is developed, this is the only feasible method of NO_x control in the stack discharge.

Ash in fuel oil is generally not of significance, since the levels are below present-day particulate emission limits. However, with oil firing, it is unlikely that the stack discharge will meet a Ringleman #1 requirement (opacity less than 20 percent), since the ash that does exist is extremely fine and seems to have a higher opacity than coal ash. Because many oil-fired plants are in use today, some treatment of the flue gas may be required to meet this standard.

Oil-fired power plants usually have considerable storage capacity in the form of above-ground tankage in close proximity, except in densely populated metropolitan areas where the tankage may be remote. Many oil-fired plants are also equipped to burn natural gas, since gas firing produces a minimum amount of air pollution; the incremental cost of the gas firing equipment is nominal, and gas may be available at attractive prices on an interrupted basis.

The most complex boiler plant is one that burns coal. The majority of utility coal-fired boilers designed and built since 1940 are fired with finely-ground coal carried into the furnace in an air suspension. The coal burns almost instantly, leaving the ash and a small amount of carbonaceous material as a waste product. At combustion temperatures the ash is in a molten con-

dition, and the furnace must be sized large enough to cool the ash to temperatures low enough that it will not stick to the tubing and housing of the convection section. This temperature varies with the chemical nature of the ash, and is in the range of 1900°–2000°F. Some of the ash drops to the bottom of the furnace and is removed periodically through a bottom hopper to an ash disposal area. This bottom ash is the coarser material and some agglomeration takes place, so that it tends to consist of relatively large particles that do not tend to become airborne. The ash carried out of the furnace with the flue gas is called fly ash and is extremely fine, with particle sizes ranging from 250 microns down to below 1 micron. The coarser ash is in some cases removed by centrifugal-type mechanical collectors, which are most effective on particles above 50 microns in size. After leaving the mechanical collectors, the flue gases pass through electrostatic precipitators, which are more effective in removing the finer particles. In most modern installations, however, the mechanical collectors are omitted. Design efficiency of electrostatic precipitators has increased in the last 15 years from a range of 85–90 percent to a range of 97.5–99.5 percent. Current and projected air quality regulations require an efficiency in the range of 99 ± 0.5 percent to meet either grain-loading or opacity requirements. Fly ash from the hoppers below the mechanical collectors and the precipitators is removed periodically and either sluiced to a disposal area in a wet condition or pneumatically transported to storage for later handling to a disposal site.

A variation of pulverized coal-firing is the cyclone furnace, which burns coal ground to $\frac{1}{4}$ inch. The air–coal mixture is injected tangentially at high velocity into the cylindrical, water-cooled section of the furnace, on which a thick layer of molten slag is carried at all times. The coal floats on the slag during the combustion process, and the hot flue gases pass into a secondary furnace. This approach is of particular interest for coal with low ash-fusion, or softening temperatures. With cyclone firing, mechanical collectors are not required, since only the finer fly ash leaves the cyclone furnace. Electrostatic precipitators used with cyclone furnaces are identical to those used for pulverized coal firing. Although other fuels, such as black liquor, wood chips, bark, bagasse, paper waste, garbage, and petroleum coke have been used or suggested as utility fuels, and have found a place in industrial and municipal steam plants, they are unlikely to be of major importance to electric generating utilities and do not present any unique problems as compared to oil/coal firing.

The steam turbine-generator plant is not affected by the choice of fuel, being the same for any of the fuels discussed above. The steam exhausted from the turbines at very low pressures, ½ to 3½ inches Hg abs., is condensed in a large surface condenser for reuse as feedwater to the boiler. The cooling water for the condenser is normally water from an adjacent river, lake, or salt-water body, and is returned to that body in a manner to enhance natural heat dissipation to the atmosphere by evaporative, conductive, and radiative effects. The temperature rise through the condenser is in the range of 15°–30°F, and it is becoming increasingly difficult and expensive to design plants for rises much below 15°F. In recent years, many plants have been forced to adopt either dilution methods, which decrease the "average" temperature rise, or closed cooling water systems, which depend on cooling towers or cooling ponds for dispersal of the rejected heat to the atmosphere. In fact, today there are increasingly few places where direct open cycle cooling can be considered for new construction. About 45 percent of all energy in the fuel shows up as rejected heat in the cooling water discharged from the plant (about 15 percent of the energy exits in the flue gas). Heat rejected in this manner into surface waters has been the subject of much discussion and regulation.

Another form of thermal power station is the combustion gas turbine. This device is similar to an aviation turboprop engine, except that the driven device is an electric generator instead of a propeller. Sizes of single machines range from 5,000 to 50,000 kW. By using multiple driving units connected to a single generator, capacities up to 250,000 kW have been achieved.

The most common form of gas turbine plant is the simple, open-cycle, unregenerative unit. These are relatively simple and inexpensive, quick to start and easy to control, but require gas or highly refined oil as fuel. Accordingly, such units are normally considered as peaking units and are not considered for base-load operation. Because of the nature of the fuel used, the exhaust gas discharge does not contain particulate matter, and SO_2 is either very low or absent. In addition, the combustion process used in gas turbines can be controlled to maintain low levels of NO_x formation. There is no aquatic thermal pollution problem with gas turbines, since the waste heat, which is contained in the combustion flue gases, is discharged directly to the atmosphere, without any requirement for steam condensing water. A major environmental problem of gas turbines is noise; both air inlet and exhaust gas outlets must be carefully arranged and provided with efficient silencers.

It is possible to design a combined cycle plant that uses exhaust heat from the gas turbine in a steam generator. Several variations of combined cycle plants have been proposed, and a few such plants are in operation. However, except where low-cost gas or distillate oil is available, the higher-priced fuel required by the gas turbine makes such schemes of only marginal interest at present. Considerable research and development work has been done to adapt the gas turbine to heavy oil or bituminous coal firing. However, erosion, and hence maintenance costs, continue to be a major impediment to this scheme.

Nuclear Plants

A nuclear power plant is similar to a fossil-fired power plant except that the fossil-fired boiler is replaced with a nuclear reactor system. This change eliminates fossil fuel handling and storage facilities, electrostatic precipitators, flue gas stack, and ash-disposal facilities. The turbine portion of the plant, however, is very similar. Due to lower plant efficiencies, nuclear power plants under consideration today and in the foreseeable future will reject a greater amount of the total heat generated into circulating water discharge. The increment ranges from 25 to 50 percent of that discharged by the average fossil-fired plant. This increase, while significant, does not change the fundamental approach to design of the circulating water systems.

Substitution of the nuclear steam supply system (NSSS) for a conventional boiler eliminates environmental problems associated with the latter, but introduces a set of different conditions.

For a typical (light water) reactor plant, there is a small quantity of gaseous, liquid, and solid waste from the nuclear power plant that is radioactive, or potentially radioactive, which must be handled in accordance with AEC regulations, principally 10CFR20, and the technical specifications for the particular plant. In addition, material shipped from the site by common carrier must comply with ICC regulations.

Solid waste material consists principally of contaminated paper and clothing and spent fuel. The latter is shipped to a chemical facility for reprocessing, while the former is shipped to a site designated by the AEC as a permanent disposal area.

Aqueous wastes are disposed of by treatment to reduce activity to acceptable levels, followed by dilution into the local surface waters, or by reuse as condensate makeup. The radioactive residue from this treatment is barreled and disposed of in the same manner as solid wastes.

Gaseous wastes are retained in the plant long enough to allow all short-half-lived isotopes to decay sufficiently, then discharged to atmosphere by dilution in the plant ventilation air. While nuclear plants in the United States were initially designed for gaseous releases on the order of 10 percent of the maximum permissible concentrations (MPC) established by 10CFR20, plants recently licensed for construction, or under consideration, are designed for releases two to five orders of magnitude below MPC by use of longer holdup systems. At these levels, radiation effects on the environs will be a small fraction of natural background levels in all parts of the United States and are technically insignificant.

Advanced reactors under development, such as sodium-cooled (breeder) reactors and high-temperature, gas-cooled reactors utilizing helium, would have essentially no radioactive liquid waste from normal operation, but may have small amounts resulting from maintenance which can be packaged and shipped from the site. Gaseous wastes could be concentrated and bottled for storage and later shipment more readily than for a light water plant, because they would be more concentrated and easier to separate from the diluting gas (usually helium).

Particular consideration is given in nuclear power plant design to special circumstances that may arise as a result of abnormal operation or from an accident. Typically, the nuclear steam supply system is located in a containment building, or structure, that is designed to contain all the energy that can be liberated as a result of a failure of the primary coolant system. For pressurized water reactors, this is a large, cylindrical concrete pressure vessel designed for an internal pressure of 50–60 psig. The concrete walls and inner liner of this structure also provide shielding against radiation that might be released inside the containment in such an event.

Other types of reactors employ similar structures of appropriate size and shape and design pressure to achieve the same objective. AEC regulations permit a somewhat higher radiation release during an abnormal incident than during normal operation. To minimize the impact of such a release, a nuclear plant is located in the center of a relatively large plot of land. Distance from the nuclear unit to the site boundary is known as the "exclusion distance," which is typically 2,000–3,000 feet.

Geothermal Steam

Geothermal steam utilizes the principle that hot water rising along a fault is subjected to the pressure of its own weight, resulting in an elevated boiling point. For example, at the depth of 1,000 feet, water does not boil until it reaches 215°C. By drilling wells deep into such a channel, the hot water is suddenly relieved of its overlying pressure and flashes into steam. This steam can be piped from the well to a generating plant, where its thermal energy is converted to electrical power. At the time the hot water and steam are rising to the surface, cold water, due to its greater density, flows downward, where it also becomes heated, resulting in a huge natural heat-convection system above the buried volcanic rock.

Geothermal wells may produce dry steam, wet steam, or only hot water. Dry steam (superheated), which is the preferred type, contains more heat energy than wet (saturated) steam and will produce more power at the same pressure. However, judging from present worldwide geothermal developments, wet-steam-producing wells are much more abundant than dry-steam systems (perhaps 20 to 1).

The feasibility of geothermal power production has been demonstrated on a worldwide basis. In 1961, about 420 megawatts of geothermal electric power were in production throughout the world. The figure is now over 675 MW and is likely to rise to about 860 MW by the end of 1971 and to over 1,000 MW in the next few years. In Iceland, it is estimated that 60–70 percent of the population will obtain heat for their houses from geothermal sources within the next decade. In 1969, Italian geothermal power plants totaled almost 400 MW of installed capacity. Considerable developments have also been reported in New Zealand, Japan, the Soviet Union, and Mexico. The Geysers field in California has a proven capacity of 1,300 MW, with 82 MW of current generating capacity.

Geothermal steam provides the potential for a pollution-free form of energy. Most of the problems associated with the combustion of fossil fuels are eliminated when the combustion and steam generation stages of the conventional power production process are bypassed. However, waste water disposal, as a result of the high saline content of the geothermal resource, gas emissions resulting from noncondensables such as hydrogen sulfide and carbon dioxide in the effluent, subsidence, and seismology are areas to be carefully evaluated in a geothermal project. In addition, geothermal power production requires large land areas, since the plants have a low yield per well and must be separated to prevent interference among wells. Also it is restricted to those locations in which the raw resource is available. In operation, the chief liabilities are the tendency to deposit scale and to corrode normal materials of construction, because of the high acid content of the condensate.

In 1967, the United States Geological Survey, following a special inventory of all lands in 13 western states, designated federal lands in five states, within an area comprising 1,051,000 acres, as having "current potential value for geothermal resources." The areas—which have been withdrawn from sale or entry—are located in California (838,000 acres), New Mexico (140,000 acres), Nevada (38,000 acres), Montana (18,000 acres), and Idaho (16,000 acres). It is estimated that the potential geothermal resources of these states are between 5 and 10 percent of the energy contained in the total fossil fuel resources of the area.

Environmental Impacts

Air Quality

Until very recently, the principal development contributing to air quality by reducing particulate emission from fossil-fired power plants was improvement in collection efficiency, first with mechanical collectors, then with the addition of electrostatic precipitators. Precipitators being constructed today reduce plant dust loadings to 0.03–0.04 grains per cubic foot.

This represents essentially total removal of all fly ash particles larger than 10 microns and removal of 80 percent or more of particles in the 1–10-micron range. It is questionable whether further significant improvement would best be made by increasing sizes or by resorting to another form of collector device in series with the precipitators, such as filter bags or scrubbers. Considerable trouble has been experienced with present-day precipitators when low-sulfur coals are used. Collection efficiency falls off rapidly as sulfur content is reduced, and, below one percent, very special (and expensive) efforts must be taken to achieve satisfactory performance.

The main function of a stack is to disperse the flue gas. Under most atmospheric conditions the flue gases diffuse downwind in a conical pattern, and at some distance from the point of emission the surface of the cone of diffusion intersects with the earth's surface. The taller the stack, the further this point is from the origin, and the more diffusion (dilution) takes place. Very tall stacks also afford protection from "trapping," which exists during periods of strong inversion, particularly in river valleys where power plants are most apt to be located. Tall stacks become a distinguishing feature on the landscape and can be seen from some distance. In addition, as they may be a potential hazard to aircraft, their height and markings are regulated by the FAA. Because of their effectiveness, stacks must be considered an important part of the abatement system for the foreseeable future.

During the last decade there has been a pronounced shift in emphasis from particulate removal to an examination and then regulation of the overall problem of air quality, including limitations on SO_2. Since all coal and oil contain sulfur in varying amounts, the selection and/or treatment of fuels represents one possible form of SO_2 control.

In the eastern reserve of the United States, coals with sulfur contents much below 1 percent do not exist in any significant quantity, and there has been a strong demand for low-sulfur coal both for utility and metallurgical use. The tremendous increase in exporting metallurgical-grade coal to Japan in the last three years has compounded this problem. Although the basic technology for producing a de-ashed, desulfurized (gasification) coal has existed for many years, the processing is expensive and has not yet proven economically feasible. Another expensive but technically feasible approach to reducing sulfur levels well below 1 percent is to change over coal-fired units to oil-fired units and purchase low-sulfur oil. As described earlier, the source of low-sulfur oil is either special treatment of regular crudes or more conventional treatment of special low-sulfur crude oils. Today there is also greatly increased interest in direct firing of low-sulfur crudes. Natural gas, an otherwise ideal solution, is in short supply and is not available on a firm basis for boiler firing except in a few isolated areas.

In the western reserve, coals tend to be of lower quality with higher moisture content, but they are also generally lower in sulfur than eastern coals. Typical sulfur contents range from 0.25 to 0.7 percent, with many seams averaging 0.3 to 0.4 percent sulfur. There is increasing demand in the midwest for western coal, at least on an experimental basis and possibly on a permanent commercial basis. However, it should be recognized that boilers are, to an extent, "tailored" to a particular coal, and it is not always practical to use coal with significantly different properties.

The alternative to reducing sulfur content in the fuel is to remove SO_2 from the flue gas. A wide variety of processes to accomplish this are under development, including lime-water scrubbers, dolomite or MgO scrubbers, catalytic oxidation and absorption, sodium bisulfite scrubbers, and fluidized bed absorption. Many pilot plants have been put into operation and several full-size (up to 150 MW) demonstration units are under construction or test. Roughly a third of the processes suggested for SO_2 removal contemplate recovery of the SO_2 in a pure form usable for sulfuric acid production. The remainder reject the SO_2 as a solid combined chemically with Ca or Mg and usually oxidized to $CaSO_4$ or $MgSO_4$. This material is disposed of by sluicing or hauling to the fly ash disposal area and may increase the volume of solid waste by as much as a factor of 2 to 4, depending upon the ash and sulfur fractions. While the recovery processes seem attractive, the cost of equipment and chemicals is increased considerably, and, furthermore, the potential supply of sulfur from this source greatly exceeds the demand for sulfur in most areas. In addition, the quality of recovered material may not be as high as that of other sources, and transportation costs to the market can make the recovered product even more expensive.

Any of the aqueous scrubbers mentioned above have the added advantage of removing some of the fly ash. In fact, plans have been made to substitute a scrubber for an electrostatic precipitator in at least one station and are being considered in others. Typical values for scrubber SO_2 removal efficiency are 40–60 percent with limestone and 90 percent with magnesia, while particulate removal efficiencies of up to 99 percent have been demonstrated with two-stage units. Of some concern is the fact that the flue gases leaving the scrubber are saturated with water vapor at the operating temperature and hence, even with reheating of 50°F or so, may result in a "wet stack" characterized by a dense white plume.

Because oxides of nitrogen have been identified as an important ingredient in smog formation, they are of special interest. To date, no feasible way has been demonstrated to remove a significant quantity of NO_x from flue gases. Since the oxides of nitrogen are very much less soluble in water than SO_2 (or CO_2), relatively little is removed in an SO_2 scrubber. The only effective means of NO_x control demonstrated to date is by combustion control. Nitrogen reacts chemically with oxygen at the very high temperatures encountered in the flames in the furnace (above 3000°F) but disassociates rapidly at somewhat lower temperatures. Thus the indicated method of control is to burn fuel at the lowest practicable flame temperature and maintain intermediate temperatures for the longest possible time in the furnace. This has been proven to be efficacious for gas-firing and to a lesser degree for oil-firing. Thus far, these techniques have not been uniformly successful with coal-firing.

Water Quality

The major water quality consideration resulting from the operation of thermal turboelectric power plants is the discharge of heated water from the condenser circulating water system. The alternatives to decreasing the temperature rise in this system are to increase flow, mix dilution water with the condenser discharge water, and distribute the discharge over a large area by the use of diffusers in the discharge conduits. None of these alternatives changes in any way the total amount of heat energy that must be dissipated; rather, they spread the heat increment over a larger volume.

Water quality standards being considered today increasingly leave little choice but to use either a cooling tower or cooling pond in many locations. Cooling towers may be either mechanical draft or natural draft. Those in use today are almost all direct-contact, air–water heat exchangers, where cooling takes place by evaporation of water into the air. For each 10°F of cooling, approximately 1 percent of the water evaporates. The water passing through the tower tends to approach the wet bulb temperature of the air. Hence, the return water temperature is a function of ambient atmospheric conditions and may vary considerably from day to day.

Mechanical draft towers achieve air flow by use of large induced draft fans mounted on the top of the tower, discharging vertically to atmosphere.

Natural draft towers use a tall concrete chimney, usually a hyperboloid-of-revolution, up to 500 feet base diameter by 500 feet high. Some water droplets are entrained in the air stream and discharged from the exhaust. This water is known as drift loss. Because of the evaporation of water passing through the tower, dissolved solids in the tower makeup water are gradually concentrated and must be controlled to prevent precipitation and scale formation on the heat transfer surfaces. This control is by a combination of chemical addition and blowdown. Lower concentration factors of 5 : 1 to 10 : 1 are common. Consideration must be given today to disposal of the blowdown, because of both temperature and mineral content. At least one tower is in operation at this time that has no blowdown other than the drift losses.

Dry-air cooling towers have also been proposed for condenser water cooling. In these towers, the hot condenser water is circulated through a series of pipes (a heat exchanger) over which air is blown. The usual arrangement includes the air intake around the periphery of the tower base and air exhaust at the top. Such a direct air cooler is limited by dry bulb temperatures, which results in reduced summer cooling capacity, the low heat transfer efficiency of the process (about one-sixth that of wet cooling towers), and the subsequent requirement for large volumes of air and surface heat exchanger area. A dry tower in England has experienced great difficulty in accommodating these requirements. The exhausted air is heated about 15°–20°F, but system evaporative losses are eliminated.

Cooling ponds are artificial lakes, created to cool water by a combination of evaporation from the surface and long wave radiation to the sky. Both cooling towers and cooling ponds have the potential of causing or increasing local fogging under certain atmospheric conditions. This matter is under investigation in several sections of the country and valuable new information is being accumulated.

It is possible to design a cooling pond as part of an open-cycle cooling water system, which returns water to the natural surface water at least partially cooled. This concept is now being employed in some instances and is under study in others. It is applicable only where there is a large area of relatively flat land available for such use. Much interest has been shown in the use of salt-water cooling towers for estuarine and ocean sites. While small salt-water cooling towers have been used to some extent throughout the world, there is little technical information available on the subject, and the environmental impact, even of small units, is unknown. Principal concern over the use of salt-water cooling towers centers around the fact that seawater contains about 3½ percent by weight of salts, principally NaCl, and the evaporative losses from the tower would increase this percentage. Thus, both the drift loss and the blowdown from a large plant have great potential for ecological damage.

The evaporative efficiency of cooling ponds can be increased by admitting water to the pond through a multiplicity of spray nozzles located two or more feet above the surface. Small spray ponds have been used for many years, but large installations are uncommon, and the technical data to determine the expected increase in evaporative efficiency are not available as yet. One fairly large installation has recently been put into service and so far has demonstrated only that maintenance due to clogging of the spray nozzles is a major problem.

In study and selection of the alternative cooling water systems mentioned above, the capital and operating costs for each system are important considerations. Also important, however, is the performance of the turbine under various conditions of backpressure. The lower the backpressure, the greater the energy that can be extracted from the steam up to some limit that is a function of turbine design and physical size. Cooling towers and cooling ponds generally will not cool the circulating water in hot summer months to the temperature of surface waters. Therefore, some efficiency of energy conversion is lost when a cooling tower is used. Further, waters from the depths of rivers and lakes tend to be colder than the water at the surface. Therefore, from the standpoint of maximum generating capability and thermal efficiency, the colder subsurface waters are the best choice, surface water the second choice, and cooling towers or ponds the third choice. The cost penalty in a typical case would be in the range of 3 to 7 percent in going from the best case to the worst case.

Radiation

Both pressurized and boiling water reactors have small amounts of gaseous waste products containing radioactive isotopes. Of principal concern are the noble gases and halogens from fuel leakage and activation products contained in the water used for core cooling. The standard treatment system is to hold up these gases for a period of time long enough to allow decay of short-lived isotopes. Holdup systems range from a relatively simple pipe, where holdup is the transit time in the pipe, to a compressor tankage system, to a trapping system using charcoal beds at ambient or cryogenic temperatures, to a cryogenic separation system that bottles all isotopes for shipment to permanent storage.

Other sources of potential gaseous radioactive discharges are plant ventilation air exhausts and radwaste concentrator vents. These sources are typically very low level and can be discharged after passing through filters to remove iodine and particulate matter.

Gaseous wastes from liquid metal or helium-cooled reactors would be similar in nature to those from water reactors, and would be treated in similar types of equipment. In either case, the carrying gas is helium and the separation of radioisotopes is somewhat easier.

Radioactivity contained in liquid wastes from light water reactor plants can be processed by dilution in the circulating water or by concentration in a demineralizer or evaporator. Suspended solids can also be removed by filtration. Bottoms from the evaporation and exhausted demineralizer resins or concentrated demineralizer regenerative solutions are drummed for offsite shipment to permanent storage sites designated by the AEC.

There are no normal process liquid wastes from helium- and sodium-cooled reactors. Incidental liquid wastes mentioned below occur in all types of nuclear plants, and provisions for handling these are included in helium- and sodium-cooled reactor plants.

Other sources of liquid wastes are laboratory drains, floor drains, equipment drains, laundry drains, and decontaminating solutions from maintenance operations. Provisions are included in each nuclear plant for holdup, monitoring, and release or treatment.

Spent fuel is periodically removed from the reactor, stored temporarily for 90–120 days, and then shipped offsite to a fuel reprocessing plant. Shipping containers must comply with regulations formulated by the ICC and AEC. Shipment can be by truck, rail, or barge; rail shipment is the most common. There is no significant environmental impact from shipping spent fuel, either at the site or in transit.

Coal contains a small amount of natural radioactivity, particularly uranium, thorium, and radium. Some of this is released with the flue gas, and the resulting radioactivity has been the subject of some study. The levels are generally on the order of less than 1 percent of maximum permissible concentrations, as established by the Federal Radiation Council.

Land Use

Important considerations in selecting a thermal plant site are availability of cooling water, transportation of fuel to the site, adequate room to build and operate, fuel storage, disposal of solid waste products, and availability of transmission rights-of-way from the site to the load center(s). For large base-load generating stations, minimum land required is for gas-fired plants and nuclear power plants (not including exclusion area). Coal-fired plants require the most area because of the space required for coal and ash facilities. Transmission requirements are independent of generation method. It is customary, but not mandatory, to locate a transmission switchyard adjacent to the power plant, and land use considerations take this into account. A high-voltage switchyard may occupy as much land as the plant itself, less fuel storage facilities or exclusion area, and must be fenced in, so supplementary use is not feasible. Transmission beyond the switchyard normally utilizes overhead lines carried on steel towers. The right-of-way is normally wide enough to afford access to the line for construction and maintenance and to afford a degree of protection in event of mechanical failure of one of the components. Vegetation along the right-of-way is controlled to prevent interference with, or close proximity to, the conductors. This is accomplished by selective cutting or topping of trees. Some supplementary use can be made of right-of-way, such as agriculture, grazing, or nature trails and other recreational facilities.

In metropolitan areas, utilities are frequently required, either by economics (land value) or regulation, to install transmission lines underground. In these cases, the transmission lines consist of heavily insulated cables installed in a pipe or conduit or a bus duct system. Insulated cables for voltages to 345 kV have been developed, but are extremely costly and have much higher losses than overhead lines. Several bus duct systems are under development, including some that operate under pressures up to 100 psig to improve dielectric conditions and improve conductor cooling. Such systems may be used for direct burial and underwater crossing.

The most economical voltage at which bulk power is transmitted increases with transmission distance and increasing capacity. Today, major systems use 220 kV, 345 kV, and 500 kV as primary voltages, with some experimental and development work at voltages up to 1,500 kV. A major system is in operation in Canada at 735 kV. Some 765 kV is already in operation in the United States, and it may well become the next primary voltage for general use in the United States. For very long lines, for deep underwater crossings, and also for stability reasons, dc transmission is of considerable interest. The Pacific D.C. Intertie, connecting California with the Pacific Northwest, is already in operation. This line operates at ±450 kV and has a design capacity of 1,400 MW. Direct-current transmission opens up new possibilities for remote or offshore siting and also for improving stability and reliability of existing transmission grids, although these applications will no doubt be limited.

Aesthetics and Noise

Recently, increasing attention is being given to aesthetic values in the design of modern power plants. Appearance of the plant is considered, and attention is given to the architectural plan to make the plant as attractive as possible, consistent with the dimensional and volume relationships of the required structures. Nuclear power plants lend themselves particularly to architectural enhancement because of their inherent cleanliness and the greater variety of shapes and structures that are possible. Supplementary land use for most coal- and oil-fired plants is limited because of the associated fuel handling and storage facilities and ash disposal areas. However, many nuclear plants are designed to have a parklike atmosphere, with minimum disturbance of natural vegetation within the exclusion area. Only the innermost parts of the facility are enclosed by fencing, and the remaining land is dedicated to some public use. Supplementary use of exclusion area land includes picnic areas, day camps, wildlife sanctuaries, and boating facilities.

Attention is given in power plant design to sources of noise that may disturb users of adjacent property. Noise may come from operating equipment, such as transformers, draft fans, cooling tower fans, or high-speed pumps. Each of these can be made less noisy by proper specification and design, or by acoustical treatment. A source of considerable annoyance is intermittent discharges from safety valves and blow-off valves. These discharges from safety valves are now normally provided with mufflers, which effect a 30–50-db reduction in noise levels. Solutions to acoustic problems are relatively successful, as evidenced by the fact that generating stations have been accepted as close neighbors in urban areas.

HYDROELECTRIC POWER PLANTS

Methods of Generation

Flow

A hydroelectric generating plant is one that uses the potential energy of a water column to drive a hydraulic turbine coupled to an electric generator. In most cases the water column is created by building a dam, the height of the column being roughly equal to the height of the dam. In mountainous country, it is often possible to extend the column by diverting the water to a much lower elevation through large conduits known as penstocks. Most hydroelectric projects are multipurpose and may, in addition to power generation, also provide flood control, water conservation, recreation, navigation, and low-flow augmentation.

The two forms of dams used predominantly on hydroelectric projects are the thin arch and gravity types. In the thin arch dams, resistance to the overturning moment is provided by buttresses tied in to the rock structure on either side of the dam. Hence this type of dam is applicable only to narrow chasms in mountainous country. Gravity dams, either concrete or earthfill, resist overturning moment by the weight of material in the dam and the breadth of the base. While it is possible to generate power with only a few feet of available head, modern hydro projects generally are designed for heads of 25–150 feet for gravity dams, 150–1,000 feet for thin arch dams, and up to 1,500 feet where the natural head is available.

The location of a dam on a river is usually controlled by natural features of the area, such as suitable soil conditions and terrain. Further, it is clearly not possible to generate more power from a river than the flow of water and the natural drop in elevation provided by nature. Much of the economically exploitable flow in the United States has already been developed, and many otherwise potential hydrolectric sites are now precluded by virtue of other, more valuable land use.

Hydroelectric generating stations combine readily with steam generating stations in a utility system. Since the energy component of hydro-power is free, it is used to the maximum (consistent with other uses) in preference to fossil or nuclear fuel. Hydro stations can be programmed to operate either as base-load stations, with steam carrying the peaks, or as peaking plants, with steam carrying the base load. In California, for example, in the last 25 years the use of hydro has shifted from the former to the latter as the load grew beyond the availability of water to carry the base load. In the United States today, only the Pacific Northwest area is serviced predominately by hydro power.

Pumped Storage

A special variation of a hydroelectric station is a pumped-storage project, wherein a natural or artificial reservoir is located at a higher elevation than the principal body of water, and the hydraulic turbines are operated part of the time as pumps to pump water to the upper reservoir. When passed back through the hydraulic turbines, power can be generated. This, in effect, becomes a storage battery for a utility system and permits use of other generating capacity available in periods of low load (such as nights and weekends). For example, a typical pump storage facility might pump for 10 hours at night and generate for 4–5 hours during the morning and afternoon peak load periods. It is recognizable from the description given above that a pumped storage plant does not generate any net energy and, in fact, delivers back about ⅔ of the input because of mechanical, hydraulic, and electrical losses.

Environmental Impacts

Hydroelectric plants have a considerably different impact on the environs than the thermal stations previously discussed. There are no waste products to discharge to the atmosphere or water. Land use factors are much more significant because of the large areas of land involved and the permanence and scale of the effects. Maintaining the area in an undeveloped state, with an undisturbed flow in the river, may be considered a more worthwhile use of the land. There is no thermal pollution; however, the river temperatures below the tailrace are often 10°–20°F lower in the summer months than the normal river water temperature, because the water is removed from the storage pond at considerable depth. Furthermore, it has recently been shown that high-turbulence outlets tend to produce very high dissolved nitrogen contents in downstream waters, which have caused some fish kills.

LONG-RANGE ALTERNATIVES

Many new schemes have been proposed for central station power generation, including coal gasification or liquefaction, fluidized bed combustion, fusion reactors, solar energy, tidal energy, and the use of magnetohydrodynamic generators (MHD).

Briefly, gasification and liquefaction of coal are simply methods of treating coal to purge it of impurities, such as ash or sulfur, that would otherwise require treatment at the flue gas stage. Gasifying coal permits the treatment of much smaller quantities of gas than exist at the flue gas stage, where air has been added to permit combustion. However, since 100 percent of the heating value of the coal is not sensible in the gas, certain losses are accumulated.

Fluidized bed combustion is a process for burning coal more efficiently. By this process, the steam is heated by conduction in a very compact configuration, thus decreasing thermal losses. There is also the potential to reduce the SO_2 and NO_x components of the flue gas emissions.

Fusion reactors would use an economical fuel of widespread availability (deuterium) and would have few radioactive products of concern. In addition, it has been suggested that such reactors operating as they do at very high temperatures, could be coupled with an efficient conversion cycle, resulting in a decrease in thermal pollution.

The use of energy from ocean tides entails the harnessing of naturally occurring, abundant energy resources. Since no combustion is involved, air pollution could be practically eliminated, and the only thermal pollution would result from the displacement of heat from one sensible form to another.

Magnetohydrodynamic generators could be used as additions to the conventional Rankine (steam) cycle, resulting in greater efficiencies than might otherwise be achieved, and causing a reduction in unwanted by-products. The technological status of these concepts is discussed further below.

Coal Gasification

Coal gasification has the potential to greatly reduce sulfur emissions resulting from the use of coal as a fuel for combustion.

One of the most highly developed processes for coal gasification is the Lurgi gravitating-bed gasifier technique. The gasifier is a pressure vessel surrounded by a water jacket, where some of the process steam is generated. Coal is injected at the top of the gasifier, while air and steam at pressures of between 15 and 25 atmospheres and at temperatures below the ash fusion point are introduced through a rotating grate at the bottom. The countercurrent contact of the coal with the gasification agents results in the coal being dried, devolatilized, and gasified completely. The fuel gas is then scrubbed free of dust, alkalis, and sulfur, after which it is burned in a pressure combustion chamber, generating superheated high-pressure steam to drive the steam turbine. In addition, the combustion gases emerging from the pressurized combustion chamber can be used to drive a gas turbine.

An important advantage that coal gasification can provide to power plant operation lies in the control of atmospheric emissions. Since the gasified coal is at a high pressure, it is considerably more concentrated than the flue gases that result from ordinary combustion and is, therefore, commensurately less difficult to scrub free of contaminants. The flue gases that result from pressure combustion are essentially free of all dust, ash, and sulfur.

Coal gasification techniques are continually being im-proved and hold promise for the future. Fifty-eight pressure gasifiers are successfully in operation for the production of town gas and synthesis gas from a variety of different coals in Europe, South Africa, and Asia. The 170-MW Kellerman Power Station in Germany, which combines pressure gasification with gas turbine processes, was scheduled for startup by the middle of 1971.

Currently, the cost of coal gasification does not make it an attractive alternative for firing utility boilers in the United States. However, a shortage of natural gas, coupled with the advanced development of gas turbine technology, larger units (which require greater exhaust volumes and are hence more difficult to scrub), and more stringent emission-abatement standards for fossil-fired plants, could result in the extensive utilization of gasified coal for utility operations.

Coal Liquefaction

Coal liquefaction is a fuel treatment process that removes impurities from fuels prior to combustion. In the process, coal is heated in multistage fluidized beds, undergoing pyrolysis. The products are char, volatiles, and vapors, which can be condensed to a high-quality fuel oil. This synthetic crude oil is essentially free of the oxides of sulfur and much of the particulates that ordinarily accompany coal burning.

Work on such a coal liquefaction scheme has been conducted since 1962. Currently, a plant to liquefy 36 tons of high-quality coal per day is in operation in Princeton, New Jersey. In view of the fact that this facility is only 0.1 percent of the size required for a large utility operation, it appears that it will be at least 10 years before coal liquefaction becomes feasible.

Fluidized-Bed Combustion

A field receiving considerable attention in regard to coal-fired boilers is the use of fluidized-bed combustion. In this process, coal is injected into a bed of inert granular material, such as ashes or crushed rock, that is being supported by the upward flow of combustion air from the plenum chamber. The bed supports the coal during combustion, and hence a coarser grade can be utilized, with resulting lower combustion temperatures and greater burning times than conventional schemes. In addition, the steam tubes are immersed in the fluidized bed, resulting in heat transfer by actual conduction, which is considerably more efficient than the heating by convection and radiation that takes place in a conventional boiler scheme. This results in considerable economies in the size and cost of the boiler. A concept for an 8 million-lb/hr steam flow fluidized-bed boiler would require about one sixth of the space necessary for a conventional pulverized-coal boiler.

The fluidized-bed combustion process can also provide partial flue gas emission control without external scrubbing. The use of a limestone regeneration cell has resulted in an 85 percent reduction in SO_2 emissions when burning 4.5 percent sulfur coal. Also, fly ash carbon reduction of 80–90 percent has been demonstrated in a single pass through a simulated carbon-burnup fluidized bed. Hence, the fly ash can be simply removed from the cycle and discharged to the ash silo. Low carbon losses can thus be achieved without incurring the high dust loadings caused by multiple fly ash reinjection. In addition, the lower combustion temperature results in decreased production of oxides of nitrogen.

Numerous problems remain to be solved before a prototype fluidized-bed boiler unit for utility operations is developed. Most of the work done to date has utilized fluidized-bed columns and has involved the study of heat transfer rates. Experience with an 18-sq-ft atmospheric steam pressure boiler with embedded steam tubes led to the substitution of a series of smaller combustion chambers for the tubes. This unit, with a steam-generating capacity of 5,000–8,000 lb/hr had difficulty with carbon burnout, resulting in high carbon monoxide losses.

Other problems concerned with generating steam at lower temperatures and pressures than those of conventional boilers, the optimum coal-feeding scheme, and channeling, which can lead to bed failure, have been encountered.

Fluidized-bed boilers have considerable potential for replacing the conventional, pulverized-coal-fired boilers. The advantages, in terms of cost, size, and emissions, are not expected to be realized until 1978–1980. Then, fluidized-bed boilers may prove especially attractive for utilities burning high-sulfur coal in densely populated areas, where size can be of considerable concern.

Nuclear Fusion

Nuclear fusion has received considerable attention as the ultimate power source. It combines the advantage of being free of air pollution with the ancillary benefit of an almost inexhaustible fuel supply.

The fusion reactor uses a fuel that occurs abundantly in nature, deuterium. Deuterium, or heavy hydrogen as it is called, appears as one of every 6,500 atoms of ordinary hydrogen in water. It can be easily and inexpensively extracted, with no deleterious effect upon the water. Although the fusion reaction of two deuterons is preferred, it takes place too slowly, and hence the deuterium–tritium reaction is the one most com-

monly considered. The reaction produces more energy per pound of nuclear fuel than is available from any other fuel.

Difficulties in the fusion reaction arise from the fact that it requires a collision between charged nuclei. The particles require high energy to overcome their electrostatic repulsion, necessitating temperatures on the order of 10^8°C, plasma (ionized gas) densities of about 10^{15} particles/cc, and confinement times of from 0.1 to 1.0 second. Accomplishing this confinement holds considerable challenge, since huge magnetic fields, on the order of 100,000 gauss, would be required. In addition, the vacuum wall surrounding the plasma (which will be at 5000°–10,000°C), the reflectors for heat removal at 1800°F, and the potassium–steam binary vapor cycle proposed to accomplish greater energy conversion efficiencies all involve techniques of design and fabrication that require a great deal of development.

Confinement, which has been achieved at temperatures of about 5×10^6°C, densities of 5×10^{13} particles/cc, and for periods of 0.02 seconds, must still be improved by a factor of 50. It may well be achieved in the mid-70's, after which time vacuum wall, reflector region, and other problems will remain to be solved. The 1980's may see the first prototype reactor and the late 1990's, or the beginning of the 21st century, should see the arrival of the first commercial nuclear fusion reactors.

Solar Energy

This abundant resource can be harnessed and used with a minimum impact upon the environment, since it involves the utilization of an otherwise naturally dissipated source. Household hot-water heaters powered by solar radiation have been commercially competitive for many years in Israel, Japan, and Australia. The production of potable water from salt or brackish supplies is the oldest application of solar energy to a technological process, dating back to 1872. In France, a solar furnace with an energy output of nearly 1 megawatt is being built. The reflecting surface is almost 3,000 square meters and the concentrator is about 2,500 square meters. It will be able to achieve temperatures as high as 6000°F.

Proposals have also been made for using solar energy for generating power in technologically advanced countries. The most publicized system would be two geostationary satellites positioned so that at least one would be illuminated by the sun at all times, 22,300 miles from earth, and parallel to the equatorial plane. Organic, photoconductive, solar cells would collect the sun's radiation and convert it to electricity, which would then be amplified and transmitted to the earth as microwave radiation, where it would be rectified and transmitted through superconducting lines.

Although such a scheme certainly would eliminate the forms of pollution indigenous to conventional generating methods, it is far from achievable with today's technology. The required collectors, amplifiers, transmitters, and rectifiers are only in the preliminary theoretical stages. The solar cells that have been built are much too inefficient and large for such an application. In addition, the microwave beam proposed would be a form of pollution itself if not adequately controlled.

In summary, existing technology for the use of solar energy is limited in application, due to the large areas of land required to collect the radiation and the transient availability of it. A football field of thin-film solar cells situated in a sunny section of the country would take 10 years to match one hour's output from a 2,000-MW conventional power plant, and the initial charges for the cells alone would amount to almost $2 per kilowatt-hour. To supply the energy requirement of Arizona in 1990, for example, would require an area of about 190 square miles of solar energy conversion surface. It is likely to be a long period, the length of which it is impossible to forecast, before solar energy will begin to fulfill its potential for supplying the needs of the earth for an abundant source of pollution-free energy.

Tidal Energy

Another inexhaustible source of naturally occurring energy is found in the tides of the ocean, which are derived from the energy of the earth's rotation. The basic scheme involves the capturing of water available at high tide in huge retention basins, such as a natural bay with a man-made dam. The basin is held at a maximum level until power is needed or the sea is sufficiently below the level of the pool, at which time it is released through hydraulic turbine-generators to the sea, creating electricity.

The technologic feasibility of tide harnessing has been demonstrated. The Rance River Plant in France, with a difference of from 9 to 14 meters between high and low tide, produces over 544 megawatts. Using a reversible turbine, power is tapped from the waters as they rush upstream at high tide and as the waters recede toward sea. The Soviet Union has also built an experimental plant on the Kislogubskaia to investigate the use of tidal power. The small coastal bays of the Soviet Union hold a power potential exceeding 8.2 million megawatts. The Passamaquoddy Bay on the United States–Canada border is considered to have potential for a huge tidal power plant.

The principal deterrent to greater development of tidal energy is economic. The Rance River project, which was started in 1963 and online in 1967, cost close to $100 million. The expenses involved in the construction of impounding structures, as well as the transmission costs entailed in providing the power to the usually distant load centers, and the intermittent, varying power levels available, make tidal power economically unattractive at this time. It is doubtful that it will become an extensively utilized resource in this century, and it will, by its nature, always be restricted to high tide areas.

Magnetohydrodynamic Generation

Magnetohydrodynamic (MHD) generation of power is based on the same principle as conventional generation. However, instead of a solid conductor (turbine rotor) moving across a magnetic field, a jet of ionized fluid is forced through it. By placing electrodes in this fluid stream, direct current electricity at relatively high potential, e.g., 2,000 volts or more, can be obtained. It is the substitution of this ionized fluid for the armature of a conventional turbogenerator that characterizes MHD generation.

Since high temperatures are required to make most fluids, and especially gases, sufficiently conductive, MHD is generally thought of as a topping cycle for conventional steam cycles. In the open-cycle system, the working fluid consists of the gaseous products of fossil fuel combustion, seeded with an easily ionizable element, such as potassium or cesium. Such seeding lowers the ionization temperature of the air from about 10,000°F to 4000°F—a temperature attainable by burning coal with heated air. The hot gas, with its increased electrical conductivity, accelerates through the MHD channel, inducing the direct-current electricity, which is extracted by means of electrodes in the wall. To generate a practical voltage, the speed of the ionized gas in the MHD channel must approach 2,000 mph. Even under these conditions, a superconducting magnet is usually suggested to create the 40,000 to 80,000 gauss field required, since, even with the seed, thermal ionization at these temperatures results in fairly low conductivities.

The exhaust gas, at a temperature of 3000°–4000°F, exits from the topping cycle into the air preheater and then to a conventional steam generator for a.c. power generation, resulting in an overall plant efficiency of 50 to 60 percent.

MHD generation can significantly reduce the amount of thermal discharge from a power plant. Assuming a conventional plant efficiency of 40 percent and an MHD-topped plant efficiency of 50 percent, the conventional plant will reject 60 percent more heat to the condensers and to the environment. Particulate matter from an MHD plant must be removed by electrostatic precipitation at a very high efficiency. This is necessitated by the economic requirement that 99.9 percent of the ionization seed be recovered. In addition, formation of nitric oxide is greatly increased in a MHD plant, due to the higher flame temperatures being utilized. Although it has been suggested that the greater concentration of nitrogen oxides might facilitate the scrubbing of sulfur oxides from the flue gas, present data on the overall environmental impact of MHD generation, when compared to that of a conventional plant are, at best, inconclusive.

The Soviet Union currently has the world's largest open-cycle MHD power station. The plant operates on natural gas and develops 75 MW. However, only 25 MW is produced by the MHD generator; the remainder results from a conventional bottoming steam cycle. The largest MHD installation in the United States provides 20-MW power bursts of three minutes' duration to drive an electric wind tunnel.

Numerous problems must be overcome before MHD can become a viable alternative to central station generation. The ability to economically recover the ionization seed has not yet been demonstrated, nor has the oxides of nitrogen problem been successfully remedied. Oxygen enrichment, required to achieve high temperatures, results in a strongly oxidizing atmosphere, necessitating the use of special materials. The thermal stresses resulting from an MHD duct with 4000°F fluid inside, surrounded by superconducting magnets operating at −450°F, are expected to be considerable. Channel insulation, electrode fabrication, and superconductive magnets are also formidable engineering challenges.

MHD generation is not now at the point of utilization, and its potential for fulfilling utility system needs during the decade of the seventies and beyond is highly speculative.

APPENDIX B: SITE ACCEPTANCE AND APPROVAL—HISTORICAL BACKGROUND

BACKGROUND

In the early beginnings of our nation there were no public utilities of any kind and, therefore, no siting or other public utility regulatory problems. The population was small and widely scattered, and families were nearly completely self-sufficient. Water was obtained from wells, springs, or nearby bodies of water. An ample supply of wood served as fuel for cooking and heating, and illumination was obtained from homemade candles or sperm whale oil lamps.

In 1790, the United States had a population of less than 4 million, of whom only 131,000 lived in towns or villages of 8,000 or more. By 1830, the population had increased to nearly 13 million, but still over 93 percent of the people lived in rural areas. However, as the population continued to grow during the 19th and 20th centuries, towns and villages grew into cities, and people more and more began to concentrate in urban areas, with an increasing dependence upon public utilities for essential services such as water, gas, and electricity. The 1970 census shows our population now exceeds 200 million, of whom 73.5 percent live in urban areas.

An evolution has taken place. A nation basically agricultural in nature and made up of persons essentially self-sufficient has become a highly industrialized society that has increased 50-fold since 1790 and that is largely concentrated in or near large cities and towns. We have undergone a major change from widely scattered small groups of people to a high population density concentrated in large urban areas, which are dependent

upon public utilities for essential services such as water, gas, and electricity.

When they first came into existence, there was some opposition raised against electric and gas utility services based upon a potential danger to life and limb, but, for the most part, electric utilities had little or no regulatory restraints during their early development. However, as early as the colonial period, it was recognized that business organizations operating under monopoly conditions should be under some government regulation. Accordingly, as public utilities were formed and granted exclusive service rights within specified areas through corporate charters, local franchises, legislative acts, and municipal ordinances, they also came under regulation.

In 1869, Massachusetts established the first regulatory commission, which was to be copied in following years by other states. For the most part, however, state public service commissions did not grow in number nor in regulatory power until the beginning of the 1920's. Except for the states of Minnesota, Nebraska, Texas, and South Dakota, every state eventually established a regulatory commission with jurisdiction over electric utilities. Their authority and jurisdiction over siting of plants and transmission lines vary greatly.

The early social and regulatory restraints on electric generating plants, transmission systems, and distribution lines did not include environmental matters. They were mostly directed at restricting the number of companies permitted to operate in a particular area, under the concept that it was not in the best interest of the general public to have different water, gas, or electric companies competing in the same service area. The extensive rights-of-way through streets and other real property required for electric and gas lines and water-mains, not to mention the heavy capital costs incurred, would have made it wasteful and economically prohibitive to have had many utilities competing in the same area. It was recognized, however, that when any charter or franchise gave one single company the monopoly to perform a service essential to the general population, then that company should be closely supervised to assure the services would be properly performed at a reasonable cost.

Well into the 20th century, new utility services were not only welcomed but were eagerly sought by the public at large and by their elected representatives throughout the country, particularly in the rural areas. In the 1930's and 1940's and well into the 1950's, high-voltage transmission lines and their accompanying towers coming across the landscape signified to the numerous small farms and rural areas the end of backbreaking drudgery and hard work. It meant the electrification of the farmhouse and all the accompanying benefits. As late as 1935, only 10.9 percent of U.S. farms had electricity. By 1970, more than 98 percent had electric service, and the farmer as well as the city dweller is now able to benefit from ample supply of electric energy. New power plants meant a significant addition to the tax rolls to support new schools, modernize fire-fighting equipment, and meet other public needs. Little or no concern was evidenced as to environmental effects.

CURRENT TRENDS

As the population concentrated more and more in large cities, competition and, therefore, the price for land increased. Steam power plants are most economically located near large bodies of water. Many of the cities and large towns were initially located on navigable rivers and other large bodies of water, and, as these population centers grew in size, power companies seeking additional sites for new power plants encountered difficulties in locating available sites not already committed to other purposes.

Seeking to place power plants close to load centers for economic reasons, utilities more and more came into competition with their customers—commercial as well as residential—for the available land. As the population increased and as the individual's use of electricity continued to rise, it was necessary for electric utilities to increase total generating capacity to meet demand. Economic and technologic advantages resulted from the increasing number and size of generating units, but

also there resulted increasing undesirable effects to the immediate areas in which the plants were located. Notwithstanding greater efficiencies, as more and larger generating units were built, the total amount of pollutants and therefore impact to the local area increased. Urban areas that were initially willing to tolerate coal-fired electric generating plants opposed more and more the expansion or addition of electric generating facilities, notwithstanding the use of less-polluting fuels, such as natural gas or low-sulfur oil. City and suburban areas that had welcomed new generating plants because of additional tax revenues to the community now began to weigh this advantage against some of the undesirable effects on the community.

In recent years, while desiring and demanding substantially more electric power, the public at the same time is showing increasing concern as to the quality of the environment.

As pointed out by the Office of Science and Technology in its August 1970 report, *Electric Power and the Environment,*

The environmental concern extends not only to the more obvious factors of smoke and gases from combustion, waste heat from power plant cooling, and possible hazards from nuclear power sources, but also to esthetic and economic objections to the use of choice land for power plants and for transmission lines.

Earlier, when a new generating facility and accompanying transmission lines nearby meant the difference between having and not having electricity, the community's decision was easy—it was in the affirmative. This is no longer the situation, even in the rural areas where electric service has been available for some time. Although utilities need new power plants or additional transmission lines to provide for future growth in the area or, in the case of transmission lines, to strengthen its interconnection with other utilities for greater reliability to customers, the general public does not sufficiently understand the problem and is not sympathetic. Since his immediate needs are being met, the average person living near a site for a proposed plant or transmission line considers the new facility to be for the benefit of an area other than his own—some large city or urban area far from him. Accordingly, there is growing opposition to new generating plants and transmission lines.

Consequently, the electric utilities and society must work together to resolve the apparent conflict between the demand for electrical facilities to meet an increasing power requirement and the desire of a growing segment of the public to prevent their construction.

APPENDIX C: NORTHERN STATES POWER COMPANY APPROACH TO OPEN PLANNING

In efforts to select power plant sites acceptable to both the utility and local and national conservation and environmental organizations, some utilities have been successful in obtaining assistance from these organizations during the site selection process. There are many methods of establishing a dialogue between the utility and such groups. An approach tried by the Northern States Power Company has shown some indication of success.

BACKGROUND

Northern States Power Company (NSP) announced in January 1970 that siting and development of all future plants and transmission lines would be discussed, in advance of decision-making, with the public. This announcement was based upon the realization that new problem-solving and decision-making mechanisms are necessary to enable utilities to site future facilities in keeping with environmental, social, and political needs of the public.

Recently, many generating facilities proposed by utilities throughout the United States have been subjected to public opposition. The relationship between utility decision-making processes and patterns of public political resistance to proposed nuclear power plants and the siting of these facilities has been studied by David G. Jopling and Stephen J. Gage of the University of Texas. The results of their study of the patterns of resistance towards the Enrico Fermi proposal, Lagoona Beach, Michigan; the Pacific Gas and Electric proposal, Bodega Bay, California; and the Consolidated Edison proposal, Monticello, Minnesota, indicate that the opposition to this additional generating capacity could have been reduced if interested segments of the public had been more deeply involved in the predecision planning process.

Therefore, a new approach must be sought in siting and developing future generating facilities to prevent undue delays in construction and operation. Open planning and participative decision-making should be considered an integral part of this new approach.

Simply defined, open planning is a process that actively seeks outside inputs, ideas, and evaluations. It is participative decision-making, which helps to determine the needs of the public. Success or failure of the process is dependent on the quantity and quality of active public participation.

POWER PLANT SITING TASK FORCE

To implement its decision to use the open planning process, NSP went to every major (and some minor) environmental/conservation groups in Minnesota—perhaps some 30 groups. Many of these had been and still are locked in bitter controversy with that utility. Some are small new *ad hoc* groups, while others are well-established large groups. Each was asked to send representatives to a Plant Siting Task Force being formed by NSP. NSP tried to convey the seriousness with which it was approaching the siting effort and asked them to send people capable and willing to make an intellectual commitment and devote the time necessary. Each group agreed to participate, although many were hesitant, fearing the request was a public relations facade and would be used by the company to the group's disadvantage. But they took the risk, and a 40-member Task Force was formed. During this period, much time was spent by NSP representatives discussing the idea with editorial writers, reporters, and opinion leaders.

The Governor of Minnesota had earlier formed an Environmental Cabinet, consisting of State department heads (Department of Natural Resources, Economic Development, Pollution Control Agency, etc.). NSP invited participation of the Governor's Environmental Cabinet, and it also accepted.

These two groups started weekend meetings with NSP in March 1970. Their initial assignment was twofold. First, to recommend to NSP a location for its next generating station, scheduled for service in 1976. The constriction of time required that a recommendation be limited to four sites that NSP already owned. The second assignment, after completing the first, was to look at the long-range problems of plant siting (including transmission lines) in a totally unrestricted way and, hopefully, to develop environmental criteria that NSP could apply in the acquisition and utilization of future sites.

The initial meetings were marked by great suspicion and hostility on the part of everyone. Each week NSP, and in some cases the State, was subjected to criticism by members of the Task Force, since it included almost every major NSP critic.

The initial meetings were devoted to a series of presentations intended to bring everyone up to some minimum level of common understanding about the nature of the electric utility business. Whenever possible, NSP used consultants, people from the University of Minnesota, and other non-NSP experts in the hope of enhancing credibility.

By about the sixth week, a significant change in attitude occurred. The group began to feel that it had a chance to do something constructive and that constant hostility from the more outspoken critics was delaying any possible progress. From that point on the dynamics of the group began to change.

NSP also learned very early that it is impossible to limit areas of discussion in this kind of group. Soon, many varied subjects were being discussed, such as advertising and marketing policy and expenditures, research, rates and rate structure, and validity of demand projections, matters that had never been presented in a public setting. All meetings were open to the public and the press was invited. At least one NSP officer was in attendance at every meeting. The initial meetings were kicked off by the Chairman of the Board, and he intermittently attended other meetings in an effort to express the seriousness of corporate intent. This kind of demonstration is vital, because the overall effort has great credibility problems with the environment oriented community.

After a time, the Governor's Environmental Cabinet began to withdraw from the meetings with the citizens' group, and NSP was faced with two separate groups meeting independently of each other. The burdens of staffing these two groups were severe. Any such effort demands complete and total support from all segments of the corporation if it is to succeed. The end result of phase one, that is to say the location recommendation of the next generating unit, resulted in two reports, one from the Task Force, and one from the Governor's Cabinet. It did not surprise the pessimists that each report recommended a different location. The Governor's group recommended what had been the NSP's first choice initially. NSP had made it clear to the Task Force at the beginning what the company's preferences were. After a great deal of internal discussion, NSP elected to follow the recommendations of the Task Force. While the reason for that decision is complex, it basically stemmed from the feeling that the rationale presented by the Task Force was more sound.

PLANT SITING TASK FORCE'S FUTURE

The Task Force continues to meet at a rigorous pace, strengthened by NSP's decision and the belief that NSP is really listening. It has divided itself (by its own choice) into four study groups, each of which makes recommendations to the full Task Force. These four groups are looking at: (1) environmental monitoring programs and research; (2) identification of future plant sites; (3) development and utilization of those future sites; and (4) the whole question of future energy demands.

The major obstacle to such an effort is distrust, and the corporation must work long, hard, and patiently to overcome this obstacle (not only distrust by the environmental community, but also the distrust that exists within the corporation itself). The whole process must be based on an implicit faith in the rational man. Once this distrust is overcome (if this is possible), the problems assume an entirely new scope.

This example of open planning is at most an experiment and a transitional forum, which will be replaced in the future by a new open planning forum where the Task Force and the corporation will work with government in a semiformal gathering, but hopefully not in a hearing format. Hearings require predetermined positions and tend to stifle flexibility and cooperative planning. If open planning fails, the advocate–adversary approach of public/governmental hearings is necessary, but it should not be looked upon as the only means to an end, especially when the objective is public involvement in planning processes.

Corporate involvement in these processes dictates considerable commitment to the philosophy of open planning. It will be necessary to experiment until a method is formulated that fits the needs of the period and the problem. These needs will change constantly, and the process must be flexible enough to change also. The risks are high, but the benefits can be the lack of delays and new insights pertaining to problem-solving alternatives.

To minimize the impact of electrical generating stations and their related facilities upon the environment, and to optimize the utilization of natural resources, a hierarchy of research and development priorities is established in Chapter 5. These priorities are divided into three categories, depending upon their impact upon human health, nonrenewable resources, and productivity, comfort, and enjoyment. A discussion of these categories and the individual areas of research and development is presented in this appendix.

MINIMIZING EFFECTS ON HUMAN HEALTH

For reasons of economics and reliability, it has been and still is desirable to site electrical generating facilities relatively close to the load. This has resulted in possible exposure of portions of the population to stack gas effluents and some small amounts of radiation. Both types of emissions may have some potentially harmful effects. The research required to determine these effects and minimize them is in two parts, in order of importance: air quality, dealing with emissions from fossil-fueled plants, and radiation sources.

Air Quality

It is expected that fossil-fueled plants will continue to provide the major source of electrical generation for many years to come. Extensive research has already been undertaken to develop improvements for those plants operating today and for those planned for the future. Individual areas for continued investigation, in

priority order, are SO_2 removal, NO_x removal, particulate removal, minor stack gas constituents identification, and stack gas constituents interaction.

SO_2 Removal

SO_2 (resulting from the combustion of oil and coal, both containing sulfur as an impurity) is an irritant to the respiratory system. If rained out of the atmosphere as sulfuric acid mist, it may be harmful to the lungs. Several methods for scrubbing SO_2 from the flue gases in a power plant are under investigation, but most result in secondary problems, such as the creation of solid-waste disposal problems. New methods are required to efficiently remove SO_2 from flue gases in a manner that minimizes secondary problems. In addition, development of advanced combustion techniques that will trap the sulfur before it is converted to SO_2, such as fluidized-bed combustion, needs to be accelerated. Removal of sulfur from oil and coal before they are introduced into the boiler is another method of resolving the SO_2 problems. Oil desulfurization processes exist today; however, they are expensive and on a small scale. More work is needed to develop improved large-scale processes. New research is also needed in the area of coal desulfurization. Ultimately, coal desulfurization may be combined with one of the coal gasification processes noted below.

NO_x Removal

When a fuel is burned in air at elevated temperatures, there is some combustion of the nitrogen and oxygen in the air to form oxides of nitrogen, particularly NO_2, which contribute to production of photochemical smog. Reduction or removal of these oxides ranks second only to the removal of SO_2 in research priorities. Two approaches to this problem are under way but need much additional work. One involves modification in the combustion process—two-stage burning, gas recirculation, adjusted burner configurations, and other methods designed to reduce the combustion temperatures. These techniques have produced substantial results so far, but further improvement is needed. The second method is the catalytic reduction of NO_2 to less harmful NO. A great deal of research will be required before this method is perfected. Research is also needed on methods of scrubbing the oxides of nitrogen from flue gases, particularly if this can be accomplished simultaneously with the removal of SO_2.

Particulate Removal

Associated with the combustion of coal and, to a much smaller extent oil, is the problem of the residual solid ash. Present collection methods, such as electrostatic precipitators, are being studied to increase their efficiency, particularly with respect to the removal of smaller size particles. Much additional research is needed on all methods of removing ash from flue gases, including electrostatic precipitation, fabric filters, and scrubbing, with emphasis on those methods that concurrently reduce SO_2 and/or oxides of nitrogen concentrations.

In addition to flue gas cleanup methods, there is a growing need for methods of removing ash content from fuel before combustion. These are the coal gasification and oil gasification techniques which can also include sulfur removal processes. Although the problem of disposal of the ash is not solved when fuel is gasified, these methods preclude release of fine particles to the atmosphere.

Minor Stack Gas Constituents

An identification of all stack gas constituents is necessary for various types of coal and oil to determine which of the minor materials present may be potentially harmful. Because of their generally small concentrations, the trace constituents present a difficult removal problem that needs further research. It may be possible to remove these materials by methods being used to remove the major stack gas materials but, failing this, it will be necessary to develop additional processes.

Effect of Stack Gas Constituents

While research is proceeding on reduction and removal of the various substances present in the stack gas from fossil-fueled plants, it is also highly desirable to research the interaction of these stack gases, separately and in combination with other atmospheric constituents. The effect of the resultant materials on humans, wildlife, and vegetation should also be determined in order to provide a reasonable technical basis for setting air quality standards. Some of this work is under way, but much still remains to be evaluated.

Radiation Sources

It is anticipated that nuclear energy will provide a significant portion of future electrical needs. To increase the acceptability of this kind of generation, some areas of research should be pursued more extensively. As already noted, nuclear plants lend themselves to more esthetic architectural design than fossil-fueled plants and can be a visual asset to the surroundings. The potential problems, however, are health-related due to the release of small quantities of radioactive substances. Work is needed in the following areas.

Reducing Radioactive Waste Releases during Plant Operation

All operating nuclear reactors produce some radioactive wastes that must be contained and processed on the plant site. At present, these wastes are often released very gradually to the environment, well within applicable federal limits. However, more research is needed on methods of further reducing or eliminating these emissions in order to remove any doubt regarding the long-term effects of this type of generating facility on man and the environment.

Reducing Radioactive Waste Releases during Fuel Reprocessing

As more reactors come into operation, there will be increased use of nuclear fuel reprocessing facilities. More work is required to develop methods for reducing releases from these plants and for effectively storing the radioactive waste products during extended cooling periods following reprocessing.

CONSERVATION OF NONRENEWABLE RESOURCES

Our nation is fortunate in having a large land mass, great quantities of both fresh and salt water, and access to large supplies of fuel. However, none of these is infinite. Effective use of these resources, in order to achieve maximum benefit from them for as long as possible, must become a basic tenet of our economy.

In the case of the electric industry, which makes use of land, water, and fuel, the continuation and expansion of research in many areas can lead to more successful conservation of these resources.

As in the previous section, the individual areas fall into three groupings which, by priority, are water and marine life, fuel, and land.

Water and Marine Life

Because of their possible sensitivity to change and the irreplaceable nature of some water systems and the life contained in them, the research areas connected with discharging heat to bodies of water have been given the top priority in conservation of resources research. The major areas of research in order of priority include alternative methods of heat disposal, effects of thermal discharge, more efficient generating processes, and thermal discharge utilization.

Alternative Methods of Heat Disposal

Currently operating steam-generating plants, and those planned for the near future, discharge heat to water passing through the steam condenser. The heat must be carried away rapidly and eventually dispersed to the atmosphere. Additional research is required to enhance dispersion of this heat and avoid releasing significant amounts of heat to bodies of water, with potential damage to marine ecology. The primary method being investigated and used today is wet cooling towers which, at present are large, expensive, and can create side problems such as local fogging conditions. Additional studies are needed in the heat transfer process to allow optimization of cooling tower design. Of particular interest, and requiring much more work, are saline towers and dry cooling towers. The former are planned for plants using seawater in the condenser. Major problems in the transport and dispersion of salt drift and concentrated salt brine are still to be solved. The dry cooling tower, which uses air to remove heat from the condenser water flowing in a closed loop, as opposed to an open loop for other cooling towers, is still in the development stage. Work to bring this type of tower to commercial application will result in much greater flexibility in plant siting.

Where the terrain permits, cooling ponds are often used. Some research is still needed in such related areas as possible additives to enhance evaporation from the ponds and in the use of sprays to further increase heat dispersion. After additional studies into droplet dynamics and nozzle plugging, sprays could become an increasingly useful method in easing the heat disposal problem.

Effects of Thermal Discharge

As in the case of air pollutants, studies of the true effects of heated water on water systems and the life forms contained therein must proceed simultaneously with the studies on minimizing heat release. This is necessary to provide a balance in deciding whether a particular thermal discharge situation has a damaging, neutral, or beneficial effect on the local environment.

More Efficient Generating Processes

The most effective way of reducing problems caused by the addition of heat to bodies of water is to reduce the amount of heat released in the generation process. A number of advanced technologies are currently under investigation, but all require significant additional research. In general, the successful development of any of these is a longer range task than the more immediately needed development of adequate heat disposal techniques for present systems.

One of the advanced methods closest to practical use is the high-temperature gas-cooled reactor (HTGR), which represents an improvement over the light water reactor in both use of uranium fuel and thermal efficiency and requires only a relatively small amount of additional development in the area of fuel reprocessing to become commercially viable. The HTGR has the potential of being an economically attractive reactor system prior to the commercial acceptance of the fast breeder reactor and should receive continued support.

Another advanced method of power production is the combined gas–steam cycle plants, which pass the products of combustion through a gas turbine and then to a conventional boiler, to increase the overall plant efficiency. Some additional developmental work is also needed on this type of plant to bring it to its full potential. Particular problems still to be resolved include fuel conditioning, improved efficiency, and gaseous effluent control.

Magnetohydrodynamics (MHD), when used as a topping cycle with more conventional systems, is predicted to achieve efficiencies of 50 percent or greater, representing a large reduction in the amount of heat to be dispersed to the atmosphere. However, before a demonstration plant can be constructed, a number of areas must still be investigated, including channel lifetime at high power, ionizing-seed recovery, and, especially, control of air pollutants.

Closed-cycle plants using liquid metals or gas in connection with nuclear reactors are in the research stage but should be pursued as a viable alternative to other generating techniques.

Fuel cells, which convert chemical energy directly to

electrical energy, require little water and have no thermal release to bodies of water. In addition, there would be significantly reduced emission of problem pollutants to the atmosphere. If used as distributed generation, there also would be a decreased need for transmission lines, thus providing a savings in land use. However, sizes of fuel cell units under investigation today are too small for anything other than supplementary energy sources. For this type of generation to be viable as a major energy source, substantially larger cells must be developed. To bring this technology to maturity for optimum use in electrical systems, research is still required in a number of areas, including cell life, fuel conversion, and optimum electrolytes.

There are several other processes being investigated for increased efficiency based on working fluids other than water. These are potentially promising, but require much further investigation to decide if any of them can be made commercially practical.

Thermal Discharge Utilization

Where possible, heat discharged from the generating process should be used for beneficial purposes. Where distances permit, steam has been supplied for heating buildings or for use in chemical processes. However, such applications are the exception, since most generating plants are not located near facilities that can utilize waste steam.

Other uses of waste heat, including use of warm water for improved agriculture or fish farming, are in the experimental stage and research should be expanded. Additional methods of using heated generating plant effluent should also be developed.

Fuel

Recently there has been much comment about a shortage of gas and a potential shortage of oil. To provide more options for the future, a number of generating methods using a variety of fuels need to be developed to meet future demands. Because of the somewhat longer-range nature of developmental programs in this area as compared to other more immediate developmental needs, this area appears to have been accorded a lower priority. However, because of the importance of reducing reliance upon a limited number of fuels, research in the following areas must be completed.

Nuclear Breeder Reactor

To greatly extend the use of uranium reserves, a type of reactor that produces fissile fuel from nonfissile uranium should be developed fairly rapidly. Large-scale demonstration of the concept should be made as soon as practicable so that the drain on uranium reserves due to the proliferation of nuclear reactors can be slowed. The breeder reactor, like the gas-cooled reactor, is more efficient thermally than the present light water reactors and will aid in preserving water quality.

Fusion

Several years of research have already been invested in the concept of nuclear fusion. The major fuel for this process is to be deuterium, an isotope of hydrogen recoverable from seawater. To date, however, technical feasibility has still not been established, but work is continuing, with the expectation of success by the end of the 1970's. Currently, this method appears to have minimal impact on the environment, but much more research is still necessary to confirm this.

Geothermal Energy

In certain parts of the United States, particularly in the western states, there exist underground large quantities of naturally heated water under pressure. Occasionally this water is pure, but generally it has a high content of solid materials and dissolved salts. This type of water, or steam, where clean, has been used for many years at various locations around the world. With research into the removal of the corrosive impurities and into achieving a single, large, energy supply from a series of wells, additional use could be made of geothermal energy to supply a portion of the U.S. energy demand.

Solar Energy

Solar energy has already been used on small scale for space applications. However, a great deal of work is needed in the collection and concentration of the very low density solar energy reaching the Earth. If breakthroughs can be made, then solar energy represents as nearly an infinite source of energy as possible.

Tidal Energy

Only in very select locations is harnessing tidal energy even remotely feasible. In the future, it is possible that this source of energy will be utilized, but, for the present, only a minor effort should be expended in its development.

Land

Expected increases in future electrical energy requirements indicate the need for more land devoted to the generation and transmission of electricity. Often, however, portions of this land can also be used for other compatible purposes, and every effort should be made to do so. The land problem, while extremely important, is not as critical as the water or fuel resources problems and is accorded a slightly lower priority. Resolution of this problem is under way and would be greatly assisted by expanded study in the following areas.

New Siting Techniques

Several options, not previously thought feasible, are being considered for siting large generating facilities, mainly in connection with nuclear power plants. Some investigation into locating these plants underground has been done, but a great deal more investigation remains in developing construction techniques and in determining applicability of individual locations for this kind of siting. A second possibility is the location of plants offshore in those parts of the country with seacoasts or large lakes. The leading method is siting on an artificially constructed island, but some thought has also been given to floating plants. More research should be conducted to determine the feasibility and optimum arrangement of such plants, and in improving underwater transmission cables capable of bringing the generated electricity from the plant to the bulk power system. The location of gas turbines on barges now ready for operation should provide initial data for the development of future waterborne plants.

One area of technology that needs a new approach is the location of facilities in areas of high earthquake potential. Determinations of the precise mechanisms by which energy is released during an earthquake and the resulting action on surrounding land and buildings are needed. From these, more appropriate and more accurate design parameters for generating plants, especially nuclear plants, can be established for those portions of the plant that are critically affected.

Underground Transmission

The technology of undergrounding has not yet reached the stage where it is economically feasible to bury high-voltage transmission lines connecting facilities more than just a few miles apart. Such undergrounding has been primarily limited to relatively short lines in metropolitan areas. Currently, a large research effort is being undertaken to develop different types of cables and insulators in preparation for large-scale undergrounding of future transmission lines. In addition, d.c. transmission is also being developed as a major step in this area. Both aspects of high-voltage transmission will require significant development in the

future but, once successful, will reduce the quantities of land necessary to meet future rights-of-way land requirements.

Dispersed Generation

In areas where new communities are developing or in established areas that are still growing, it may be desirable to build generating facilities of a size to serve just the individual community. Backup can be provided by the bulk power system, but the need for transmission could be lessened.

Also, smaller units can be located at substations to improve the overall land use efficiency. The fuel cell, as mentioned previously, can be developed to serve this distributed generation function. Another possibility requiring much additional work is the large secondary storage battery, which would be charged from the bulk system during off-peak hours and discharged to provide electricity to the community during periods of greater demand.

Advanced Substation Designs

It has become increasingly evident that more compact substations could be designed to use less land. Present investigations into compressed-gas-insulated stations, if continued through successful demonstrations on major substation facilities, could result in greatly reduced land area requirements for the future. Several facilities of this type are currently being originated outside Paris by Electricité de France.

INCREASED PRODUCTIVITY, COMFORT, AND ENJOYMENT

Once work is substantially under way in guaranteeing minimal effects on health and in conservation of resources, it becomes proper to investigate ways that the production and distribution of electrical energy can be adjusted to add to productivity and an improved life style. Several approaches can be taken, including modification of utility facilities, joint ventures with other industries, and more efficient use of the electricity. Also, existing information should be searched and additional information developed necessary to determine the sociological impacts of electrical facilities and what alternatives are acceptable in various situations. For example: Are people willing to accept less electricity than they might ordinarily demand rather than have power plants in the vicinity of their residence? Many such questions can be asked and ought to be resolved through this type of study.

Modification of Facilities

Both visual and aural aspects of electrical facilities need improvement. As for most of the areas mentioned, much work is still required. Those items listed under more efficient use of land will have a significant impact on many of the visual problems. In addition, investigations must be continued in the following areas.

Architectural Improvements

Studies similar to those performed to improve the design of transmission towers should be undertaken for all electric utility structures, particularly electric-generating plants.

Noise Control

The increased use of both gas turbines and EHV transmission lines have introduced noise problems that must be resolved. Since the noise in each case results from different processes, individual noise suppression studies for each source are required. For the benefit of in-plant personnel, additional noise-control research is also needed in all generating facilities.

Joint Ventures with Other Industries

Increased efforts should be expended to site generating plants in such a way that compatible secondary uses of plant facilities and emissions can be employed where possible. For many years, lakes associated with hydroelectric facilities have been used for recreational purposes.

For other types of plants, this multiuse of land is being carried out on a small scale at present and needs further investigation. For example, additional parks, camp, and picnic facilities should be studied in connection with nuclear power plants.

An area that has received preliminary study is siting generating plants in an industrial complex. Such complexes could include fuel-processing plants, plants which utilize the low-grade heat from the generating plant, such as a waste treatment plant, and other manufacturing facilities. More serious analysis must be given to this concept in connection with present nuclear and fossil-fueled plants and with future advanced generating systems.

More Efficient Use of Electricity

As one means of slowing the growth of electrical energy demand, the more efficient use of electrical energy by the consumer should be examined. This could include studies of improved building insulation, use of heat pump environmental control systems, more efficient appliances, etc. It is unlikely that the effect on load growth of such improvements in efficiency would be major. However, any effort to promote more efficient usages will lessen the emissions to the environment and the use of natural resources.

Working Group III

Energy and Economic Growth

317

INTRODUCTION

The growing concern for the environmental effects of energy consumption and particularly the consumption of energy for the production of electricity, has raised the question of whether the projected growth in consumption of energy is necessary or desirable. Even if such growth is desirable, the suggestion has been made that various types of restraints be imposed on the growth of electricity consumption as a means of protecting the environment. The more common means suggested include the inversion of the usual declining block rate structure, rationing, or imposition of special use taxes. The extent to which such measures would effectively restrain growth in electric energy consumption is at least open to doubt, but the economic and social implications of such measures raise further questions.

Such growth restraints, if effective, would be intended to reduce the need for additional power plant sites, and curtail the increase in noxious discharges to the atmosphere, thermal discharges to water bodies, and radioactive emissions resulting from electric power generation. They would also reduce the need for additional transmission line rights-of-way. Finally, they would help to conserve our energy resources.

It is assumed by proponents of such restraints that the restraints would result in improved environmental conditions. However, improvements in environmental quality thus accomplished may be in competition with other objectives of our society, such as those for further economic growth, rising material standards of living, and particularly, the increased employment opportunities and rising income levels for those groups now among the economically underprivileged They would also conflict with some national policies specifically intended to encourage the availability of an abundant low-cost supply of electric energy.

Furthermore, it is by no means clear that restraint on electric power growth will in fact result in environmental improvements. Unless such restraints are imposed consistently on all energy forms, the mix of energy consumption may shift in ways which may worsen the environmental impact of energy use. In contrast, a shift to electricity may be environmentally desirable, such as in the case of the expansion of electrified mass transit facilities. Also, in many cases, environmental improvements are likely to require even greater use of electric energy. Such uses of electric energy as in the handling and recycling of waste material and especially in the development of individual electric transportation are likely to greatly expand electric use while improving the environment and possibly reducing total energy requirements. Exhaustion of higher quality mineral resources is also likely to require greater use of electric energy as it becomes increasingly necessary to exploit mineral resources of lower quality.

Historically, a large part of the growth in electric energy use has been the result of the substitution of electricity for direct fuel use. For the longer run, electrification of energy use is likely to continue and may provide the only means of circumventing possible energy shortages. A number of on-going and prospective energy studies will, through their supply and demand evaluations, throw more light on the question of the need to conserve our energy resources. There is some reason to believe that technological developments could increase very substantially the magnitude of our economic fuel resources within the foreseeable future. Within two decades, breeder reactors may greatly extend the energy recoverable from uranium, and in three or more decades, fusion power may provide society with virtually limitless energy resources. The most attractive means for obtaining commercial energy from fission and fusion reactions involve the production of electricity. Thus, the long-term trend of energy use toward more electricity matches the technological prospects for the very large increases in supply.

The balance of the report of the Working Group on Energy and Economic Growth deals with the issues highlighted above. Specifically, Chapter 1 examines the source of growth for electric energy. Chapter 2 traces the traditional forecasts and the components of electric use. Chapter 3 investigates means of influencing future demand and evaluates the possible impact of such influences. Chapter 4 looks at needs for further research. Chapter 5 summarizes the findings and reports on the conclusions of the Working Group.

1

SOURCES OF ELECTRIC ENERGY GROWTH

Growth in the use of electric power comes from four basic sources: (1) an increase in the population, (2) an improvement in the standard of living, (3) the development of new applications for the use of electricity, and (4) the shifts from other sources of energy to electricity.

The following sections merely highlight some of these factors. There is no attempt here to evaluate the desirability or necessity of such growth; that is done in Chapter 3. However, it is intended to reflect the types of general growth in use of electricity which generally underlie the forecasts discussed in Chapter 2.*

In order to study growth in use of electric power, it is useful to examine the pattern of consumption in 1969. Residential customers comprised nearly 90 percent of the total number of customers, but they used only about one third of the total electricity distributed. The commercial customers used about one fifth, while industrial customers used about two fifths of the energy. The remaining 5 percent of the kilowatt-hours were used for street lighting, railroads and public authorities. Since many of the factors which affect growth vary with class of service, the growth factors affecting each class will be discussed separately.

RESIDENTIAL AND RELATED USES

The number of residential customers generally parallels the population. However, as income increases, more and

more family units can set up independent households. For example, increased pension and social security payments make it possible for older people to live in separate households. Similarly, as wages increase, more young people can establish their own households and fewer of them will double up. But offsetting this to some degree is the lower average consumption of these households. Finally, as affluence increases, a greater number of people can afford a second home—e.g., an in-town apartment and a rural retreat. The result of these factors over the balance of this century is that the number of residential customers is expected to increase at a slightly higher rate than the population.

The improvement in standard of living has led to the increasing ownership of appliances. Table 1 shows the sources of energy growth over two decades for residential customers of a southeastern utility. The increases in saturation of the various appliances are clearly evident. In addition, higher standards of living are reflected in greater comfort heating and cooling and in greater conveniences with such appliances as dishwashers, disposals, and self-defrosting refrigerators. Over the three decades to the year 2000, there is likely to be a continuing improvement in the standard of living reflected in higher saturation of these appliances.

Through the years, the electric power requirements of various appliances have changed as technological changes have occurred. Some of these changes have tended to dampen the growth in electric use. The fluorescent lamp, for example, produces several times more light per watt than the incandescent lamp. The new solid-state radios and television sets use only a frac-

---

* For a more complete discussion of growth factors and possible modification of historical growth trends, see: Federal Power Commission, *National Power Survey, Vol. II,* 1970.

TABLE 1 Sources of Residential Average Annual Use [a, b]

Electric Appliance	1949			1959			1969		
	Saturation (Percent)	Appliance Annual Use (kWh)	Contribution to Annual Use (kWh)	Saturation (Percent)	Appliance Annual Use (kWh)	Contribution to Annual Use (kWh)	Saturation (Percent)	Appliance Annual Use (kWh)	Contribution to Annual Use (kWh)
Refrigerator	69	360	248	97	415	403	98	660	647
Range	31	1,350	419	63	1,350	851	80	1,350	1,080
Water heater	16	4,050	648	44	4,490	1,976	71	5,175	3,674
Space heating	2	8,860	177	18	10,710	1,928	31	11,260	3,491
Air conditioner:									
Room [c]	5	1,250	6	15	1,355	203	36	1,680	605
Central	—	—	—	2	3,500	70	9	4,100	369
Television	1	400	4	74	400	296	99	400	396
Washer:									
Automatic	8	100	8	28	100	28	49	100	49
Nonautomatic	38	50	19	45	50	23	40	50	20
Dryer	4	940	38	9	1,130	102	35	1,335	467
Freezer	7	895	63	21	900	189	38	980	372
Dishwasher	1	325	3	5	285	14	16	340	54
Miscellaneous [d]	—	—	1,132	—	—	1,323	—	—	2,376
			2,765			7,406			13,600

[a] Source: *The Methodology of Load Forecasting,* A Report to the Federal Power Commission, prepared by the Technological Advisory Committee on Load Forecasting Methodology for the National Power Survey (1969), p. III-12.
[b] Covers a large utility system in the Southeast.
[c] Saturation is defined as the percentage of residential customers having one or more room conditioners.
[d] Lighting, small appliances, supplemental heat, and other uses.

tion of the power of vacuum-tube sets. On the other hand, frost-free refrigerator–freezers use several times the power of the small refrigerator of a generation ago. But, of course, in return they provide a far higher level of service to the consumer. This simply illustrates the many technical, social, and economic forces affecting the rate of growth in the demand for electricity.

Table 1 also shows the effect of the introduction of new appliances or new applications of electricity. In 1949, electric space and water heating were in their infancy, television was nearly unknown, and the dishwasher was not yet common. In addition, the miscellaneous classification reflects the addition of the myriad of electrical devices introduced in the last two decades. In the next few decades further refinements of existing devices and development of entirely new devices are likely. Extension of temperature control to outdoor areas, more automatic cleaning devices, electric garden accessories, and possibly the electric car could become commonly used.

Finally, the shift from other energy applications can also be seen in the table. One such major shift is in space and water heating. But cooking and clothes drying are the other major fields of competitive energy sales, with some competition in cooling beginning to emerge.

COMMERCIAL USE

The number of commercial establishments has grown to meet population growth. In addition, there has been a rapid expansion of education, medical, and other service industries, including government, accompanied by commensurate growth in electricity needs. Moreover, the business population has grown in some cases to replace functions formerly done in the home—such as food preparation and auto cleaning. On the other hand, some functions formerly performed by commercial establishments have now been shifted back to the home, such as laundry and entertainment. At the same time, heating and cooling of commercial space have significantly added to the commercial consumption of electric energy.

INDUSTRIAL USE

As noted earlier, industrial use constitutes about 40 percent of the total electricity consumption. This use varies widely among industries. Table 2 shows a breakdown of electricity use among the major industrial groups and the relationship between the cost of electricity and the value of products shipped. Clearly, the

TABLE 2 Use of Electric Energy by Industry Groups in the United States [a, b]

SIC			1966 (Million kWh)				1967 Cost of Electricity Purchased as a Percent of Value of Shipments (5)
			Purchased (1)	Generated Less Sold (2)	Total (3)	Percent (4)	
20		Food and kindred products	23,574	2,573	26,147	5.4	0.4
21		Tobacco manufactures	675	—	675	0.1	0.01
22		Textile mill products	18,426	701	19,127	3.9	0.9
23		Apparel and related products	3,167	19	3,186	0.7	0.3
24		Lumber and wood products	6,485	944	7,429	1.5	0.8
25		Furniture and fixtures	2,260	80	2,340	0.5	0.5
26		Paper and allied products	24,166	23,025	47,191	9.8	1.0
27		Printing and publishing	5,056	—	5,056	1.0	0.4
28		Chemicals and allied products	92,155	21,016	113,171	23.4	1.4
29		Petroleum and coal products	15,756	4,282	20,038	4.1	0.6
30		Rubber and plastics products, n.e.c.	10,688	614	11,302	2.4	0.9
31		Leather and leather products	1,195	85	1,280	0.3	0.4
32		Stone, clay, and glass products	22,793	1,662	24,455	5.0	1.4
33		Primary metal industries	99,700	23,460	123,160	25.4	1.5
33	1	Steel rolling and finishing	(41,911)	(10,676)	(52,587)	(10.9)	—
33	2	Iron and steel foundries	(5,703)	(46)	(5,749)	(1.2)	—
33	3	Primary nonferrous metals	(40,767)	(12,603)	(53,370)	(11.0)	—
34		Fabricated metal products	12,222	117	12,339	2.5	0.6
35		Machinery, except electrical	14,939	649	15,588	3.2	0.5
36		Electrical equipment and supplies	16,903	300	17,203	3.6	0.5
37		Transportation equipment	23,077	278	23,355	4.8	0.4
38		Instruments and related products	2,223	578	2,801	0.6	0.4
39 & 19		Miscellaneous manufacturing, including ordnance	5,092	41	5,133	1.1	0.5
		Withheld	1,739	1,623	3,362	0.7	—
		Total	402,291	82,047	484,338	100.0%	0.7

[a] Source:
Cols. (1), (2), and (3): U.S. Department of Commerce, Bureau of the Census M66(AS)-4, *Annual Survey of Manufacturers—1966.* "Fuels and Electric Energy Used, by Major Industry Groups, and by Divisions and States."
Col. (4): Calculated from U.S. Department of Commerce, "Fuels and Electric Energy Used in Manufacturing," Preliminary Report, *1967 Census of Manufacturers*, Series MC67(P)-7, July 1969, and U.S. Department of Commerce, "General Statistics for Industry Groups and Industries," *Annual Survey of Manufacturers—1968*, M58(AS)-1, June 1970.
[b] ( ) Do not add to total SIC 33 due to disclosure problems.

chemical and primary metals industries consume the largest amounts of electricity.

There are several major factors influencing growth in industrial use of electric energy. One such factor has been the substitution of materials more energy-intensive in their production for less energy-intensive materials: for example, aluminum siding for wood, aluminum for copper, synthetic fibers for cotton and wool.

A second factor has been the continuing substitution of energy-consuming capital equipment for labor. A third factor has been the substitution of electric energy for other energy forms as in electric steel and glass melting furnaces.

Such developments are expected to continue to influence the magnitude of industrial electricity consumption in the next few decades. In fact, environmental requirements may encourage the replacement of other fuels with electricity. A possible partially offsetting factor may be a gain in efficiency in the use of electricity. Since electricity is more expensive per Btu than other energy forms, when it is used as a substitute it is usually done with less loss and greater efficiency at the point of use.

OTHER USES

The development of mass transportation for people and material, particularly for medium and short distances in the crowded megalopolises developing along the Atlantic Seaboard and around the southern edges of the Great Lakes, may imply a large increase in the demand for electricity. Rapid transit systems, such as are now being constructed in San Francisco and Washington, D. C., are likely to be introduced and extended

TABLE 3 U.S. Electric Utility Sales by Classes, Losses, and Generation, 1950, 1960, and 1970, and Forecasts for 1980, 1990, and 2000 (Billions of Kilowatt Hours) [a]

| Year | Sales | | | | | Losses [d] | Genera-tion [e] |
	Residential	Commercial	Industrial [b]	Other [c]	Total		
1950	70 (24.9%)	52 (18.5%)	142 (50.5%)	17 (6.1%)	281 (100.0%)	48	329
1960	196 (28.7)	115 (16.8)	345 (50.5)	27 (4.0)	683 (100.0)	72	755
1970 (prelim.)	448 (32.2)	313 (22.5)	572 (41.1)	58 (4.2)	1,391 (100.0)	139	1,530
1980 [f]	880 (31.5)	650 (23.3)	1,110 (39.6)	160 (5.6)	2,800 (100.0)	275	3,075
1990 [f]	1,590 (30.0)	1,325 (25.0)	2,015 (38.0)	370 (7.0)	5,300 (100.0)	528	5,828
2000 [f]	2,640 (29.0)	2,365 (26.0)	3,370 (37.0)	725 (8.0)	9,100 (100.0)	900	10,000

[a] Source: Sales by sector for 1950, 1960, and 1970 are from EEI publications cited in Table 4.
[b] Commercial sales tend to be the "Small light and power" category used by electric utilities, and industrial tends to be the utilities' "Large light and power" category.
[c] Street and highway lighting, other public authorities, railroads and railways, etc.
[d] Difference between generation and sales, primarily but not entirely losses.
[e] From Table 5.
[f] The breakdown into sales category follows some of the literature but should be regarded as illustrative rather than definite, particularly in view of overlapping of the categories in actual utility recording of sales.

in major cities and may be supplemented by such other devices as moving sidewalks and stairways, and conveyor belt deliveries—all powered by electric energy.

Improved street and highway lighting is the goal of traffic engineers and law enforcement agencies. The increase in efficiency of lamps used in this lighting has greatly increased illumination with less than a corresponding increase in power use. Nevertheless, improvements in general lighting are likely to increase the use of electric energy for such purposes.

The analysis by major classes of consumers of the total electricity projections made by the Federal Power Commission *National Power Survey* Regional Advisory Committees is shown in Table 3. It should be noted that some decline is projected for the share of the total consumption of electricity represented by industrial use. Residential consumption is expected to remain close to its existing share of the total. The commercial category is expected to show some moderate increase in its share to provide for an improved environment in shops, offices, and the like. A significant increase is projected in the share of other types of sales, such as street and highway lighting, other public authorities, railroads and railways. These latter uses for such purposes as sewage and water treatment and improved mass transportation are likely to enhance the public convenience and to have overall improved environmental effects.

ELECTRICITY AND ENVIRONMENTAL FACTORS

The preceding discussion has purposely not concentrated on the environmental impacts of energy utilization. In general, environmental considerations have not been an important determinant in the user's selection of energy forms or in the choice of use of energy. In the future, environmental considerations are likely to play a more important role. Chapter 3 discusses the possibility of a reduced rate of growth from this point of view.

However, it appears that rather than diminish electric energy growth, environmental considerations may in fact accelerate that growth relative to other energy forms. This is true because

1. Central station energy conversion can be done at locations remote from population centers, where the environmental impacts will be least deleterious.*
2. Central station energy conversion can control the emission of pollutants more effectively and efficiently than numerous small furnaces. Improvements in technology further reduce the environmental impact of central station electricity generation.
3. Electric energy can best tap noncombustion energy sources: nuclear fission, falling water, and ultimately, nuclear fusion.
4. Other environmental needs will require large applications of electricity: sewage treatment, effluent control, and recycling of used materials. Recycling, while increasing electricity use, may reduce the energy required for mining and processing progressively lower-grade ore deposits.

---

* Even with the location of electric generating stations at remote locations, controversies may arise over such issues as the intrusion upon a hitherto untouched natural environment and the aesthetic effects of transmission lines. Nevertheless, new technology such as dry cooling towers and underground transmission may, over the next several decades, modify the constraints on power plant site selection imposed by large cooling water requirements and resolve the transmission line problem.

2

ENERGY AND ELECTRICITY GROWTH FORECASTS

RANGE OF FORECASTS

The growth of total energy and electric power during the last two decades provides a frame of reference within which to judge the probable magnitude of energy and electric power usage as it is likely to develop in this century, the period of primary interest for this study. In the 10 years from 1950 to 1960, domestic energy consumption grew from 34.1×10^{15} Btu to 45.0×10^{15} Btu, a compound annual growth rate of 2.8 percent.

By 1970, energy consumption had increased to 68.8×10^{15} Btu, for a compound annual growth rate during the 1960–1970 period of about 4.3 percent, significantly higher than in the previous decade. (See Table 4.)

Electric power generation has grown considerably faster than total energy consumption during these periods. Total electricity generation in the United States in 1950 was 389 billion kilowatt-hours. By 1960, this had grown to 844 billion kilowatt-hours, a compound annual growth rate of 8.0 percent, compared to

TABLE 4 Comparison of U.S. Energy Requirements for Electric Power Generation with Total Energy Consumption, 1950, 1960, and 1970, and Forecasts for 1980, 1990, and 2000

| Year | Total Electric | | | Total Energy | | |
	Generation (billions of kWh) [a] (1)	Heat Rate [b] (Btu/kWh) (2)	Btu for Generation in 10^{15} Btu (1) × (2) (3)	Amount [c] (10^{15} Btu) (4)	Millions of Tons of Coal Equivalent [d] (5)	Percent Electric (3)÷(4) (6)
1950	389	14,030	5.5	34.1	1,420	16.1
1960	844	10,071	8.5	45.0	1,875	18.9
1970 (prelim.)	1,638	10,500	17.2	68.8	2,870	25.0
1980	3,202	10,500	33.6	101.8	4,240	33.0
1990	5,978	10,500	62.8	150.7	6,280	41.7
2000	10,150	10,000	101.5	223.0	9,290	45.5

[a] From Table 5.
[b] Heat rates for 1950 and 1960 are from Edison Electric Institute, *Historical Statistics of the Electric Utility Industry* and *Statistical Year Book of the Electric Utility Industry for 1969*. The heat rate for 1970 is estimated. For 1980, 1990, and 2000 heat rates assume an increasing proportion of light water reactors, peaking units, gas turbines, and pumped storage, as well as the greater application of air pollution control devices and cooling towers which result in an increase in the heat rate. This is in part offset by the retirement of older existing units by more efficient units. Energy input for hydrogeneration is assumed to be at the central steam station heat rate.
[c] Data on Btu of total energy for 1950 are from U.S. Department of the Interior *Minerals Yearbook* and for 1970 from U.S. Department of the Interior news release dated March 9, 1971. Projection from 1970 is made on an assumed compound annual growth rate of 4 percent.
[d] Total energy in column (4) converted to coal equivalent at 24 million Btu per short ton.

2.8 percent for total energy. In 1970, production of electricity in the United States reached 1,638 billion kilowatt-hours, for an average growth rate during the 1960–1970 period of 6.8 percent, compared with 4.3 percent for total energy during the same period. It should be noted that in the last decade the difference between the rates of growth of total energy and of electricity has narrowed.

There are many forecasts of future energy and electric energy requirements for the United States available in the literature. Reference to these can be found, among others, in the Edison Electric Institute's Statistical Committee compilation, *Bibliography and Digest of U. S. Electric and Total Energy Forecasts, 1970–2050,* and the Battelle Memorial Institute's Pacific Northwest Laboratories publication, *A Review and Comparison of Selected United States Energy Forecasts.*

Energy is an important input in the production of the nation's goods and services. While the comparison of energy to gross national product (GNP) does not imply a direct causal relationship between the two, it helps in understanding the possible range in forecasts.

Since 1920, there has been a declining trend in the amount of energy required to produce a unit of real GNP. This has been the result of a number of factors (e.g., the increasing technical efficiency in using energy, including the improved thermal efficiency of power plants, higher efficiency of combustion engines, more efficient fuel conversions such as gas heat versus coal heat, use of transistors instead of vacuum tubes, fluorescent lighting, and the like) as well as increasing efficiency of economic organization of production. Recently, however, the rate of decrease in the ratio of

energy to real GNP has been slowing and has even shown some signs of reversal (National Economic Research Associates, 1971).

Since World War II, the trend of electric power generation, including industrial self-generation, has been an increase of about 3 kilowatt-hours for every dollar increase in GNP when GNP is expressed in 1958 dollars.

If it is assumed that the total energy–GNP ratio will stabilize at the present level and growth in real gross national product will be 4 percent per year for the balance of the century, total energy requirements of the United States for the year 2000 would be about 220×10^{15} Btu. The United States Bureau of Mines, however, has forecast energy consumption at 168×10^{15} Btu in the year 2000 with a 4 percent per year growth in the real GNP, implying a continuing decline in the energy–GNP ratio. These two projections bracket the typical energy forecast figures.

In the case of electric power, the relative difference in estimates is even greater than the differences in the total energy forecasts. One of the higher forecasts is based on the assumption of a continuing rate of growth of 7 percent per year. Under this assumption, electric power generation would increase from 1.64 trillion kilowatt-hours in 1970 to 12.5 trillion kilowatt-hours in the year 2000. This type of forecast is frequently associated with a population of 320 million resulting in an annual use of 39,000 kilowatt-hours per person.

A considerably lower estimate of 5.6 trillion kilowatt-hours in the year 2000 results from the use of an S-shaped (Gompertz) curve of per capita consumption (developed by the Edison Electric Institute) in conjunction with a population estimate of 280 million. This population

TABLE 5 U.S. Electric Energy Requirements and Peak Demands, Actual 1950, 1960, and 1970, FPC Regional Advisory Committee's Projections for 1980 and 1990, and Extension of These Projections for 2000 [a]

| Year | Generation (Billions of kWh) | | | Peak Demand, Electric Utilities (Millions of kW) | Installed Generating Capacity, Electric Utilities (Millions of kW) |
	Electric Utility	Industrial Self Generation	Total		
1950	329	60	389	64	69
1960	755	89	844	133	168
1970 (prelim.)	1,530	108	1,638	275	340
Regional Advisory Committee Projections					
1980	3,075	127	3,202	555	668
1990	5,828	150	5,978	1,051	1,261
Extension of Regional Advisory Committee Projections [b]					
2000	10,000	150	10,150	1,800	2,150

[a] Source: 1950 data from Edison Electric Institute, *Historical Statistics of the Electric Utility Industry;* 1960 data from EEI, *Statistical Year Book of the Electric Utility Industry for 1969;* and 1970 data from EEI advance release dated April 1971 (mimeo).
[b] Arbitrary extension for illustrative purposes. These figures have not been reviewed by the Regional Advisory Committees.

figure, combined with the estimate of 20,000 kilowatt-hours per person in the year 2000, leads to the 5.6 trillion kilowatt-hour estimate.

One possible pattern of electric power consumption is that projected by the Regional Advisory Committees to the Federal Power Commission *National Power Survey*. This forecast for 1990, together with a somewhat arbitrary extension to the year 2000 leads to a projection of 10.2 trillion kilowatt-hours by the end of the period as is shown in Table 5. This forecast is the sum of the forecasts for each region, which, for the most part, were derived by the individual electric utilities in each region.

The specific level of generation expected by the end of this century is not crucial to an analysis of the power plant siting problem, especially in the shorter run of the next decade. Whether the rate of growth is 7 percent, 4 percent or something in between, the problem of obtaining power plant sites with suitable protection for the environment remains. On the other hand, projected levels of generation are by no means irrelevant over the longer term. If power generation should be at the level of 12½ trillion kilowatt-hours in the year 2000, approximately 2½ to 3 times as much new capacity will have to be built in the next 30 years than if the figure is 5.6 trillion kilowatt-hours. This slower rate of growth would permit more time for the development of the necessary technology which would substantially reduce the siting problem.

Table 4 shows the forecast of the nation's total energy supply and the share converted to electricity. The percentage supplied to electric generation depends not only on forecast of electric power and total energy but also on the assumed thermal efficiency of electric power generation. Based on the *National Power Survey* forecast and a 4 percent increase in total energy consumption, the share of energy converted to the electric form will increase from 25 percent in 1970 to 45 percent in the year 2000. (See Table 4.)

IMPLICIT ASSUMPTIONS IN FORECASTS

The general range of forecasts discussed above must, by their very nature, be based on either explicit or implicit assumptions with regard to factors affecting the economy as a whole as well as the energy industries. In order to clarify the basis on which these forecasts were made, there follows a discussion of what was assumed with regard to these factors.

Most such forecasts implicitly reflect the expectation of continuing past trends as experienced over the last several decades. This does not imply fixed socio-economic factors. Rather, it implies a rate of change in these factors continuing as it has in the recent past.

For example, the rate of introduction of new utilization devices is implicitly expected to continue in the aggregate as in the past with the same relative energy utilization. Also, the rate of growth in personal income is expected to continue at the same relative rate as it has in the past.

Population

There appears to be some indication in recent demographic data and in the development of social movements such as Zero Population Growth of a possible reduction in the rate of growth of the population. These forecasts do not reflect any further decline in the rate of growth other than that which may have taken place during the decade of the 1960's. Insofar as a more rapid decline may take place, the forecasts for the period after 1990 may be overstated.* Prior to that time, the probable effect on electricity demand of a slower population growth would be small since the population which will form the households during this period has already been born. In fact, during this period, a reduction in births may increase discretionary income (i.e., personal income available after taxes and expenditures for essentials) per household and thereby possibly raise actual energy use above that forecast.

GNP and Standard of Living

The forecasts assume a growth in the GNP of about 4 percent per year reflecting the current national economic objectives of the Congress and of the President and the Council of Economic Advisors. Were the objectives to change, there could be an effect on growth in energy demand depending on the character of the accompanying changes in the economic structure. For example, were policy to shift in favor of producing more leisure time per capita rather than material goods, the rate of growth of GNP could decline. Were that leisure time to be spent in activities less intensive in energy or electric energy use than otherwise would have been, then the forecasts are overstated.

The forecasts also contemplate a continuing increase in the standard of living of the general population. Should the rate of growth in population change, then, the same rate of improvement in living standards would be sustained with a commensurate change in the rate of GNP growth. A reduction in this rate of improvement in living standards would, in general, cause the forecasts to

---

* Even with such a decline, population will increase significantly by the end of the century. According to an interim report of the Commission on Population Growth and America's Future, *Population Growth and America's Future*, if family size should decline to a two-child average and remain there to the year 2000, population will still increase to about 280 million.

be too high. On the other hand, an acceleration in the redistribution of income toward lower income groups would probably cause the forecast to be too low, since their use of energy is likely to grow as a result of rising levels of income to a greater extent than in the case of those already at higher levels of income and energy use.

Technology

The forecasts clearly imply no supply limitation in terms of electric generating and transmission capacity. This means, therefore, that to whatever extent technological innovation is necessary to make these forecasts compatible with environmental goals, it is assumed that the technology will be developed in time and at a price to meet the needs.

Similarly, natural resources are implicitly forecast to be in plentiful supply. Specifically, sufficient fuel is assumed to be available, whether it is coal, oil, gas or uranium. Resources such as air and water for cooling, land for siting, capital for investment and labor for construction and operation are also assumed not to be limiting factors.

The forecasts do not imply use of these resources in the identical way they are being used today. In fact, they imply technological and utilization progress to avoid a significant increase in their costs relative to the general price level. Progress is assumed to be sufficient to overcome any constraint of growth in electricity demand which might otherwise be imposed by resource limitations.

Price of Electricity

Most forecasts implicitly assume that the price of electricity, at least relative to substitute energy sources, will continue to decline slightly. Also, they assume that the relative price of all energy sources to the prices of all goods and services will remain roughly constant.

To the extent that the relative price movements are greater or lesser than those expected, and to the extent responsiveness to price is operative over the range of expected change, then these forecasts may be overstated or understated.

Shift in Energy Use

Over the past two decades total energy use has grown at 3½ percent annually while electric use has been growing at 7½ percent. This means electricity has been capturing a growing share of the energy markets. Since most forecasts predict continuing current rates of growth for both total and electric energy, this implies a continuing growth in the proportionate share of the market for electricity. To the extent that these shifts in energy use patterns in favor of electricity do not continue, then the electric forecasts are overstated.

Growth in Use

The forecast of electric energy use reflects a larger rate of growth than the rate of population increase. This implies a continuing growth in electricity use per capita. In part, this growth stems from the shift in energy mix discussed above; in part, from growing saturation of existing appliances; and in part, from the development of new utilization devices. These growth factors include continuing growth in air conditioning, electric heating, higher standards of lighting, automation in factories, new electricity-using manufacturing processes and new electricity-intensive industries, such as recycling firms and new forms of electric transportation.

To the extent that such growth in saturation and/or development of new uses does not take place at the same rate as over the past decade, the forecast rate of growth will be too high.

Utilization Efficiency

With abundant and relatively inexpensive energy, past history does not reflect extremely high utilization efficiency in the various energy applications in the home, store, office, or factory. The relative Btu cost of electric energy is higher than that of other fuels. Therefore, when electricity is used in competitive applications, it is typically done with less losses or greater efficiencies at the point of use. For example, because of the higher cost of electricity, electrically heated homes are usually better insulated and have less air infiltration than others. Electric steel furnaces have lower losses than other types.

To the extent that an increasing relative price of fuels or a heightened environmental ethic leads to a more conservative use of energy or a willingness to employ additional capital to conserve energy, energy forecasts will be too high.

3

MEASURES TO RESTRICT GROWTH IN ELECTRICITY DEMAND

The previous chapters have discussed the nature of electric consumption growth, have reviewed the range of forecasts both for total energy and electricity, and, finally, have reviewed the assumptions under which these forecasts were made and the gross effect of varying these assumptions. This might be termed the traditional approach to determining power plant site requirements.

If it is deemed desirable or necessary to limit the number and size of plant sites, then clearly, some external force would have to be added to reduce electric power consumption from what it otherwise would be. The following sections are examinations of the external forces that might be applied.

These external means are first described, and then an assessment is made of the likely extent of their impact on both electric and total energy requirements. Finally, the operative time span required for the impact to be felt is generally assessed. This is especially important since plant sites must usually be selected as much as 10 years before the commercial service date of the plant. The effect on the economy and the social system of applying various means of limiting energy growth is also discussed and the feasibility of applying such means is appraised.

As these following sections bring out, control of the economy is extremely difficult, and even focusing that control on a single element is highly complex. Any policy instrument may have many side effects, including some which are difficult to anticipate. It may lead to unintended consequences and may even have effects contrary to those intended.

TAXES AND STANDARDS AND THEIR IMPLICATIONS

An increase or decrease in the cost of fuels and electricity resulting from the imposition of additional taxes will have a direct impact on the price of energy to the ultimate consumer. In the regulated sectors of the energy industry, such as electric utilities, this result would be obtained directly through the regulatory process which seeks to relate the price of the product to the consumer to the cost of supply. In considering the imposition of taxes as a means of retarding energy growth, care must be taken to impose such taxes equitably among suppliers of all the various forms of energy. Such an approach is necessary to avoid distortions in resource allocation effects and to make possible the choice among the several forms of energy on the basis of an equitably competitive market.

The extent that the levying of taxes raises the price of one type of energy compared to competitors should reflect the degree to which that particular supplier imposes on the environment so that his supply costs may properly reflect the total costs to society of making that energy form available. The consumer of that energy form would then be required to pay a price that more fully reflects the cost to society of making that energy available to him.

Discriminatory treatment among the several suppliers of energy would result in a distortion of the market for energy. To the extent that prices do not properly reflect the environmental impacts resulting from the use of each of the energy forms, this discriminatory treatment could well have detrimental rather than beneficial effects on the environment.

In considering methods to discourage growth in energy consumption, the entire range of energy markets needs to be considered. The imposition of taxes or onerous environmental standards, if confined to only one of the energy forms, cannot achieve the desired effects. It would tend to shift the demand to other forms which would make possible avoidance of these standards and taxes. For example, measures that would increase the cost of central station electric energy generation without at the same time imposing similar requirements on self-generation of electricity would encourage consumers, particularly large industrial consumers, to undertake their own electric generation as a means of avoiding the additional cost burdens they would otherwise have to bear if they were to continue to purchase electric energy from central stations. Shifts to other fuels would also be encouraged. For example, in many applications, oil or gas can substitute for electric energy, and measures which would increase the cost of electric energy relative to the cost of these other fuels would shift demand in their favor at the expense of electric energy. It is unclear whether this would have a beneficial or detrimental environmental impact. Restrictions on the use of coal or oil for electric generation and severe limitations on the use of nuclear energy may create pressure to permit increased use of natural gas for electric generation. This could create sufficient demand pressure on natural gas prices to encourage consumers in other sectors of the economy, where such stringent limitations may not be imposed, to convert their facilities from gas to oil or coal, and the net result might well be increased rather than diminished environmental degradation. The concentration on a single energy form without adequate consideration of the ramifications for the total energy market raises the risk that efforts to improve the environment will be in vain. Environmental policies, therefore, as they affect the use of energy, need to be established with the entire range of energy markets clearly in view. This is not to say that measures to improve the environment should not have a differential impact, but the impact of these measures should be proportionate to the burdens that each of the energy forms imposes on the environment.

There is reason to believe that the cost of electricity-consuming devices is a more significant determinant of consumer demand for electricity than the price of electricity itself. Therefore, an effective way of limiting electric power growth would be to impose severe taxes on such electricity-consuming devices as ranges and ovens, clothes dryers, and air conditioning equipment. The magnitude of the impact of such taxes on the use of such electric-energy-consuming appliances will vary inversely with income. Thus, in effect, these taxes would deny the benefits of such conveniences in the home to those sectors of the population with the lowest incomes, who would therefore be tasked to pay the price of environmental protection. One of the major social goals at the present time in the United States is to raise the standard of living and the quality of housing for the least privileged segments of our society. Quite apart from income limitations, one of the characteristics making housing substandard is the inadequacy of electrical wiring, which limits the ability to install electricity-consuming devices, irrespective of the willingness of the occupant to purchase such devices. Part of the task of upgrading housing in the United States has been to provide housing capable of supplying the conveniences made possible by appliances, which necessarily involves providing the wiring which will permit these appliances to operate. Much of the public housing in New York City forbids the installation of air conditioners, but there is little reason to think that our society will or indeed should continue to deny the occupants of such housing the opportunity to enjoy the benefits of air conditioning. It, too, has a beneficial environmental effect which needs to be considered.

It should also be kept in mind that for almost 40 years it has been a national objective to make available to the consuming public an abundant supply of electric energy in whatever quantities may be required at the lowest possible prices. The federal government has provided loans at extremely low interest rates and has expended additional large sums for the construction of power production facilities with tax forgiveness in an effort to achieve this objective. Restrictions on growth in electric energy and taxes which would raise the price of electric energy to inhibit such growth would be directly contrary to this objective. While our society may desire to modify or even abandon this objective, it should not do so without full recognition of what is being done and an affirmative decision that the choice is proper.

PRICES AS A GUIDE TO RESOURCE DECISIONS

Whatever their source, price changes operate through essentially the same economic process. Consequently, this section discusses the effects of changes in energy prices and, especially, of electricity rates, including those that result indirectly from changes in costs and those that are imposed directly as a means of affecting growth in demand. Before evaluating these effects and comparing the price effects of alternative policy approaches, it is helpful to develop some basic economic concepts and to define two terms that have a special technical meaning to economists: "internalization of costs and benefits" and "price elasticity of demand." Both of these terms are potentially confusing and are often misused in ways that cloud discussions of environmental policy. The Working Group therefore believes it is important to clarify these concepts before discussing the effects of changes in energy prices.

Resource allocation decisions are made in our economy both through political and governmental processes and through the free market as constrained by legal rules. Though in practice prices are frequently an imperfect and incomplete guide to resource decisions, the free market performs well as a means of social choice in situations where all important benefits and costs are "internalized," that is, where those who benefit (e.g., consumers) voluntarily pay a price that compensates those (e.g., producers) who voluntarily incur the necessary sacrifice (costs) in return for compensation (income). In practice, market compensation does not, and in many instances cannot, cover all side effects, but the market performs well as the economy's workhorse wherever uncompensated side effects are not perceived as unreasonable. These side effects involve unpaid-for benefits as well as uncompensated costs. The main strengths of the market as a resource allocation device are twofold. First, it operates through voluntary action; consequently, it harmonizes differences in individual values and responds flexibly to changes in these values. Second, it makes it unnecessary to predict directly all of the ramifications of a policy change, such as the introduction of a new standard to preserve environmental quality. By voluntarily acting in response to prices and costs, buyers and producers help to discover the allocation that accords with individual values. This is important because the side effects of price changes and environmental regulations in a complex industrial economy are far-reaching and almost impossible to predict.

It may not be possible to internalize all of the benefits and costs relating to energy and the environment. But it is essential to remember that the aim of such policy instruments, standards, criteria, taxes, and price policy is to move in the direction of increased internalization. For instance, fuel quality regulations aim at internalizing the cost of avoiding air pollution while leaving buyers of electricity free to choose the extent to which they wish to pay the increased cost or relinquish the benefits of some electricity use.

This concept of internalization is relevant to the practical evaluation of policy instruments affecting prices to deal with electricity supply and the environment, because the logic of internalization implies that the policy action should penalize or reward the immediate source of the benefit or cost. Measures aimed at dealing with environmental effects of an electricity supply alternative (e.g., radiation) should be specific to that alternative. The choice among alternatives can then be made, giving appropriate weight to the environmental protection costs imposed by each alternative. A consistent and effective set of policy measures dealing with the economy's most serious environmental problems—whether they arise from the smelting of metal, the generation of electricity, from chemical processes or any other source—would significantly affect prices and costs throughout the economy. We do not know how the price of electricity relative to the fuels or the prices of the fuels relative to other goods in the economy at large would be affected. Conceivably, use of electricity might be encouraged or discouraged on net balance, and there are likely to be other effects involving changing costs of the fuels and electricity in different geographical areas, different seasons of the year, and different hours of the day.

If prices of all of these goods, including electricity, are reasonable reflections of their cost to society, then they will serve as a guide to allocating scarce energy and environmental resources, and there is no clear additional benefit from policies with the direct purpose of discouraging the use of electricity or other energy. For this reason, such policies are undesirable so long as environmental effects of energy supply can be reasonably contained by cost–price instruments and so long as it is not necessary to conserve energy supplies because of prospective resource shortages. However, the pressure of industrial growth and material consumption on the environment in the future may become so great that direct growth-discouraging policies, including price increases to discourage demand, could become necessary. Prior to adoption of such growth-discouraging policies, it will be necessary to consider carefully their broad political, social, and economic implications, on both the international and domestic fronts. Such policies may confront the nation with grave social conflicts and serious questions of social equity which may transcend the environmental issues.

What would be the effects of an increase in energy prices, or in electricity rates relative to other sources of energy? It is useful to discuss the effects on energy use and other economic effects.

PRICE ELASTICITY OF DEMAND

Price changes take effect largely through the *relative* prices of competing alternatives. For instance, a 10 percent increase in all energy prices would not change the competitive balance among energy forms, but it might induce energy users to conserve energy by, say, using more insulation, installing storm windows, or buying smaller automobiles for gasoline economy, provided that there are no offsetting increases in prices of insulation and other alternatives that directly or indirectly compete with energy. Some shifts to other alternatives, such as insulation, could possibly shift the locus of energy use from the residential consumer to the manufacturer of insulation materials, with little or no net gain in environmental quality. Indeed, the net environmental effect, including effects other than from fuel combustion, could be negative. Where the balance would lie is by no means easy to determine.

The effect of a price change—in electricity or in other forms of energy—depends on both the size of the price change and the responsiveness of demand to changes in relative prices. This response is conventionally measured by a coefficient of price elasticity, which is the percent change in quantity demanded caused by a 1 percent change in the price of the good in question. Price elasticity measures a *partial* effect, i.e., the effect of price alone when all other prices, income, population, and other factors affecting demand are held constant.

We do not attempt to present quantitative estimates of elasticity here; there have been a few experimental statistical studies (for example, Fisher and Kaysen, 1962) but there is no consensus among informed experts on the magnitude of price elasticities. However, some useful things can be said about the general order of magnitude involved and the factors on which elasticity depends.

In appraising elasticity effects, the following considerations should be kept in mind:

1. The most important single factor affecting elasticity is the availability and cost of competing alternatives. Elasticity is likely to be substantial only in cases where potentially active alternatives are clearly present.

2. The larger the class of goods for which elasticity is being measured, the smaller the elasticity, because substitution among goods in that class washes out. Consequently, the demand for electricity is more elastic than the demand for all energy, and most of the elasticity of demand for electricity stems from substitution between electricity and other energy forms. Equally important, the elasticity of industrial demand in a particular utility's service area (say for aluminum reduction) is much higher than the national or world elasticity for the same type of load, because of the influence of power rates on plant location.

3. Elasticity is positively related to the portion of the buyer's costs or budget accounted for by the good in question. For instance, a customer is likely to care about a 10 or 15 percent increase in his annual space heating costs, but the electricity cost of watching an hour's TV program is so small that the user will ignore

it. Because of this factor, the elasticity relevant to drastic rate increases might be higher than that for moderate increases from the present low rates.

4. For a buyer to respond to price incentives, he must be aware of the impact on his electric bill of his electricity-use decisions. Residential customers typically have little knowledge of how much energy most of their appliances use per billing period and do not relate their monthly bills to the amount of use. Consumers are more aware of the cost impact of energy-using appliances that are separately metered or are on special rate, such as water heaters. Consumers are also likely to be more sensitive to the cost of appliances which consume large amounts of energy (such as electric stoves).

5. Price responses require time to take effect. Initially, the buyer can respond only by varying the intensity of use of his existing electric devices, but with the passage of time he has increasing flexibility as old electricity-using equipment is replaced and as energy-using plants, equipment, and appliances designed for optimization with the new prices account for an increasing share of the total. The direct effect of a given increase in rates can be expected to take years to work itself out. If rate increases do not recur periodically, then continued growth in nonprice factors such as income may cause consumption to resume its old rate of increase after the direct price effect is largely worked out.

6. Energy-saving responses to price increases, such as increased use of insulation for walls and water pipes, increased use of storm windows, or greater efficiency in power-using devices, are subject to technical limits which are fairly readily identified upon investigation. Furthermore, the cost of the energy required to operate an appliance is expended in relatively small sums over a long period of time. The impact of accumulated savings over an extended period is likely to be reduced when compared with the large capital outlays required to achieve such savings.

7. A relatively low price elasticity in combination with a large price increase can have a significant impact on the rate of growth in demand. For instance, a rate increase of 10 percent, in the case in which elasticity is as low as 0.35, could still cut in half a year's load growth of 7 percent.

8. Large price increases for most goods usually have some demand effect even though it is frequently small, indirect, subtle, and hard to observe.

It is generally accepted among experts that the elasticity of demand for energy as a whole is very low for moderate price increases and would require years to take effect. For electricity, elasticities may be higher, but they arise mainly from substitution between electricity and other energy forms. Aside from some major uses such as space and water heating, residential electricity rates appear unimportant to the decisions to purchase and use appliances. The purchase of appliances probably depends more closely on income, appliance prices, including installation costs such as for special wiring or piping, and the terms on which appliances are marketed and financed.

For most industrial and commercial uses, space and process heating excepted, power is a minor proportion of total cost, and there is no energy source competitive with electricity. In these cases, price elasticities of demand for electricity and energy at the national level are probably extremely low. In heavy power- and fuel-using industries—such as metals, chemicals, and petroleum refining—power is an important cost element, and there is some competition among energy sources, but conservation of energy in these industries is subject to technical limits.

For a fairly wide range of electricity price increases, it is doubtful that the need for additional electric capacity will be significantly curtailed, because the overall demand effect is likely to be small. Utilities with a responsibility to prepare in advance to meet expected loads cannot give such dubious effects the benefit of the doubt in planning new capacity. This is especially the case at a time when regulatory agencies are urging electric utilities to increase their reserve capacity margins to achieve improved standards of reliability.

Over longer periods of time, the adjustment possibilities are greater, and the rate of growth in electricity use might be significantly slowed.

RATE REGULATION AND PRICE POLICY

There is no simple, direct link between the overall level of an electric utility's costs and its rate structure, which determines the rates paid by particular customers. Public utility rate design is a complex technical field, and neither expert opinion nor utility practice is uniform. Consequently, within the general legal framework of cost of service under which the industry operates, there is room in rate structure design for discretion by regulatory commissions and publicly or cooperatively owned power systems. Regulated investor-owned utilities also have some discretion for adjusting rate structures, though their rate policies are subject to regulatory approval. The only requirement is that total revenues cover total costs, including a return on investment sufficient to maintain financial integrity.

The costs of meeting present and expected environmental quality criteria are contributing to a change in electricity cost trends from their historical pattern of decline (in "real" or constant dollar terms) to one of increase, quite possibly at a substantial pace. These trends have led to increased support for rate structures using "marginal" or "incremental" cost as the basis for rate setting rather than the prevailing approach largely based on "embedded" or historical cost. In addition, some have suggested changes in rate structures specifically designed to discourage consumption of electricity. For this purpose, the increasing block rate proposal has been advanced, usually for residential rates. It would reverse the customary declining block rate structure by imposing rates that increase with increased customer use. The rate for small quantities would be low, and the incremental rates would be progressively higher for higher levels of customer consumption.

Marginal cost pricing has been proposed by economists in various forms for a long time, and incremental cost approaches to pricing have long been used by electric utilities in power pooling, for off-peak loads, and in setting some industrial rates. The development of practical, workable marginal costing and pricing guidelines for an entire rate structure is much more recent and has not as yet been adapted to the conditions of the American electric power industry. The main rationale for its use is to increase the efficiency of power supply and to improve resource allocation in the economy by relating rates to the effect of the buyer's decision and the impact of his consumption on overall system costs.

Historically, the declining block rate structures used in the electric power industry have generally reflected the approximate level of the incremental cost of serving the increased use per customer. One consequence of the past pattern is that changes in major rate schedules were comparatively infrequent, involving only moderate changes in rate level. The revenue–cost balance has generally assured stable earnings close to the "fair rate of return" adjudged by the regulatory system to attract necessary capital, prevent confiscation of investor property, and protect consumers from monopoly exploitation.

There is good reason to believe that the trend in the cost of power system expansion may change from one of decrease to increase in real terms, so that the cost of expanding a power system may become substantially higher than the average cost of old and new equipment that forms the cost base of present power rates. This development has led to the advocacy of marginal cost rates, especially for on-peak uses, as a means of limiting the use of electricity to those who are willing to pay for the cost of that expansion.

OTHER EFFECTS OF RATE STRUCTURE REFORMS

Incremental Cost-Based Reforms

Rates based on long-run incremental costs might or might not have substantial effects on load growth; the direction of effect would be to flatten the load curve. The main problems associated with such reforms are ones of practical implementation: cost measurement and estimating cost impact in the light of uncertainty and shifting load patterns. There are also adjustment problems for affected ratepayers. However, the incremental cost approach, properly understood, is nothing more or less than an attempt to make prices reflect the cost of supply alternatives as viewed by system planners and operators as they economize in power supply decisions. Such rates may or may not imply large differences between on- and off-peak rates, depending on the complexities of supply and demand for each affected system.

Increasing Block Rate

Increasing block rates and other rate reforms unrelated to power supply costs would distort economic incentives to energy users and suppliers.

Legal standards limiting utility earnings to a fair rate of return and to nondiscriminatory rate structures bearing a reasonable relation to cost of services may not be consistent with increasing block rates. Implementation of this approach might require special taxes and frequent rate adjustments to keep total revenues in line with cost of service if nonprice factors continue to generate growth in consumption per customer over time.

Utilities must reconcile their various obligations to stockholders, ratepayers and the broad public interest. Successful reconciliation of these responsibilities is most likely to occur if managerial incentives are clear and well directed. A vigorous, dynamic management that intelligently and successfully pursues its opportunities for gain while recognizing its public responsibilities and obligations is greatly to be preferred over a static, confused, and disheartened management. The latter frame of mind is encouraged when utilities are forced to adopt practices—such as the proposed rate structure—that are intended to deny opportunities for growth. For this reason, should it be in the public interest to penalize power supply alternatives that adversely affect the environment, the penalty should apply directly to the particular supply alternative having the adverse environmental effect, such as the selection of fuel for electric generation, rather than to power consumption on the whole.

An enterprise that seeks growth and engages in competition with a reasonable chance of success has an interest in reducing costs and finding ways to obtain growth and cost reduction while incurring the costs of maintaining environmental quality. A firm that is invited to lose in the competitive race for the sake of the environment has little incentive to seek imaginative solutions to environmental problems or to keep costs down.

The increasing block rate proposals raise an additional, more specific incentive problem: In itself, the proposal provides no consistent set of incentives indicating which power supply alternatives do not have adverse environmental effects.

There would also be problems of adjustment hardships. The burden of the increased rates would fall heavily on those customers whose homes are already equipped with electric water heaters and electric heating facilities. Buyers who chose electricity on the basis of the existing rate system, or whose dwellings were already equipped with electric water heating or space heating facilities, would suddenly be faced with a large increase in their heating bills. Even if they switch to oil or gas, they must still pay the high cost of conversion.

RATIONING AND END USE CONTROLS AND EFFECTS

Direct controls on the use of energy have been suggested as a means of slowing the growth in energy use and indirectly reducing adverse environmental side effects arising from the provision of energy. Such controls could conceivably take many forms. Examples include regulations limiting the horsepower of automobiles, prohibition of "frivolous" electrical appliances or limiting their number by means of permits, and restriction of new capacity additions by heavy power-using industry in areas where the side effects of energy supply on the environment are particularly severe. Direct end use controls could also be used for the purpose of shifting energy consumption from one energy form to another that poses a less severe environmental problem. In considering rationing devices, we are not concerned with temporary load shedding or temporary load control during severe capacity shortages. Rather, we are concerned with direct controls aimed at affecting the long-term growth in energy use.

Rationing to curb "nonessential" uses requires a public judgment of what constitutes a nonessential use, as well as means of implementation and policing that are effective at a reasonably low cost.

Because electricity is demanded at the flick of a switch and it is costly to meter appliances separately, direct controls on the residential use of electricity are costly to effect and difficult to police. Consequently, if any methods of end use control and rationing are adopted, they are likely to take the form of limitations on the ownership of electricity-using devices or, less probably, of kilowatt or kilowatt-hour quotas to large industrial users.

In the United States, direct controls on the use of goods and services have been a rarely used policy device, being used as expedients to deal with special shortage situations, as in wartime, and to deal with products, such as drugs, posing unusual problems to the user or to society.

The prime reason that rationing has usually been confined to crisis situations is that it is a cumbersome, inflexible, and costly device that requires special administrative and enforcement machinery to prevent cheating (such as a black market in appliances), to evade appliance rationing, to allocate whatever is being rationed, and to prevent corruption in administration.

The Working Group considers the other policy instruments discussed in this report to be generally superior to rationing or other direct controls as long-term methods of allocating scarce energy resources and of reconciling environmental values with energy supply and consumption.

In many instances rationing or end use controls on electricity would simply shift energy use to more costly energy alternatives considered inferior by the user. Like other means of affecting consumption, rationing can affect the impact of energy supply on the environment only by its effect on the rate of growth in total supply: by itself it provides no guides concerning which power supply alternatives should be chosen, nor does it give potential energy users the freedom to decide for themselves how much energy they wish to use if they must pay the costs of abating environmental effects of energy supply.

It has been suggested that the growth in the demand for all energy and especially electricity could be retarded by higher prices imposed by various additional forms of taxes either directly upon each of the sources of energy or upon energy-consuming appliances and equipment and by the increased costs of meeting stringent environmental standards. In the case of electricity, rate structure changes which would raise the price of increasing consumption have been suggested.

POPULATION CONTROLS AND EFFECTS

It has been suggested that one way of reducing energy demands is to reduce the rate of growth of population. Chapter 2 discussed this issue in relation to the forecast over the next 30 years. In that discussion it was pointed out that, if there is any effect, a reduced rate of growth in population might increase energy demands for the first two decades. Thereafter, as long as increased discretionary income effects on energy demand do not offset the reduced number of households, there would be a net decrease in the energy demand over what it would have been without population control.

LIMITATIONS ON NATIONAL PRODUCT GROWTH

Finally, it has been suggested that a reduction in growth of the total economy would result in a comparable reduction of growth in energy demand. While this may be correct, all other things equal, the implementation of such a policy would be at best very difficult and could have highly undesirable social and political consequences. A better alternative would appear to be to impose standards by law or regulation which would attain environmentally acceptable productive processes for all goods and services and to encourage research in technology which would make possible a reconciliation of economic growth and an acceptable environment. The cost of all goods and services would then more fully reflect their benefits and costs to society. Insofar as such a policy would increase the real price for goods and services (the price in terms of resources required), some limitation on the rate of growth in national product would be achieved. However, such limitation would be in direct relation to achieving the desired environmental consequences.

As the previous chapter indicates, there is a general understanding of the interrelation of energy, and more specifically electric energy, and the operation of the economy. However, there is a lack of quantitative understanding of these relationships. A great deal of research is required before it will be possible to provide the necessary and appropriate economic evaluation of many of the proposals designed to mitigate the environmental impact of electric supply. The most pressing needs for further information are in the areas of price elasticity, efficiency improvement in energy utilization, economic impact of higher energy price, nonprice effects on demand, and net environmental effects of the shifts in consumption patterns. These are discussed below.

PRICE ELASTICITY

There have been few studies of the effect of changes in price on the level of electricity demand (see Fisher and Kaysen, 1962, for an example). Those which have been done have been limited by inadequate data and shortcomings in understanding of electricity user behavior, relevant technology, rate structures, and other pertinent factual material. The evidence to date indicates relatively little responsiveness of electricity demand in many consumption areas to relatively small changes in price level. On the other hand, other areas of consumption are believed to be quite responsive to price change, and very large changes in price may have an effect in all consumption areas. The use of electricity

for air conditioning, for example, appears to be relatively insensitive to price. In the case of space heating applications, total energy use appears to be relatively insensitive to price, but price, especially the price of electricity relative to that of other fuels, appears to have a significant effect on the form of energy selected. Similarly, price changes may have little impact on the use of electricity for lighting industrial plants but may have a large effect on the form of energy used for the certain industrial processes.

Generally, changes in the price of electricity are likely to have their greatest effect on demand for those uses in which the cost of fuel or electricity is large over the life of the using equipment and for which substitution is possible between electricity and one or more other fuels. Research is needed to determine the extent of such price effects in each category of use and the time lag of the impact on use. A combined approach—utilizing the quantitative tools of econometrics and the detailed knowledge derived from surveys and marketing experience for an integrated analysis of the factors affecting the demand for electricity and fuels—would be very useful for the formulation of sound policies.

EFFICIENCY IMPROVEMENT

Opportunities for achieving greater efficiency in electricity use may be stimulated by higher price levels, but we know little about the extent to which such opportunities may eixst. The most obvious means for reducing electricity consumption is by turning off lights in unoccupied space, reducing the amount of cooling or heating of space, limiting the use of labor-saving equipment, etc. These and similar changes in human habits may involve substantial losses of convenience or comfort.

More likely is the possibility of increasing capital investment in utilization devices to increase their efficiency of operation. This could involve better insulation of refrigerators and ovens, added insulation and double-glazing of windows in homes and commercial equipment to reduce losses and improve efficiency of conversion, etc. Changes in industrial technology or in the organization of production may also make possible reductions in electricity consumption per unit of product. For example, the aluminum industry has, in the past decade, achieved some reductions in the consumption of electricity per pound of aluminum.

At present we have no systematic knowledge of where such opportunities may exist nor the degree of price rise which might induce such efficiencies. Research is needed in this area as well as in the motivational aspects of achieving the acceptance of such changes. It should be noted that in other economies there is more careful husbanding of electrical resources. This may be solely a relative price effect, or it may involve social habits as well.

ECONOMIC IMPACT

Changes in the consumption of electricity cannot be looked at in isolation from their effect on the total economy. Such changes, whether brought about by price or other means, will change the balance of the internal economy. They may, for example, encourage the substitution of plastics for aluminum, cause shifts in employment trends, or affect geographic location of facilities.

In addition, such changes will also affect the economy in relation to international markets and may be in conflict with other policies. For example, it is currently proposed again to permit accelerated depreciation by industry. The purpose is to stimulate the economy and particularly capital investment in plant and equipment in order to expand and modernize the industrial plant of the United States. Such investment is likely to include a large share of electricity-using equipment and thus stimulate further expansion in the demand for electricity. Studies to determine the extent to which rising prices for electricity would discourage such substitution of capital equipment for labor, and the extent to which this would impair the competitive position of United States industry in world markets, would be desirable. More than a decade ago, the European Economic Community, in their concern for possible energy short-

ages and high prices of energy in Western Europe, undertook studies of the effect of such prices on the competitiveness of various European industries. Clearly, further information on the regional, national, and international economic impacts is needed before implementing policies to limit electric growth. Further, economic effects of limiting electric growth may be in conflict with other national economic goals. Research is needed to clarify these effects and help establish priorities.

Energy markets are presently in a state of flux, in no small part as a result of fuel quality regulations and other restrictions and requirements imposed by environmental policy considerations. Interfuel competition, for example, is clearly being affected by SO_2 and particulate restrictions. Research into changing patterns of energy supply and demand as they relate to environmental policy measures is especially important to appraise the impact of these policies on fuel-producing areas, fuel transportation requirements, and the like. This research is also important if decisions in power supply planning are to be based on reasonable assumptions concerning fuel supply, and if fuel and environmental policies are to be properly integrated so as to permit the economy to function without impairment from fuel supply inadequacies.

Research into the workability, economic impact, and relative merits of alternative control policies, such as effluent taxes, standards (e.g., fuel purity versus emission standards), and various forms of rate structures, would also be useful in formulating effective policies.

NONPRICE EFFECTS

There are a number of factors other than price that affect the use of electricity and that are subject to national or social goals. For example, population growth will in the long run influence demand for electricity—so will the size of family income. Both are subject to influence through national government policies. Research into the effect on energy demands of changes in population and income would help to focus strategies on the proper objectives.

Similarly, land-use planning has become increasingly important as land resources have dwindled. Development of housing strategies—single-family versus multiple-family, cluster cities versus a megalopolis, etc.—all affect the intensity of energy utilization. Research in these areas would help to develop planning consistent with energy objectives.

ENVIRONMENTAL EFFECTS

The premise of most proposals to limit electric energy use has been to improve the environment. There exist too few data on the environmental effects of today's production of electricity and on its potential impact after the implementation of technological improvements which are currently possible.

But even more than this, there is virtually no information as to the total environmental consequences of the alternatives to electric use. It has been suggested that a reduction in electrical use might be translated into increased consumption of an alternative fuel. Whether this is environmentally desirable is a complex measurement of relative efficiencies, effluent emission, and effluent effects. Similarly, substitution of insulation to reduce energy consumption has been suggested as a desirable environmental goal. Yet there is virtually no data to show that the production of insulation is environmentally less detrimental than the alternative utilization of energy. Clearly, if a strategy leading to environmental improvement is desired, it must be based on means which achieve those ends.

These suggested areas for further economic research are not, by any means, intended to be exhaustive. They do indicate, however, areas for research in which our currently available information appears to be inadequate.

5   SUMMARY AND CONCLUSIONS

A cessation of growth in total energy requirements or of electric energy in the absence of revolutionary changes in the structure of our society is difficult to visualize in the foreseeable future. Although there is a growing awareness of the need to control population growth, such growth, possibly at a reduced rate, is likely to continue for the balance of this century at least. This growing population will have to continue to be housed, clothed, and fed with increasing quality while at the same time resorting to lower quality raw material resources. Life styles may well change. An increasing share of the population may have to be supported by a diminishing proportion of the population at work as the workweek may be shortened, as years in school increase, and as the retirement age is perhaps reduced. Increasingly, the choice may well be made for non-material forms of consumption with increasing shifts towards services. But the extent to which this would curtail the need for increasing quantities of energy is uncertain. These will require increased productivity and in essence this will mean a substitution of energy-consuming equipment for human labor. Even if the rate of growth diminishes, the general order of magnitude of the power plant siting problem will remain.

At the present time there appears to be a conflict between economic growth and environmental preservation, but this conflict need not be a permanent condition. This is not to say that the need to improve the environment is not a pressing issue. However, if we were to succumb to the temptation to impose improvised, only partly thought-through measures, these could have a detrimental impact not only on the environment but on our society as a whole. The problem we face is one of imposing the requirement that currently available technology be employed to the fullest extent feasible to minimize the detrimental environmental impact of energy consumption. This can be done by imposing environmental standards which will reduce effluents to acceptable levels on all energy applications. We note that further research is needed to help determine acceptable levels from the technical point of view.

Where this is not possible with existing technology, it is important that pressure be maintained to undertake the research effort needed to develop the improved technological means of resolving our need for energy and our desire for a better environment. The other Working Groups will document the specific areas where this is necessary. To the fullest extent, the cost of minimizing the impact on the environment should be internalized in the cost and price structure of the electric power industry as well as all the energy industries. To the extent that this raises costs for the energy industries or changes the relative costs among them and thereby restrains growth and changes the pattern of energy demand, it will do so in a way that reflects the desires of the public as expressed in the free market.

REFERENCES

Battelle Memorial Institute, Pacific Northwest Laboratories. *A Review and Comparison of Selected United States Energy Forecasts* (prepared under contract for the Office of Science and Technology). Washington, D.C.: U.S. Government Printing Office.

Commission on Population Growth and America's Future. *Population Growth and America's Future*. Washington, D.C.: U.S. Government Printing Office.

Edison Electric Institute, Statistical Committee. *Bibliography and Digest of U.S. Electric and Total Energy Forecasts, 1970–2050*. 1970.

Federal Power Commission. *National Power Survey,* Vol. II. Washington, D.C.: U.S. Government Printing Office, 1970.

F. M. Fisher and C. Kaysen. *A Study in Econometrics: The Demands for Electricity in the United States*. Amsterdam: North Holland Publishing Company, 1962.

National Environmental Research Associates, Inc. *Energy Consumption and Gross National Product in the United States: An Examination in a Recent Change in their Relationship*. March 1971.